# CISM COURSES AND LECTURES



The series presents lecture notes, monographs, edited works and proceedings in the field of Mechanics, Engineering, Computer Science and Applied Mathematics.
Purpose of the series is to make known in the international scientific and technical community results obtained in some of the activities organized by CISM, the International Centre for Mechanical Sciences.

INTERNATIONAL CENTRE FOR MECHANICAL SCIENCES

COURSES AND LECTURES - No. 425

# EMERGING METHODS FOR MULTIDISCIPLINARY OPTIMIZATION

EDITED BY

JAN BLACHUT
THE UNIVERSITY OF LIVERPOOL

HANS A. ESCHENAUER
UNIVERSITY OF SIEGEN

Springer-Verlag Wien GmbH

This volume contains 123 illustrations


Originally published by Springer-Verlag Wien New York in 2001
SPIN 10837205

In order to make this volume available as economically and as rapidly as possible the authors' typescripts have been reproduced in their original forms. This method unfortunately has its typographical limitations but it is hoped that they in no way distract the reader.

ISBN 978-3-211-83335-3 ISBN 978-3-7091-2756-8 (eBook)
DOI 10.1007/978-3-7091-2756-8

# ACADEMIC YEAR 2000

## The Oswatitsch Session

## in Memory of

## Univ.-Prof. em. Dr. phil. Dr.-Ing. E.H. mult.

## Klaus Oswatitsch, Vienna, and his Work

Klaus Oswatitsch was born in Marburg/Drau, in 1910. He attended the University of Graz, where he also graduated as Doctor of Philosophy. Between 1938 ad 1946 he worked as a scientific assistant of Ludwig Prandtl in Goettingen at the former Kaiser-Wilhelm-Institute for Flow Research (now Max-Planck-Institute). In 1943 he received his *venia legendi* from the University of Goettingen.

Oswatitsch's fundamental research in the field of fluid dynamics, which brought him worldwide reputation and fame, was stimulated by Ludwig Prandtl. For the first time ever, Oswatitsch calculated the condensation processes in moist air in Laval jets. His ideas on implementing nucleus formation and the subsequent growth of drops into the fundamental equations of fluid dynamics are still, after more than 50 years, the starting point for any investigation into this highly topical field of research. Furthermore, he discovered the principle of the collision diffuser, and proved it both experimentally and theoretically. In the diffuser, a supersonic flow can by generated without substantial loss by a sequence of diagonal collisions with a final weak perpendicular collision. A further important work is his formulation of the fundamental law of air resistance as an integral of the entropy flow.

Between 1946 and 1956 Oswatitsch continued his scientific work in England, France and Sweden, where he concentrated on investigations of transonic and hypersonic flows. The laws of similarity, the equivalent statement, and the surface rule as well as his integral equation method are pioneering works that have entered into many practical applications.

Oswatitsch established the Institute of Gas Dynamics of the DFVLR (German Air- and Spacecraft Research and Experimental Association) in Aachen, and later in Goettingen. From 1960 to his retirement, he was Full Professor and Head of the corresponding institute at the Technical University of Vienna, Austria, The wide field of his activities is illustrated by his and his coworkers' works on, e.g., multidimensional nonlinear characteristic procedures, novel supersonic jets, separation of flows, etc. His immense productivity is proved by more than a hundred original papers, articles in books, and textbooks that have become world-standard. Science occupied the greater part of Prof. Oswatitsch's life; the large number of his students clearly shows that he was always prepared to generously share his knowledge. Everybody who has had the privilege of knowing him personally will always hold him in high esteem as a highly stimulating senior scientist, but also as a friend with a strong personal and caring interest in his coworkers' lives.

Prof. Oswatitsch received a large number of honours and awards. Amongst others, he held three honorary doctorates, was a member and honorary member of numerous academies and societies, carrier of the Prandtl Ring, and recipient of many honorary medals.

## *PREFACE*

*The continuous pursuit of ever more demanding end-user and environmental objectives is driving the requirements for sophisticated engineering tools. One of the prime tools, which offer the possibility of facing the challenge successfully, is optimization. Optimization does not belong entirely to a specific field but it spans across a range of human activities and it embraces disciplines as far afield as pure and applied mathematics, numerical analysis, computer and natural sciences, continuum mechanics, material sciences, economics, etc. Being a universal tool, optimization is subjected to constant advancements due to progress being made in the interrelated fields. The above observations have all been visible throughout a week long series of lectures and accompanying discussions at the Centre International des Sciences Mécaniques, CISM in Udine, Italy.*

*Effectiveness of optimization is frequently measured by its ability to solve complex, real-world problems. It appears that in most cases of applied optimization, mainly mathematical programming and optimality criterion methods have been favoured for solving the resulting non-linear constrained optimization. However, the treatment of problems with larger number of degrees of freedom and/or with larger number of design variables is a very time-consuming process. Additionally, in many cases of gradient calculations in the first and second order methods (sensitivity analysis), it transpires that these methods are not sufficiently accurate to be able to find the exact optimal solutions. The remarkable increase in computational power which has been achieved during the last decade offers substantial speed-ups of computations. In this context, many of the "old" methods (e.g., 0th order methods) are being re-discovered and many more are being developed and researched. In this constantly changing environment it is essential to keep pace with these related developments and CISM has provided us with an ideal opportunity to accomplish this.*

*The course has provided a balanced training in both theoretical and practical aspects of optimization. The material has been presented by six lecturers according to the following list:*

***J. Blachut** (The University of Liverpool, UK) has presented non-gradient techniques. This covered modern heuristic search methods based on the study of nature. This fast growing area of research is quickly moving from the Operations Research and Computer Sciences area into a wider engineering domain. Three methods have been selected for the course, i.e. simulated annealing, tabu search and genetic algorithms. Unprecedented advances in computing hardware bring into the forefront some other zero order techniques like dynamic programming, random search methods and hybrids, as well. Potential of these methods has been illustrated by case studies relevant to underwater and pressure vessel applications where stress, stability and collapse need to be considered subject to severe limitations on geometric imperfections.*

***H. Eschenauer** (University of Siegen, Germany) has introduced various aspects of multidisciplinary optimization through his long time experience in the field. Topics of his lectures included technology oriented product development within a design process, structure of the modular multidisciplinary optimization concept, optimization strategies and algorithms. The latter included approximation* *procedures, e.g. design-of-experiments methods, modelling, simulation and practice oriented real-world applications. The importance of correct problem definition and modelling for optimization computations has also been emphasised. Case studies included satellite antenna, automotive components, power generation switch gears, pressure vessels, etc. The complex nature of these examples showed how important is the role of supporting disciplines, in order to be able to create a mathematically correct model for analysis and meaningful optimization.*

***P. Hajela*** *(Rensselaer Polytechnic Institute, Troy, USA) has lectured on some cutting edge optimization technology including various speculative methodologies. His six lectures included: application of genetic algorithms in multidisciplinary optimization, GA based modelling of immune networks with applications in decomposition and multicriterion optimization as well as GA's and computational intelligence. Discussion of the neural networks' application for function approximation and the modelling of fuzzy environment in order to extract the most useful and practical information has showed where this technology is currently moving.*

***K. Marti*** *(Federal Armed Forces University Munich, Germany) presented an elegant mathematical approach to realistic problems which nowadays we are faced with, i.e. stochastic optimization. These subjects, of rapidly growing importance, have covered: theory of statistics and probability, stochastic optimization techniques, stochastic optimization methods applied to robust designs, optimal design under stochastic uncertainty and fuzzy-strategies.*

***A. Schoofs*** *and* ***K. Rijpkema*** *(Eindhoven University of Technology, The Netherlands) have lectured on design-of-experiments methodologies and response surface methods. They continued with optimal experimental design and they covered the following topics: point selection techniques, orthogonal arrays, D-optimal and DACE designs, design sensitivities in response surface methods and fractional factorial experimental designs. Finally, issues in design-of-experiments application to computer simulations were also covered.*

***V.V. Toropov*** *(University of Bradford, UK) has considered local, global and mid-range approximations, multi-point approximation methods based on response surface fitting, kriging and screening experiments. Next, genetic programming methodology for selection of global and mid-range approximations have been introduced together with the use of simplified numerical models as approximations (corrections and tuning). A range of illustrative examples has supported the main features of his lectures and these included civil, aerospace and materials engineering cases.*

*The course has delivered an up-to-date overview of major advances, emerging trends, projected and industrial applications in the field of multidisciplinary optimization The main thrust of the course has concentrated on the current status of the field, on a search for commonalities, and on innovative and promising methods. Elements of speculative approaches to optimization have also featured in the delivered lectures. An emphasis has been placed on the multiobjective approach. It has been demonstrated that this approach carries a fair amount of generic knowledge across a wide range of engineering disciplines. Due to the above nature of the course, it has attracted specialists from a wide class of*

*backgrounds, i.e. from mathematicians and physicists, to aerospace, civil and mechanical engineers. Participants who subscribed to the course have come from industry, academia and academies of sciences. Some of the participants were supported by the generous UNESCO-ROSTE sponsorship. This has certainly facilitated access to state-of-the-art research findings in the above mentioned subject.*
*It is our pleasure to present to the reader the contents of 35 hours of lectures which we hope will contribute to the ongoing progress of optimization.*

*The editors wish to express their gratitude to the Board of the International Centre for Mechanical Sciences CISM, and in particular to its Rector, Professor S. Kaliszky and to the Secretary General of CISM, Prof. G. Bianchi, for making this meeting possible, to the lecturers for their devoted efforts, and to the participants for their attention and useful discussions during the course. Further thanks are due to Professor C. Tasso and the members of the CISM secretariat, and to those that have assisted the editors in preparing the manuscripts and in secretarial tasks, namely Michael Wengenroth (Siegen).*

*Jan Blachut*
*Hans Eschenauer*

## CONTENTS

# LIST OF SYMBOLS

**Note:** The following list is restricted to the most important symbols, notations, and letters in the book, and listed individually for each chapter.

Scalar quantities are printed in roman letters, vectors in boldface, tensors or matrices in uppercase, boldface type.

## Chapter 1

### Indices and Notations

| | |
|---|---|
| $a, b, c$ | Latin indices valid for 1,2,3 in tensor notation |
| $\alpha, \beta$ | Greek indices valid for 1,2 in tensor notation |
| CCD | Central Composite Design |
| OA | Orthogonal Array |
| $\dot{(..)}$ | time derivative |
| $(\ )\|$ | covariant derivative |
| $\in$ | an element of |
| $a \wedge b$ | *a* and *b* are valid |
| $\sum$ | sum |

### Latin Letters

| | |
|---|---|
| $\boldsymbol{A}$ | coordination matrix |
| $\boldsymbol{A}_f$ | sensitivity matrix |
| $a_{f_{ij}}$ | elements of the sensitivity matrix |
| $\boldsymbol{b}$ | vector of coefficients estimated by regression analysis |
| $C_{ep}^{ijkl}$ | elastic-plastic tensor |
| $E$ | expected values |
| $\boldsymbol{e}_i$ | *i*-th unit vector |
| $F_i$ | stochastic objective function |
| $\boldsymbol{F}$ | system matrix |

| | |
|---|---|
| $f, \boldsymbol{f}$ | objective function, objective function vector |
| $f^i$ | volume forces |
| $f_\Gamma$ | arbitrary functional |
| $G_v$ | optimization functional |
| $G_j$ | stochastic constraint |
| $\boldsymbol{g}$ | vector of $p$ inequality constraints |
| $h(r)$ | thickness |
| $\boldsymbol{h}$ | vector of $q$ equality constraints |
| $J_\Gamma$ | domain functional |
| $m$ | mass |
| $n$ | rate of revolutions<br>value of desired reliability |
| $p$ | preference / substitute objective function |
| $p^i$ | components of surface forces |
| $\boldsymbol{p}_j$ | pseudo load vector |
| $Q^i$ | dissipated heat quantity |
| $q^i\big\|_i$ | heat flow quantity |
| $R_{VS}$ | strength constraints |
| $R_{FV}$ | failure safety |
| $\Re^n$ | set of real numbers |
| $r_B$ | radius of a small hole (bubble) |
| $\boldsymbol{r}$ | load vector |
| $s^i$ | subsystem |
| $T$ | temperature field / kinetic energy |
| $U^*$ | complementary energy |
| $u_i$ | components of the displacement vector |
| $\boldsymbol{u}$ | state variable vector |
| $\boldsymbol{u}_B^{i(m)}$ | boundary state variable vector |
| $\boldsymbol{u}_I^i$ | internal state variable vector |
| $V$ | volume of a structure |

| | |
|---|---|
| $W$ | weight |
| $\boldsymbol{X}$ | matrix of a sequence of experiments |
| $X$ | feasible domain |
| $\boldsymbol{x}$ | vector of $n$ design variables |
| $y$ | second order polynomial approximation |
| $\boldsymbol{y}$ | analysis variable vector |
| $z$ | noise factors |
| $\boldsymbol{z}$ | transformed variable vector |

**Greek Letters:**

| | |
|---|---|
| $\alpha_D$ | drop-off angle |
| $\beta_a$ | reliability index |
| $\Gamma$ | boundary of a domain |
| $\gamma$ | tape shear angle |
| $\gamma_{ij}$ | strain tensor components |
| $\delta s$ | distance between the old and the new boundary |
| $\varepsilon$ | step size factor |
| $\eta$ | aeroelastic efficiencies |
| $\kappa_B$ | curvature |
| $\boldsymbol{\lambda}_i$ | auxiliary variable vector |
| $\mu_y$ | mean value |
| $\rho$ | density |
| $\sigma_{\hat{y}}^2$ | variance, standard deviation |
| $\tau_r$ | reference stress |
| $\tau_{rr}, \tau_{\theta\theta}$ | radial and tangential stresses |
| $\phi$ | arbitrary coupling function, AIRY's stress function |
| $\Omega^p$ | domain |
| $\omega$ | angle velocity |

# Chapter 2

**Indices and Notations**

| | |
|---|---|
| GAs | genetic algorithms |
| NS | neighbourhood size |
| SA | simulated annealing |
| TS | tabu search |
| TT | tabu tenure |

**Latin letters**

| | |
|---|---|
| $Ci$ | concentric sphere number 'i' and of radius ri |
| $D$ | shell diameter |
| $E$ | potential energy |
| $F$ | cost function |
| $g$ | constraints |
| $L$ | length of a barrel, also length of cylindrical flange |
| $L_o$ | initial length of cylinder |
| $N_k$ | epoch length or number of random moves at a given temperature level |
| $nit$ | number of iterations |
| $n_1, n_2, n_3$ | parameters describing generalised ellipse |
| $Pr$ | acceptance probability of the Boltzman distribution |
| $p$ | external pressure |
| $p_{bif}, p_{coll}, p_y$ | bifurcation buckling, collapse and yield pressures |
| $R_o$ | mid-surface radius of a shell |
| $R_s$ | radius of spherical portion in a torispherical dome |
| $r$ | knuckle radius, random number or meridional radius of barreled shell |
| $T$ | dimensionless parameter in the acceptance criterion (traditionally called temperature) |
| $T_o, T_f$ | initial and final temperatures for a given cooling scheme |
| $t$ | wall thickness of a shell |
| $\boldsymbol{u}$ | random unit vector |
| $W_H, W_T$ | weight of hemispherical and torispherical dome, respectively |
| $X_t, Y_t, X_c, Y_c$ | tensile and compressive strength parameters for composites |
| $\boldsymbol{x}$ | design vector |
| $\boldsymbol{x}'$ | feasible neighbour of a given solution $\boldsymbol{x}^*$ |

**Greek letters**

| | |
|---|---|
| $\alpha$ | parameter in a geometric cooling schedule |
| $\delta$ | ratio of step-wise change in beam's stiffness |
| $\lambda$ | scalar step length |
| $\theta_i$ | lamination angle of ply number '$i$' |
| $\sigma_{yp}$ | the yield point of material |

# Chapter 3

**Latin Letters**

| | |
|---|---|
| $A = A(X)$ | vector of cross sectional areas |
| $(A_0, b_0)$ | fixed matrix for the representation of the feasible domain $D$ of the design vector $X$ |
| $a$ | total $(v\text{-})$ or $(m+v)$ vector of model parameters including the external load parameters in optimal structural design |
| $C$ | equilibrium matrix |
| $c_g, c_f$ | weighting or scale factors related to expected cost functions or failure probabilities |
| $cov(M)$ | variance/covariance matrix of a random vector $\boldsymbol{M}$ |
| $c_{pj}$ | ratio of plastic capacities of $j$-th element |
| $D$ | feasible domain of the design vector $X$ described by simple nonstochastic constraints, like box constraints |
| $E$ | expectation operator |
| $E_j$ | modulus of elasticity of $j$-th element |
| $F$ | vector of internal loadings (forces, moments) |
| $F = F_0(\sigma_y, X)$ | right hand side of the linearized yield condition |
| $G_0 = G_0(a, X)$ | objective function like volume, weight or more General cost for the manufacturing of the structure |
| $G^{max}$ | upper cost bound |
| $H$ | matrix arising from the linearization of the yield constraints |
| $h(\omega)$ | random vector in stochastic optimization problem |

| | |
|---|---|
| $I$ | identity matrix |
| $LP$ | Linear Program |
| $L_j$ | length of $j$-th element |
| $M_i = M_i(a, X)$ | $i$-th limit state function or safety margin |
| $M = M(a, X)$ | vector of all limit state functions |
| $M^{min} = M^{min}(a, X)$ | minimum of all limit state functions |
| $M_{pj}^y, M_{pj}^z$ | bending plastic capacities of $j$-th element |
| $N_{pj}$ | principal axial capacity of $j$-th element |
| $p$ | $(\nu -)$ vector of model parameters besides external loadings |
| $p_f = p_f(X)$ | probability of failure |
| $P(...)$ | probability of a certain event |
| $q = q(X)$ | generalized variance function |
| $q^-, q^+$ | cost factors |
| $Q$ | covariance matrix |
| $R$ | $m$-vector of external loadings |
| $s^*$ | minimum failure costs |
| $SLP$ | Stochastic Linear Program |
| $SOP$ | Stochastic Optimization Problem |
| $T$ | transposition of a vector or matrix |
| $\dot{T}(\omega)$ | random vector in two-stage stochastic optimization problem |
| $T(\omega)$ | random matrix in two-stage stochastic optimization problem |
| $T_{pj}$ | twisting plastic capacity of $j$-th element |
| $U$ | feasible domain of the dual (kinematic) LP |
| $U_{j\nu}$ | finite subset of $U$ |
| $u, \tilde{u}$ | dual variables in the kinematic LP |
| $w = (u, \tilde{u})$ | dual vector |
| $V(...)$ | variance of a random variable |
| $vec(\omega)$ | indicates the randomness of the corresponding expression "*vec*", |
| $\overline{vec}$ | expectation of the random variable *vec* |
| $W$ | recourse matrix in two-stage stochastic optimization |

| | |
|---|---|
| $X$ | $r$-vector of design variables |
| $X_k$ | $k$-th component of the design vector $X$ |
| $y = y(a, X)$ | performance function in quality engineering |
| $y = y^+, y(\omega)$ | second stage decision variables in two-stage stochastic optimization |

**Greek Letters:**

| | |
|---|---|
| $\alpha^{max}$ | maximum failure probability |
| $\alpha_j$ | probability of the $j$-th realization (scenario) of the random parameter vector $\alpha(\omega)$ |
| $\eta$ | vector of weight or cost factors |
| $\Delta$ | discrepancy vector |
| $\Delta^+, \Delta^-$ | positive, negative part of $\Delta$, respectively |
| $\gamma = \gamma(z)$ or $\gamma(y)$ | cost or loss function |
| $\Gamma = \Gamma(X)$ | expected cost function or vector of expected cost functions |
| $\lambda$ | eigenvalue of a certain matrix |
| $\lambda^{max}, \lambda^{min}$ | maximum, minimum eigenvalue, respectively |
| $\theta, \theta_h$ | parameter or parameter vector of a probability distribution |
| $\varphi$ | probability density |
| $\sigma_y$ | vector of yield stresses |
| $\sigma^U, \sigma^L$ | vector of yield stresses in compression, tension |

# Chapter 4

## Indices and Notations

| | |
|---|---|
| $\wedge$ | estimated value |
| $-$ | stochastic variable |
| $\dot{(..)}$ | time derivative |
| $(..)'$ | derivative |

| | |
|---|---|
| $(..)^T$ | matrix transpose |
| $(..)^{-1}$ | matrix inverse |
| $E(..)$ | expected value operator |
| $P(..)$ | probability operator |
| $V(..)$ | variance operator |

**Latin Letters**

| | |
|---|---|
| $\boldsymbol{D}$ | damping matrix |
| $B$ | normalized average-square bias |
| $bce_f(\boldsymbol{x})$ | equality behavior constraint |
| $bci_j(\boldsymbol{x})$ | inequality behavior constraint |
| $C(\boldsymbol{N})$ | optimality criterion for experimental design |
| $d(\boldsymbol{x}, \boldsymbol{N})$ | normalized variance of the response estimator in x |
| $\boldsymbol{D}$ | design space |
| $dce_j(\boldsymbol{x})$ | equality design constraint |
| $dci_j(\boldsymbol{x})$ | inequality design constraint |
| $\boldsymbol{D}_f$ | feasible design space |
| $EISE$ | Expected Integrated Standard Error |
| $F$ | objective function<br>F-test value |
| $\boldsymbol{F}$ | multiple response design matrix |
| $f$ | regression model function |
| $\boldsymbol{f}$ | vector of regression model function |
| $\boldsymbol{H}$ | hat matrix |
| $\boldsymbol{I}$ | identity matrix |
| $J$ | averaged mean squared error |
| $\boldsymbol{K}$ | stiffness matrix |
| $\boldsymbol{M}$ | mass matrix |
| $MSE$ | Mean Squared Error |
| $MS_E$ | Mean Error sum of Squares |
| $MS_R$ | Mean Residual sum of Squares |
| $\boldsymbol{N}$ | set of design points |

| | |
|---|---|
| $r_i$ | studentized residuals |
| $\boldsymbol{S}$ | diagonal variance matrix |
| $\boldsymbol{s}$ | search vector |
| $SS_E$ | error sum of squares |
| $SS_R$ | residual sum of squares |
| $S_{yy}$ | total sum of squares |
| $t_{\alpha \quad 2}$ | student-t criterion value |
| $\boldsymbol{u}$ | displacement vector<br>multiple response vector |
| $\mathrm{v}(.)$ | scaled prediction variance |
| $\boldsymbol{V}$ | variance-covariance matrix |
| $V$ | normalized average variance |
| $\boldsymbol{W}$ | weight function matrix |
| $\boldsymbol{x}$ | design variables |
| $\boldsymbol{X}$ | design matrix |
| $\boldsymbol{y}$ | response vector |

**Greek Letters:**

| | |
|---|---|
| $\alpha_q$ | stepsize |
| $\beta$ | column vector with regression-parameters |
| $\delta_{ij}$ | Kronecker delta |
| $\varepsilon$ | model error |
| $\theta$ | physical constants |
| $\lambda_{max}$ | largest eigenvalue |
| $\mu$ | mean |
| $\rho$ | correlation matrix |
| $\sigma$ | standard deviation |
| $\xi$ | input quantities |
| $\omega$ | eigenfrequency |

## Chapter 5

### Latin Letters

| | |
|---|---|
| $A_i^k \leq x_i \leq B_i^K, B_i^k \leq B_i$ $(i = 1,\ldots,N)$ | move limits |
| $A_i \leq x_i \leq B_i \quad (i = 1,\ldots,N)$ | side constraints |
| $\boldsymbol{a}$ | vector of tuning parameters |
| $C(\boldsymbol{x},\boldsymbol{a})$ | correction function |
| $F_0(\boldsymbol{x})$ | objective function (to be minimized) |
| $F_j(\boldsymbol{x}) \leq 1 \quad (j = 1,\ldots,M)$ | normalised constraint functions |
| $\tilde{F}_j^k(\boldsymbol{x}) \quad (j = 0,\ldots,M)$ | approximation functions |
| $f(\boldsymbol{x})$ | response function corresponding to a simplified numerical model used as an approximation of the function $F(\boldsymbol{x})$ |
| $\boldsymbol{g}(\boldsymbol{x}) = [g_1(\boldsymbol{x}),\ldots,g_k(\boldsymbol{x})]^T$ | vector or regressors |
| $k$ | iteration number |
| $M$ | number of constraints |
| $N$ | number of design variables |
| $P$ | total number of plan points |
| $Q(S_i)$ | measure of quality of the approximation represented by the tree $S_i$ (in the genetic programming algorithm) |
| $R(\boldsymbol{x})$ | correlation function |
| $\boldsymbol{r}(\boldsymbol{x})$ | vector of correlations between the response at a point $\boldsymbol{x}$ and plan points |
| $w_p$ | weight coefficients for a plan point $p$ |
| $\boldsymbol{x}$ | vector of design variables |
| $Z(\boldsymbol{x})$ | Gaussian random function with zero mean used in kriging |

### Greek Letters

| | |
|---|---|
| $\Phi(S_i)$ | fitness function for the tree $S_i$ |
| $\sigma^2$ | variance of $Z(\boldsymbol{x})$ |

## Chapter 6

**Indices and Notations**

| | |
|---|---|
| $[T]$ | transition or causality matrix in neural networks |

**Latin Letters**

| | |
|---|---|
| $A_c$ | precision of representing numbers in binary coded GA's |
| $AI$ | autorotational inertia of blade |
| $b_i$ | binary valued index to define matching bits |
| $b_j$ | prescribed value of $j$-th constraint |
| $C_T$ | rotor thrust coefficient |
| $d_i$ | nodal displacements |
| $E$ | sum-squared errors |
| $E$ | Young's modulus |
| $e_j(t)$ | error at $t$-th time step in $j$-th output |
| $F$ | objective function |
| $F$ | activation function for a neuron |
| $F_z$ | maximum peak-to-peak values of the scaled shear force |
| $f(X)$ | vector of objective functions |
| $f(d)$ | membership function used in counterpropagation neural networks |
| $f_i$ | $i$-th objective criterion |
| $f_i^{min}, f_i^{max}$ | minimum and maximum values of $i$-th objective criterion in fuzzy design |
| $G_j$ | fuzzy feasible set |
| $g_j$ | $j$-th inequality constraint in optimization problem |
| $HP_h$ and $HP_f$ | horsepower required during hover and forward flight |
| $HP_a$ | available horsepower |
| $h_k$ | $k$-th equality constraint |
| $h_s$ | individual contribution of a Kohonen neuron in counterpropagation networks |
| $I$ | sectional moment of inertia |

| | |
|---|---|
| $I_i^*$ | ideal design point |
| $k_g$ | maximum permissible bending stresses in beam structure |
| $L$ | beam length |
| $l$ | length of bars |
| $M_y$ | maximum peak-to-peak values of the scaled flap bending moment |
| $M_z$ | maximum peak-to-peak values of the scaled lead-lag bending moment |
| $m^i$ | tuning mass at $i$-th location along blade span |
| $P,Q$ | applied loads |
| $p_c$ | probability of crossover |
| $p_m$ | probability of mutation |
| $R$ | Tsai-Wu structural failure criterion |
| $T_j$ | bias constant of $j$-th neuron |
| $t_i$ | flange and web thicknesses of blade box-structure |
| $U$ | match score in immune network simulation |
| $u_i$ | distance metric in multicriterion design |
| $W$ | number of interconnection weights in the neural network |
| $W_y, W_z$ | beam section moduli |
| $w_{ij}$ | interconnection weight between $i$ and $j$ neurons of two successive layers |
| $w_i$ | weighting index of $i$-th objective criterion |
| $X$ | design variable |
| $\mathbf{X}$ | a vector |
| $X^L$ | lower bound on $X$ |
| $X^U$ | upper bound on X |
| $x_i$ | input signal to neuron $i$ |
| $Y_j$ | output of $j$-th neuron |
| $Y_j^T$ | target value of the output of $j$-th neuron |
| $Z_j$ | weighted sum of inputs to neuron $j$ |

**Greek Letters**

| | |
|---|---|
| $\beta$ | sample size in immune network simulation |
| $\delta$ | twist shape parameter |
| $\delta_r, \delta_s$, *and* $\delta_o$ | distance measures used in counterpropagation neural networks |
| $\varepsilon$ | bound on generalization error |
| $\eta$ | learning rate |
| $\lambda_c$ | taper ratio |
| $\mu$ | membership function in fuzzy logic |
| $\rho$ | density of material |
| $\mu_g, \mu_f$ | membership functions of constraints and objective in fuzzy design |
| $\sigma$ | rotor solidity |
| $\sigma_{all}$ | allowable stress for material |
| $\sigma_{buck}$ | static stresses due to buckling |
| $\tau_R$ | point of taper inception along span |
| $\Omega$ | rotational speed of blade |
| $\theta_t$ | blade twist at root |
| $\theta_1$ and $\theta_2$ | ply orientations in wing box |

# CHAPTER 1

# Multidisciplinary Optimization Procedure in Design Processes - Basic Ideas, Aims, Scope, Concepts -

H.A. Eschenauer

FOMAAS, University of Siegen, Germany

**Abstract.** An important goal of engineering activities is to improve existing and to develop novel technical designs, structural assemblies, and components. The "best-possible" or "optimal" structures / systems are the ones that correspond to the designer's desired concepts, meeting at the same time the multidisciplinary requirements and/or specifications referring to manufacturing, assembling, operation, etc. In comparison with the "trial-and-error" approaches still used in the engineering environment and provided with considerable uncertainties, the determination of optimal solutions especially for large-scale and complex structures by means of continuously improved algorithms and strategies, so-called "emerging methods" is more reliable and efficient. These procedures will be a need in the design process in future, and they are already increasingly entering industrial practice. In the introductory chapter, three applications in Topology, Product and Process, and Robust Multicriteria Optimization are described.

## 1.1 Introduction

Up to the present, traditional design processes comprise a sequential procedure, ranging from the project definition phase over conceptual and detailed design to production planning, maintenance and repair. Here, the freedom of decision of the involved disciplines in the single project phases depends on the initial decision making. Current research attempts to achieve a mutual integration of the different disciplines in the entire product development. This type of design process, which is called Concurrent Engineering (CE), intends to ease the inter- and multidisciplinary cooperation (Lawson and Karandikar, 1994, Parsaei and Sullivan, 1993). An essential feature of this process is the existence of a large number of objectives, demands and constraints in the initial design phase; moreover, in most cases several sources of stochastic uncertainties, noises, etc. (stochastic variations of the material, external loads, manufacturing errors, modeling errors, etc.) must be taken into account already at the project definition phase. These facts have led to Multidisciplinary Design Optimization (MDO) emerging as a supporting tool for Concurrent Engineering. MDO comprises, as fundamental components, design-oriented analysis methods, approximation concepts, mathematical systems modeling, decomposition, search algorithms, optimization procedures, and the user as human interface (Sobieszczanski-Sobieski, 1989, Eschenauer, 1992).

The field Multidisciplinary Structural or Design Optimization (MDO) is extremely important as a bridge between all engineering disciplines to Structural and Continuum Mechanics (solid mechanics, load-bearing structures, fluid- and thermodynamics, material theory) and to applied and numerical mathematics and communication sciences (field theory, finite methods, mathematical programming, computer architecture, parallel processing, etc.). The rapid development of the computer technology connected with the increase of efficiency of optimization algorithms enables us to find more rapidly and as exactly as possible "optimal" or at least "best-possible" designs for all kinds of structures and systems (Eschenauer, 1992, Yoshimura and Takeuchi, 1994).

The typical motivations for the application of modular MDO-concepts in engineering processes can be divided into the following categories, including corresponding applications in industrial practice:

1. *Quality improvement and cost reduction*

   Examples:

   - Saving of weight and energy
     (Air-/spacecraft, automotive industry, rolling-mill industry)
   - Improvement of accuracies
     (Telescopes, rolling mills, robotics)
   - Larger reliability of constructions
     (Turbosystems, high voltage plants, vessels, apparatus)
   - Improvement of comfort conditions; reduction of sound radiation
     (Transportation and traffic systems, tool machines)

2. *Reduction of development time (Rapid Prototyping) for new products - Tool for decision-makings*

   Examples:

   - Selection of advanced materials
     → Stochastic and discrete optimization
   - Finding the best-possible initial designs
     → Topology optimization, expert systems
   - Linking of product and process optimization
     → Dynamic optimization
     → Design of Experiments
   - Increase of efficiency by MDO
     → Establishing parallel clusters for distributed computing

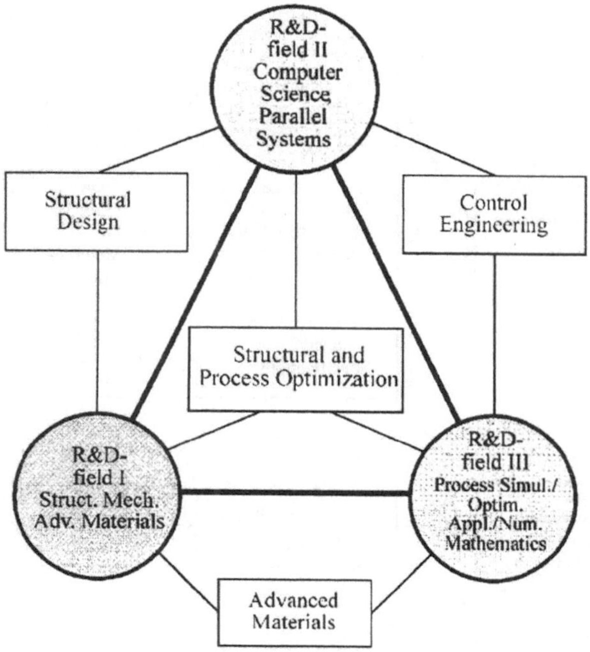


**Figure 1.1** MDA-triangle for Structural and Process Optimization.

One of the CE's goals is to include the various disciplines to a higher degree in the concept phase in order to extend the expertise available and to provide further freedom of decision, where reliability and robustness of decisions with respect to parameter variability, noise, etc. must be a central issue. If the design is influenced by stochastic quantities, the basic structural optimization problem with random parameters must be replaced by an appropriate deterministic substitute problem involving failure, survival probabilities and/or expected cost, losses taking into account random parameter variability. Thus, it is necessary that a large number of experts work in parallel in order to establish a clear physical/technical as well as decision theoretical basis for subsequent decisions. As an example, Fig. 1.1 shows the MultiDisciplinary Analysis (MDA)-triangle for Structural and Process Optimization (Eschenauer, 1998).

## 1.2 The Technology-Oriented Product Development (TOP)-Methodology

After general contemplation concerning the design process in a systems sense, the so-called Technology-Oriented Product Development TOP and the principles of optimization procedures are presented.

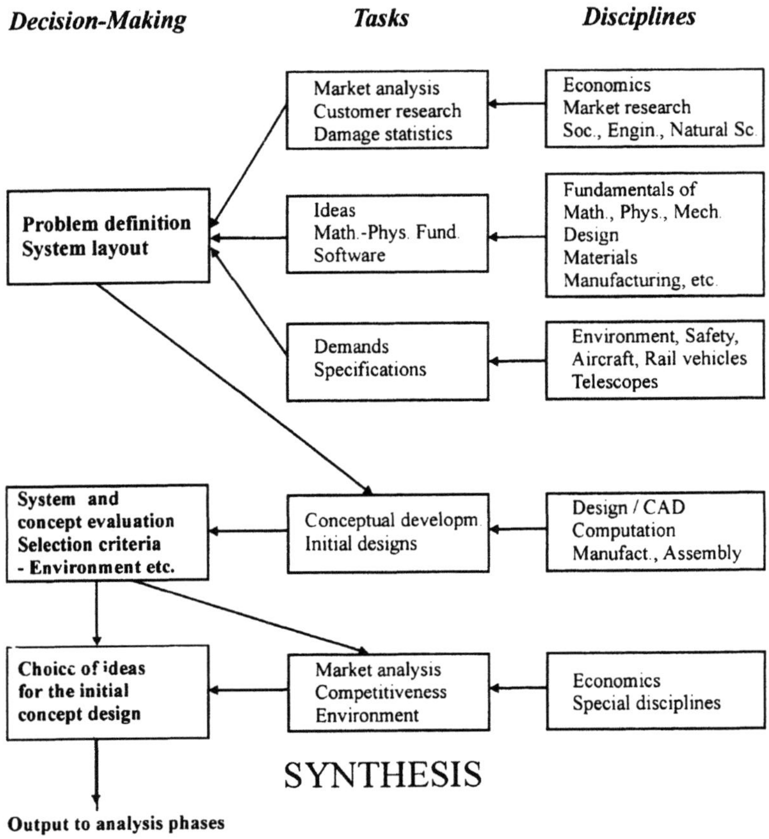


**Figure 1.2** Decision-making, tasks, and disciplines at the beginning of the TOP-Synthesis.

The TOP-process for developing new machines, products, plants, etc. is subdivided into a Synthesis and an Analysis phase with different decision-making levels. First, the synthesis consists of the problem definition and system planning, the concept and system evaluation, and rough initial concepts designs (Fig. 1.2). Already in this early stage of the design process, optimization tools should be applied in order to obtain information on the global and local behaviour of the new product or system (Papalambros and Chirehdust, 1990).

Fig. 1.3 shows the single phases in the TOP-Analysis, where each phase can be presented as a control loop under a systems engineering point of view, at the end of which decision criteria for the designer have to be established for the successive phase.

Each phase is a control process under a systems engineering point of view, at the end of which decision criteria for the designer have to be established for the successive phase. In the case of the project definition phase these are the system specifications for the development phase.

**Structural Model:**
*Load cases*
*Environmental condition*
**Optimization Model:**
*Objectives* $f(\boldsymbol{x})$
- *Primary* $f_1(\boldsymbol{x})$ *Cost*
- *Secondary* $f_2(\boldsymbol{x})$ *Accuracy*

*Constraints* $\boldsymbol{g}(\boldsymbol{x}) \leq 0$
- *Deformations*
- *Stresses*
- *Poitning error*
- *Eigenfrequencies*
- *Eigenmodes*
- *etc.*

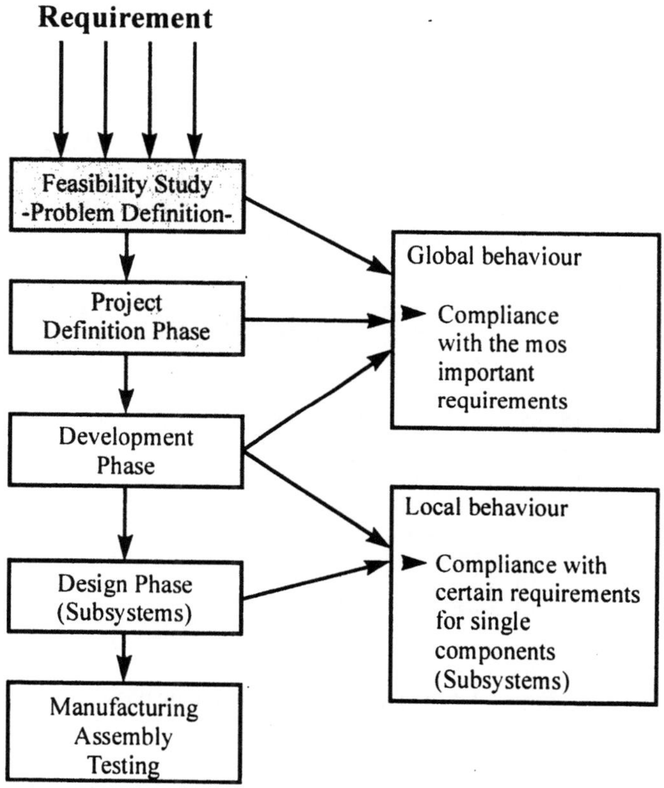


**Figure 1.3** Single phases in the TOP-Analysis.

Fig. 1.4 shows, as an example, a loop for the project definition phase with the most important steps:

(1a) Requirements, conditions → preliminary specifications
(1b) Non-technical criteria (costs, time, schedule)
(2) Establishing design alternatives (concept variants)
(3) Structural modeling → structural analysis
Optimization modeling → structural optimization
Search for appropriate computation methods
(4a) Collection and evaluation of all other project studies - system analysis
(4b) Comparison of nominal and actual status with the requirements
(5) Selection of the "best-possible" design - decision-making process
(6) Establishing the final system specification - start "development phase"

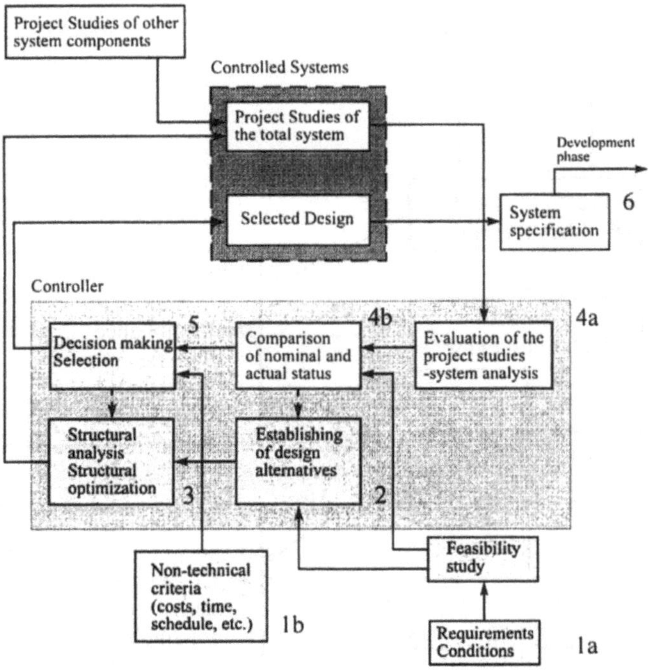


**Figure 1.4** Loop of the project definition phase

The process within the single loops can be made considerable more efficient by using optimization procedures in the stages 2), 3), 4) as it will be shown later in the applications. Figs. 1.5a,b) present the influences on the design features and the design costs according to VDI 2235 (1987). The two curves in Fig. 1.5a) demonstrate the quantitative tendencies on how the knowledge about design increases in the course of the design process, while the freedom of design becomes smaller and smaller. Similarly, characteristics for the possibility of influencing the costs and the specific estimation of costs in the different phases are presented in Fig. 1.5b). The "best-possible" or "optimal" design alternative applies to that structure which corresponds to the designer's desired concept (Eschenauer et al., 1991).

The solution of optimization problems requires software systems which are easy to use, provide sufficient efficiency, and are available for practical application. Several optimization algorithms should be linked to structural analysis procedures in a suitable manner by means of optimization model processors.

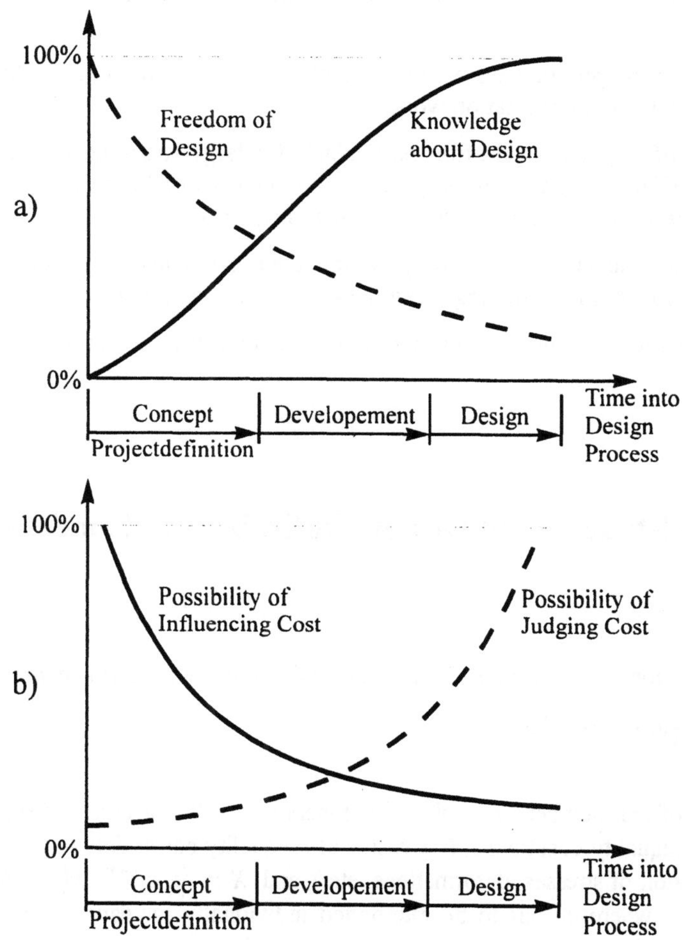


**Figure 1. 5** Influences on (a) design features and (b) design costs in the course of a design process (VDI-Guideline 2235, 1987).

In general, a software system should meet the following requirements:

- Use of different methods for structural analysis such as finite and analytical methods; possibility of selecting the suitable optimization algorithm for an optimization problem from a number of efficient methods,
- application of automatic design and evaluation models (pre- and post-processors) for a wide range of standard problems in optimization modularis; simple integration of special optimization models when required,
- modularised architecture with standard interfaces between the different program modules,
- extension of the program system by integrating additional modules without the need for comprehensive implementation work,
- reduction of the numerical effort particularly for large structural optimization tasks by applying efficient algorithms (e.g., sensitivity analysis of FE-structures, solution methods for linear and nonlinear systems of equations, etc.),
- efficient data management for large optimization problems (large-scale systems), utilization of modern programming techniques (parallel computing),
- support facilities for input data generation and output documentation.

## 1.3 On the Structure of Design Optimization Concepts

### 1.3.1 Definitions

A Scalar Optimization Problem (SOP) is generally defined by the following expression:

$$\underset{x \in \Re^n}{Min}\{f(x) | h(x)=0\,;\, g(x) \le 0\} \tag{1.1}$$

with $\Re^n$ the set of real numbers, $f$ an objective function, $\boldsymbol{x} \in \Re^n$ a vector of $n$ design variables, $\boldsymbol{g}$ a vector of $p$ inequality constraints, $\boldsymbol{h}$ a vector of $q$ equality constraints (e.g., system equations for the determination of stresses, deformations, etc.), and $X := \{\boldsymbol{x} \in \Re^n : \boldsymbol{h}(\boldsymbol{x})=\boldsymbol{0}\,;\, \boldsymbol{g}(\boldsymbol{x}) \le \boldsymbol{0}\}$ the "feasible domain", where $\le$ has to be interpreted individually for each component. The main problem of structural optimization is that the objective function(s) and constraints are in most cases nonlinear functions of the design variable vector $\boldsymbol{x} \in \Re^n$.

In practice, a so-called Vector or MultiCriteria Optimization problem (MCO problem) must be solved. Such optimization problems with multiple criteria functions are often signified by an objective conflict; i.e., none of the possible solutions permits a simultaneous optimal fulfilment of all objectives, and the individual solutions of the single objective functions are different (Stadler, 1988, Eschenauer et al., 1990).

Mathematically, this problem reads as follows:

$$"\underset{x \in \Re^n}{Min}" \{f(x): h(x)=0\,; g(x) \leq 0\}. \tag{1.2}$$

Apart from the objective function vector $f(x)$, all symbols are the same as in (1). In addition to the usual objective weight W, there are other problem-dependent goals that are of great importance nowadays. These are:

- $f_1(x)$ costs (manufacturing, material, assembly, etc.)
- $f_2(x)$ structural requirements (deformations, stresses, stability, eigenfrequencies, dynamic responses, aeroelastic efficiency, flutter speeds, etc.)
- $f_3(x)$ surface accuracy (optics, astronomy, material manufacturing, etc.)
- $f_4(x)$ quality loss function (environmental conditions, imperfections, etc.)

A vector $x \in \Re^n$ is then and only then called functional-efficient or Pareto optimal or p-efficient for the problem (1.2) if no vector $x \in \Re$ exists with the characteristics

$$f_j(x) \leq f_j(x^*) \qquad \text{for all } j \in \{1,...m\} \tag{1.3}$$

$$f_j(x) < f_j(x^*) \qquad \text{for at least one } j \in \{1,...m\}.$$

For all non-Pareto-optimal vectors, the value of at least one objective function $f_j$ can be reduced without simultaneously increasing the functional values of the remaining components. Solutions of nonlinear multicriteria optimization problems can be obtained by transformations into Scalar Optimization problems by means of a substitute problem of the form

$$\min_{x \in X} p[f(x)] \tag{1.4}$$

if there exists an $\tilde{x} \in X^*$ with the characteristic

$$p[f(\tilde{x})] = \min_{x \in X} p[f(x)]\ . \tag{1.5}$$

The function p is called preference function or substitute objective function. It is important to prove that the solutions $\tilde{x}$ of all substitute problems are functional-efficient with regard to X and to the set of objective functions $f_1,...,f_m$ i.e., a point $y = f(x)$ really lies on the efficient boundary $\delta Y^*$.

A very efficient method is the trade-off method (constraint-oriented transformation):

$$p = f_1;\ g_j \leq f_j;\ j = 2,...,m \tag{1.6}$$

with

$$p = f(x) \quad x \in X \qquad \text{Primary objective;}$$

$$g_j \leq f_j;\ (j = 2,...,m) \quad \text{Secondary objectives.}$$

More details of the MCO-methods are described in Stadler (1988) and Eschenauer et al. (1990).

### 1.3.2 The "Three-Columns-Concept"

In structural optimization, it is essential to note that the application of optimization theories in the design process depends on the theoretical aspects of the technical problems. When dealing with the structural optimization problem, it is recommendable to proceed according to the "Three-Columns-Concept" consisting of the columns Structural Model, Optimization Algorithms, and Optimization Model (Fig. 1.6) (Eschenauer, 1989).

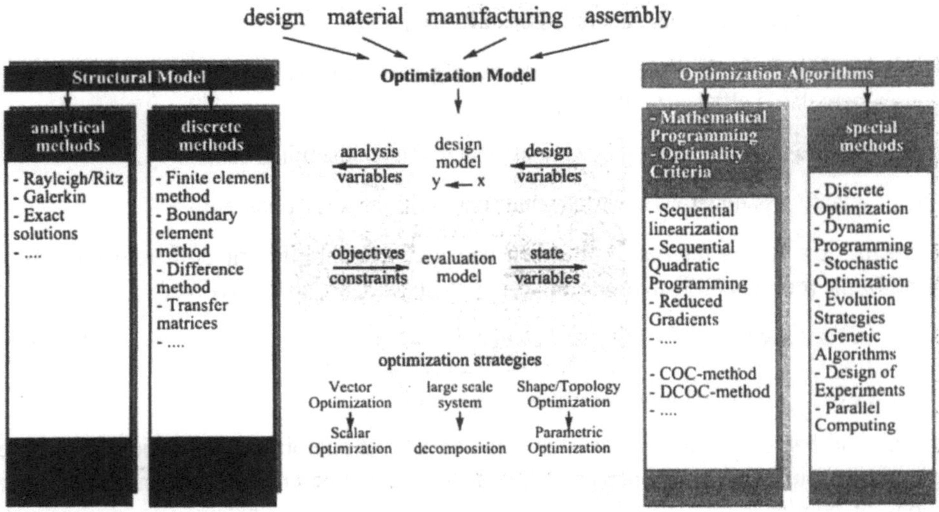


**Figure 1.6** Three-Columns-Concept for solving optimization problems.

**Column 1: Structural model**

Any structural optimization requires the mathematical definition of the physical behaviour of the structure. In the case of mechanical systems, this refers to the typical structural response subject to static and dynamic loading such as deformations, stresses, eigenvalues, etc. Furthermore, information on the stability behaviour (buckling loads) has to be determined. All state variables required for the objective function and constraints have to be provided. The structural calculation is carried out using efficient analysis procedures such as the Finite Element Method (FEM) or Transfer Matrices Methods (TTM). In order to ensure a wide field of application, it should be possible to adapt several structural analysis methods.

**Column 2: Optimization model**

When treating structural optimization problems, the most important task is the unambiguous linkage of structural model and optimization algorithms. The generation of the so-called optimization model requires creativity, and the designer also has to be familiar with the structural and process behaviour of the component or process to be optimized. For that purpose, the objective function as well as the structural requirements must be defined and then formulated mathematically as objective functions and constraints. If several objectives exist, the problem must be solved by means of an optimization strategy (in the present case by a Multiobjective Optimization). Furthermore, it is important to choose suitable design variables in order to achieve a sensible design modification, a fact that also requires the designer to be familiar with the component behaviour. In correspondence with the above-mentioned items, the optimization model is subdivided in design model, evaluation model, optimization strategy and sensitivity analysis.

**Column 3: Optimization algorithms**

The optimization algorithms solve the actual minimization problem. Since the convergence of a certain algorithm is not secured, a library of Mathematical Programming (MP)-algorithms is supplied. This library contains, among others, an SLP-algorithm (Eschenauer et al., 1990, Kneppe, 1986, Vanderplaats et al., 1990), a GRG-algorithm (Abadie and Carpentier, 1969, Bremicker, 1989), a hybrid SQP-GRG-algorithm (Bremicker, 1989, Parkinson and Wilson, 1986, Powell, 1982), and the Method of Moving Asymptotes MMA (Svanberg, 1987). For special problems, a single algorithm or a combination of algorithms can be chosen. Besides that, Design-of-Experiments (DOE)-methods, Response Surface and Robust Design methods are integrated. Furthermore, tests with genetic algorithms have been carried out (Schoofs et al.,1992).

On the basis of the "Three-Columns-Concept" and on the requirements mentioned before, the software system SAPOP (Structural Analysis Program and Optimization Procedure) due to the optimization loop (Fig. 1.7) was developed (Bremicker and Eschenauer, 1989, Eschenauer et al., 1993).

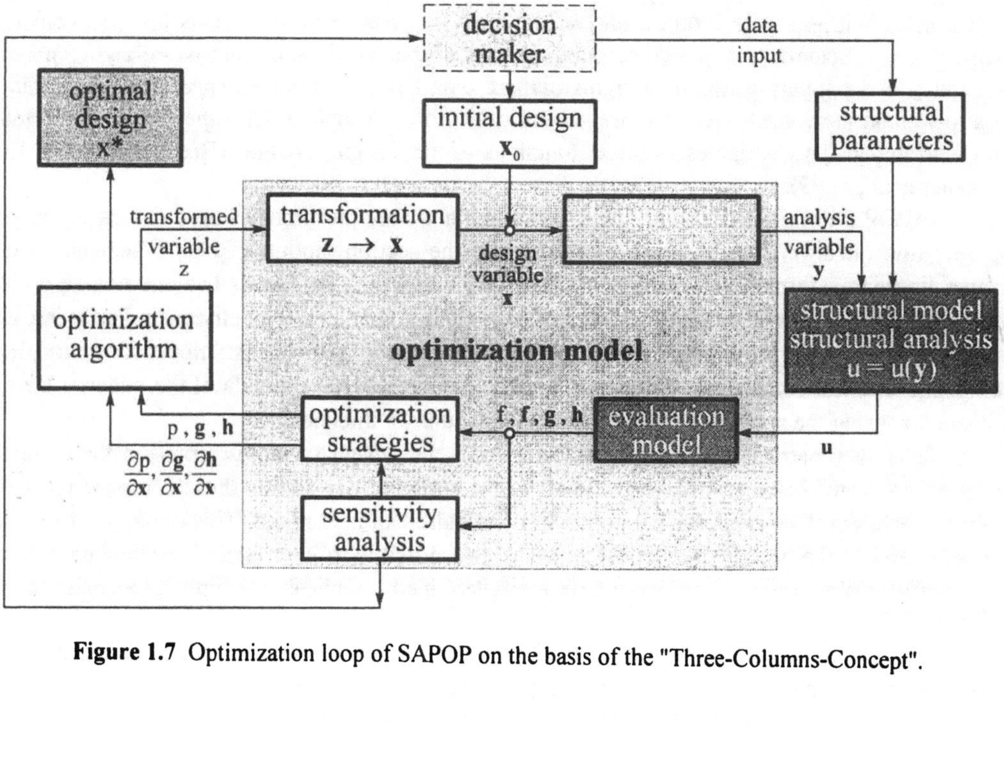


**Figure 1.7** Optimization loop of SAPOP on the basis of the "Three-Columns-Concept".

### 1.3.3 Models of the Optimization Loop

- *Design Model*

The task of a design model is to calculate the analysis variables $y$ from the design variables $x$ by using a unique mapping rule:

$$y = f(x) \quad , \quad x \in \Re^{n} \quad , \quad y \in \Re^{n_y} \ . \tag{1.7}$$

The analysis variables are a subset of the structural parameters which are required to describe the physical behaviour. In structural optimization problems, the analysis variables are usually sizing quantities (e.g., thicknesses, cross-sections, moments of inertia, etc.), geometrical dimensions, shape parameters or material quantities. In particular when using discrete structural analysis methods (FE-methods, finite differences methods, etc.), it is important that all structural elements as well as the node topology are determined by the design model. One part of the design modeling can be carried out by linear mapping

$$y(x) = Ax + y_0 \ , \tag{1.8}$$

where $A$ is the coordination matrix and $y_0 \in \Re^{n_y}$ is a constant vector. Due to a special structure of the coordination matrix, various design models can be realized. Variable linking is achieved if just one element in each row has a value "one", and if all other elements are zero. The columns of the matrix, however, contain several "one"-elements, Each analysis variable corresponds to just one design variable, whereas a design variable can be allocated to several analysis variables. The addition of a constant element to a design variable allows to consider a constant part of an analysis variable. With variable fixing a part of the analysis variable vector equals just this constant element, i.e., all elements of the corresponding rows of the coordination matrix are zero. Another possibility is the superposition of design variables to an analysis variable, where several elements of a row attain the value "one". A more general case of this design model is the linear transformation with the analysis variables as linear functions of the design variables (Bremicker, 1989, Eschenauer et al., 1993).

In SAPOP a design model with variable linking and variable fixing is used for cross-section optimization (sizing). The moments of inertia and the section moduli of beam elements result from a linear transformation whereby each row of the coordination matrix has one non-zero element differing by a corresponding similarity factor. This guarantees geometrical similarity in case of a variation of the cross-sections of a beam. The formulation of the design model for optimizing composite material components refers to the same principle. Here, the data of the material law is obtained by using the analysis variables *fibre angle* and layer thickness.

As far as geometrical optimization is concerned, the design variables represent the components of the vector between two given points of the structure. By adding these components –and the respective constant elements, all points of a structure can be defined. This corresponds to the design model of the additive superposition. It is also possible to interpolate intermediate points. The coefficients of the coordination matrix are determined by means of simple geometrical considerations.

For shape optimization problems, SAPOP provides various shape functions to define the surface shape of a component. They allow the transformation into a parametric optimization problem (Fleury and Schmit, 1980, Kneppe, 1986, Bremicker, 1989). Among others, the following approach functions are used: Lagrange polynomials, Bèzier curves and Bèzier surfaces, B-spline curves and surfaces, modified ellipse functions, etc. Control points are employed to determine the free coefficients of most of the shape functions. The coordinates of these control points are determined from the design variables by using the design model for geometrical optimization as described above (Kneppe, 1986).

- *Evaluation Model*

The expression *state variables* $\boldsymbol{u}$ of a structural mechanical system refers to quantities such as deformations, stresses, resulting forces, strains, eigenvalues, etc. These state variables depend on the design variables by way of the structural model. The task of the evaluation model is to formulate the objective function vector $\boldsymbol{f}$ and the constraints $\boldsymbol{g}$ as functions of these state variables.

Generally, the constraints represent limiting demands on the state variables; i.e., the quantities of a state variable $u_i$ should be bounded by an upper and lower limit. This demand provides two constraints which are normalized for numerical reasons:

$$u_i^l \le u_i \le u_i^u \quad \rightarrow \quad g_1 = \frac{u_i}{u_i^u} - 1 \le 0 \quad , \quad g_2 = 1 - \frac{u_i}{u_i^l} \le 0 \ . \tag{1.9}$$

The limiting values $u_i^u$ and $u_i^l$ can also depend on the design variables. According to definition (1.8), SAPOP calculates the deformation and eigenvalue constraints. The stress constraints are maximum bounds whose formulation requires stress criteria. Buckling constraints can be employed in order to optimize trusses. For large structures it is advantageous to reduce the number of constraints. For this purpose, SAPOP provides a "constraint linking strategy" which determines the most critical constraint value of a certain constraint type in a defined domain. Apart from the constraints already mentioned, upper and lower bounds of the design variables are considered.

- *Sensitivity Analysis*

The objective of design sensitivity analysis is to calculate gradients of the structural resonses and cost functions with respect to small changes of the design variables. The determination of the gradients of the objective function and the constraints is a highly important step in the optimization process, since these gradients are not only a prerequisite for the majority of optimization algorithms, but they also provide important information on the structural sensitivity when changing arbitrary structural parameters. The choice of an appropriate method of sensitivity analysis strongly influences the numerical efficiency and has thus impact on the entire course of the optimization. For this reason, the treatment of the fundamentals of structural optimization shall be dealt with separately with some remarks on frequently applied techniques for determining the gradients.

An overview of sensitivity methods in structural optimization is given in, among others, Adelmann and Haftka (1984), Haug and Arora (1978), Haug et al. (1986).

Introducing the state variables $\boldsymbol{u} \in \Re^{n_m}$ the vector of objective functions and/or inequality constraints can be written as follows:

$$\boldsymbol{f} = \boldsymbol{f}(\boldsymbol{x},\boldsymbol{u}) \quad \text{and} \quad \boldsymbol{g} = \boldsymbol{g}(\boldsymbol{u},\boldsymbol{x}) \quad \text{with} \quad \boldsymbol{u} = \boldsymbol{u}(\boldsymbol{x}) \; . \tag{1.10}$$

The following explanations concentrate on the objective functions $\boldsymbol{f}$ and can similarly be applied to the constraints. At first, the total differentiation yields:

$$d\boldsymbol{f}(\boldsymbol{x},\boldsymbol{u}) = \frac{\partial \boldsymbol{f}}{\partial \boldsymbol{x}} d\boldsymbol{x} + \frac{\partial \boldsymbol{f}}{\partial \boldsymbol{u}} d\boldsymbol{u} = \left( \frac{\partial \boldsymbol{f}}{\partial \boldsymbol{x}} + \frac{\partial \boldsymbol{f}}{\partial \boldsymbol{u}} \frac{\partial \boldsymbol{u}}{\partial \boldsymbol{x}} \right) d\boldsymbol{x} \tag{1.11}$$

$$\rightarrow d\boldsymbol{f} = \boldsymbol{A}_f \, d\boldsymbol{x} \; , \tag{1.12}$$

where $\boldsymbol{A}_f = \left[ a_{f_{ij}} \right]_{m \times n}$ is called the sensitivity matrix with

$$\frac{\partial \boldsymbol{f}}{\partial \boldsymbol{x}} = \left[ \frac{\partial f_i}{\partial x_j} \right]_{m \times n} , \quad \frac{\partial \boldsymbol{f}}{\partial \boldsymbol{u}} = \left[ \frac{\partial f_i}{\partial u_i} \right]_{m \times n_u} , \quad \frac{\partial \boldsymbol{u}}{\partial \boldsymbol{x}} = \left[ \frac{\partial u_h}{\partial x_j} \right]_{n_u \times n} . \tag{1.13}$$

In the sequel some of the methods for determining $\boldsymbol{A}_f$ are briefly described:

*a) Numerical sensitivity by Overall Finite Differences (OFD)*

The Overall Finite Difference (OFD)-approach implies that the entries in the sensitivity matrices are approximated by simple finite difference quotients:

$$a_{f_{ij}} \approx \frac{f_j[\tilde{x}_i, \, \boldsymbol{u}(\tilde{x}_i)] - f_j[x_i, \, \boldsymbol{u}(x_i)]}{\varepsilon \, x_i} \tag{1.14}$$

with $\quad \tilde{\boldsymbol{x}}_i = \boldsymbol{x} + \varepsilon \, x_i \, \boldsymbol{e}_i$

$\varepsilon$ step size factor ,

$\boldsymbol{e}_i$ $i$-th unit vector .

The OFD-method can be quite easily implemented and is completely independent of the structural model. The method is also applicable for tasks beyond the field of optimization. However, the OFD-analysis has the great disadvantage that for $n$ design variables $n+1$ complete structural analyses of the total structural response are required, a fact that leads to very extensive computation times in case of larger optimization problems. In addition, the occurrence of round-off errors does not allow the relative perturbation $\varepsilon = \Delta x_j / x_j$ to be chosen arbitrarily small. Based on experience, values of $\varepsilon \approx 10^{-5}$ to $10^{-3}$ are suitable.

*b) Analytical sensitivity analysis*

In order to reduce the extensive computational effort, analytical methods are increasingly employed for gradient calculation. The procedures are closely linked with the applied structural analysis procedure (cf. Olhoff and Lund, 1995), and their realization thus renders manipulations in the source code necessary. These methods have originally been developed for the FE-methods, but they can be generalized to all other structural analysis procedures which transform the differential equations of the considered mechanical system into a set of algebraic equations (transfer matrix, difference procedures, analytical solution methods, etc., see Eschenauer et al. (1990)). Here, our considerations will be limited to those linear systems that have a system matrix equation of the form:

$$\boldsymbol{F}\boldsymbol{u} = \boldsymbol{r} \quad \text{with} \quad \boldsymbol{F} = \boldsymbol{F}(x) \quad \text{and} \quad \boldsymbol{r} = \boldsymbol{r}(x)\,. \tag{1.15}$$

In (1.15) $\boldsymbol{r}$ represents the load vector and $\boldsymbol{F}$ the system matrix.

By implicit differentiation of (1.15) with respect to any of the design variables $x_j$ ( j = 1,...,n), rearrangement of terms, and multiplying the equation by $\partial f / \partial \boldsymbol{u}$, one can replace the derivatives of the state quantities in (1.11) by the following expression:

$$\frac{\partial f}{\partial \boldsymbol{u}}\frac{\partial \boldsymbol{u}}{\partial x} = \frac{\partial f}{\partial \boldsymbol{u}}\boldsymbol{F}^{-1}\underbrace{\left(\frac{\partial \boldsymbol{r}}{\partial x} - \frac{\partial \boldsymbol{F}}{\partial x}\boldsymbol{u}\right)}_{\boldsymbol{p}}\,, \tag{1.16}$$

where $\boldsymbol{p}$ is the so-called *pseudo load vector* associated with the design variable $x_j$.

With $\boldsymbol{r}$ known and $\boldsymbol{u}$ obtained by solution of (1.15), computation of the pseudo load vectors $\boldsymbol{p}_j$ ($j$ = 1,..., $n$) in (1.16) only requires that the design sensitivities $\partial \boldsymbol{r} / \partial x_i$ and $\partial \boldsymbol{F} / \partial x_j$ of the load and the system matrix are known.

Eq. (1.16) can be treated in two different ways:

- *Design Space Method as a direct method*

First, we compute the pseudo load vector $\boldsymbol{p}_j$ associated with each $x_j$ as described above, and then determine from (1.15)

$$\boldsymbol{F}\frac{\partial \boldsymbol{u}}{\partial x} = \left(\frac{\partial \boldsymbol{r}}{\partial x_j} - \frac{\partial \boldsymbol{F}}{\partial x_j}\boldsymbol{u}\right) = \boldsymbol{p}_j \tag{1.17}$$

the corresponding gradients of the state quantities $\partial \boldsymbol{u} / \partial x_j$. Since the form of (1.17) is analogous to that of (1.15), each of these gradients ($j$ = 1,..., $n$) can be solved from (1.17) using the same factorisation of the global system matrix $\boldsymbol{F}$ as was used when solving (1.15) for the vector $\boldsymbol{u}$. Thus, (1.17) has to be solved for $n$ right hand sides $\boldsymbol{p}_j$ for each load case $\boldsymbol{r}$ in (1.15).

Finally, the gradients $\partial f / \partial \boldsymbol{u}$ and $\partial f / \partial x$ are calculated in order to establish the total differential according to (1.11).

- *State Space Method as an auxiliary variable method*

Here, we introduce an auxiliary variable vector $\lambda_i$ the transpose of which we define as the following product according to (1.16):

$$\lambda_i^T = \frac{\partial f}{\partial \boldsymbol{u}} \boldsymbol{F}^{-1} \quad (i = 1,\ldots,m) \ . \tag{1.18}$$

This yields the equation system

$$\boldsymbol{F}^T \boldsymbol{\lambda} = \left(\frac{\partial f}{\partial \boldsymbol{u}}\right)^T \tag{1.19}$$

for $\boldsymbol{\lambda}$ that is to be solved for the $m$ right hand sides $\frac{\partial f}{\partial \boldsymbol{u}} \quad (i = 1,\ldots,m)$.

Generally, analytical and semi-analytical methods of sensitivity analysis are able to reduce the computational time to a mere fraction of that necessary for the OFD-approach. This is mainly due to the fact that the system matrix of the former methods has to be factorised or inverted only once for a sensitivity analysis.

Depending on the number of equation systems to be solved in (1.17) and (1.19), the direct method is preferred in those cases, where fewer design variables than constraints are defined in the optimization model, whereas the auxiliary variable method should be applied to problems with a prevailing number of design variables. For further information on this topic, see Haug and Arora (1978) (*Active Set Strategy*).

- *Semianalytical methods*

The pseudo-load vector $\boldsymbol{p}$ necessary for the analytical procedures is formed from the derivatives of the system matrix $\boldsymbol{F}$ and the load vector $\boldsymbol{r}$. In several sensitivity analysis applications size optimization of FE-structures with element matrices which are simple polynomial functions of just one design variable is carried out, i.e., the derivative of these matrices can be carried out analytically on the element level.

For more general problems such as shape optimization problems it is difficult to establish a general analytical method. Therefore, it is suggested that the problem should be approached semianalytically which means that the necessary derivatives of the submatrices in (1.14) are determined by a numerical differentiation as in (1.14), e.g.,

$$\frac{\partial \boldsymbol{F}}{\partial x_i} \approx \sum_{j=1}^{n_e} \frac{\boldsymbol{F}^j(\tilde{\boldsymbol{x}}_i) - \boldsymbol{F}^j(\boldsymbol{x})}{\varepsilon x_i} \tag{1.20}$$

with $\tilde{\boldsymbol{x}}_i = \boldsymbol{x} + \varepsilon\, x_i\, \boldsymbol{e}_i$ .

- *Emerging concepts as optimization strategies*

a) Decomposition strategies

The optimization of large-scale systems or structures leads to optimization problems with a large number of design variables and constraints in combination with very complex system analyses.

The complexity of the integrated solution of those large scale problems is increased by several difficulties. The convergence rate of generally applied MP-algorithms decreases with the number of design variables. In combination with time-consuming system analyses which often execute during an optimization run, this leads to extensive computation times. The complexity is even increased in the case of multidisciplinary optimization problems due to the organizational effort in carrying out integrated system analyses (Sobieszczanski-Sobieski, 1988, 1989, Bloebaum et al., 1990, Bremicker and Eschenauer, 1989, Eschenauer and Weinert, 1993, 1994). According to the decomposition of the total system S into $n_S$ subsystems, the corresponding variable and function vectors are divided into subvectors, and the equations are decomposed into blocks of equations:

$$\begin{aligned} \boldsymbol{u} &\Rightarrow \left[\boldsymbol{u}^1, \boldsymbol{u}^2, \ldots, \boldsymbol{u}^{n_S}\right], & \boldsymbol{g} &\Rightarrow \left[\boldsymbol{g}^1, \boldsymbol{g}^2, \ldots, \boldsymbol{g}^{n_S}\right], \\ \boldsymbol{u}^i &\Rightarrow \left[\boldsymbol{u}_I^i, \boldsymbol{u}_B^{i(1)}, \boldsymbol{u}_B^{i(2)}, \ldots, \boldsymbol{u}_I^{i(m)}, \ldots, \boldsymbol{u}_B^{i(n_S)}\right] & \boldsymbol{\Phi}(\boldsymbol{x}, \boldsymbol{u}) &\Rightarrow \boldsymbol{\Phi}^i\left(\boldsymbol{x}^i, \boldsymbol{u}_B^i, \boldsymbol{u}_I^i\right) \ \forall\ i\ , \\ \boldsymbol{x} &\Rightarrow \left[\boldsymbol{x}^1, \boldsymbol{x}^2, \ldots, \boldsymbol{x}^{n_S}\right], & \boldsymbol{f} &\Rightarrow \left[\boldsymbol{f}^1, \boldsymbol{f}^2, \ldots, \boldsymbol{f}^{n_S}\right]. \end{aligned} \tag{1.21}$$

In (1.21) the subvectors $\boldsymbol{u}_I^i$ are the internal state variables of the subsystem $S^i$, and $\boldsymbol{u}_B^{i(m)}$ are those boundary state variables of the subsystem $S^i$ which describe the coupling with the subsystem $S^m$. In order to ensure compatibility of a solution found by any decomposition strategy, the condition

$$\boldsymbol{u}_B^{i(m)} = \boldsymbol{u}_B^{m(i)} \tag{1.22}$$

has to be fulfilled. Global design variables which are used for the design of more than one subsystem can easily be considered in the proposed decomposition strategy. They are defined redundantly in the subvectors of the respective subsystems. Global functions (computable on the main system level only) in the definition of the objectives or constraints can be considered as functions of an imaginary subsystem without local design variables. They are considered as external functions in all real subsystems. Fig. 1.8 shows the decomposition for two subsystems.

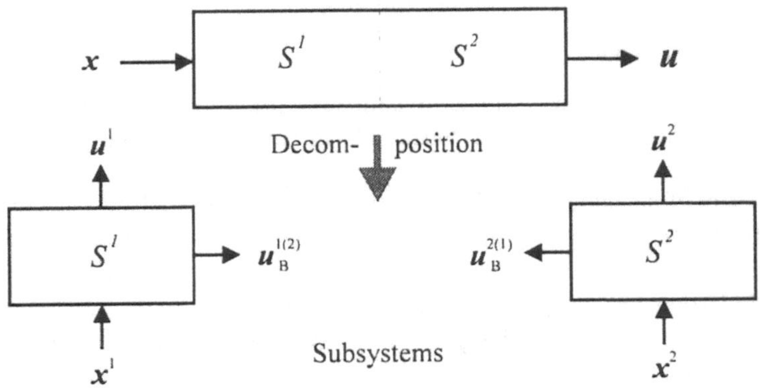


**Figure 1.8** Decomposition of a system $(n_S = 2)$ .

When the quantities decomposed according to (1.21) are inserted into (1.1), we formally obtain a suboptimization problem of the form

$$f^i(x^{i*}, u^i) = \min_{x^i} \left\{ f^i(x^i, u^i) \middle| x^i \in X^i \wedge \Phi^i(x^i, u^i) = 0 \right\}$$

$$\text{with} \quad X^i = \left\{ x^i \in \Re^{n^i} \middle| g^i(x^i, u^i) \geq 0 \right\}; \quad x^{il} \leq x^i \leq x^{iu} \tag{1.23}$$

for each subsystem $S^i$ (Fig. 1.8). Since the single subsystems are coupled via boundary state variables of common interfaces, global design variables, global constraints or nonseparable objectives, a coordination strategy has to be employed to obtain subsystem solutions which are optimal and compatible with regard to the integrated system. The subsystem solutions can be nonoptimal with regard to single subsystems.

b) Topology optimization concepts for continuum structures

Optimization of the basic topology or layout of a design with more or less complex structure has a decisive influence on the cost-efficiency of the design. It is of great importance for the development of new products to find the best-possible initial layout for given specification values (objectives, constraints) in a very early stage of the design process (conceptual and project definition phase). Thus, fundamental research was and still is done in order to establish very exact and efficient procedures for solving such problems. Over the last decade, the researchers have been occupied with two different kinds of conceptual processes, i.e., the Material or Micro-structure Technique and the Geometrical or Macro-structure Technique. It is the objective of this lecture to present an overview of these techniques as examples of emerging methods.

The topology of a structure, i.e. the arrangement of material or the positioning of structural elements in the structure, is crucial for its optimality. Traditionally, the topology of a design is in most cases chosen either intuitively or by use of already existing designs (Current-Design-World-

State). In future, however, there is a significant necessity of and interest in improving the quality of products by finding their best possible topology in a very early stage of the design process. Earlier, we have become acquainted with several research activities throughout the world concentrating on this problem, and a number of solution procedures that have been developed. Very roughly one can distinguish between two classes of approaches, the so-called geometrical or Macro-approaches, vs. the material or Micro-approaches. Both can be represented in the following abbreviated manner, and illustrated in Fig. 1.9:

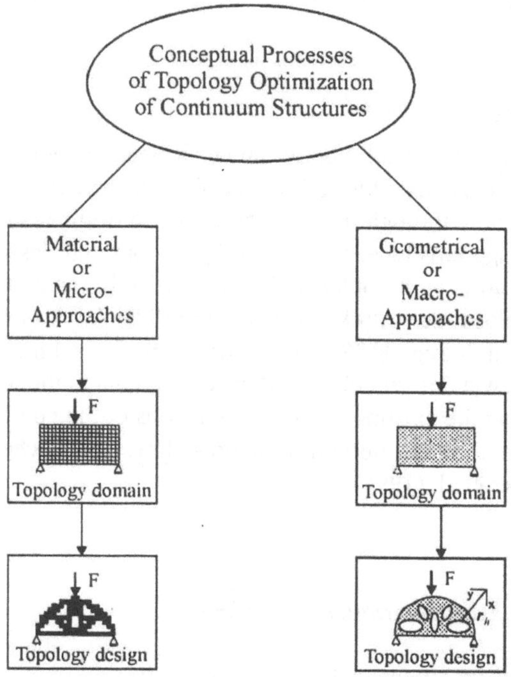


**Figure 1.9** Conceptual Processes of topology optimization.

– *Material or Micro-approaches*

Here, our design objective is to find that structural topology which renders a given design objective on optimum value subject to a prescribed amount of structural material. It is assumed that in solid form the amount of material is less than the amount that would be needed to cover the entire admissible domain for the continuum. Hence, for the initial design it is normally chosen to distribute the material evenly in some porous, microstructural form over the admissible design domain (left-hand column of Fig. 1.9). Based on the works on the homogenization of porous materials, Bourgat (1973), Bendsøe et al. (1993, 1994) have developed the *homogenization method* for topology optimization. In this context, they have referred to research activities carried out by Allaire and Kohn (1993). The homogenization method works without a 0-1-check, since it considers microcells which form the small structural areas of the topology domain. These microcells consists of massive material, and of a no-material area (void) so that a porous material behaviour can be simulated. For the computation, this porous material is then subjected to a homogenization, and thus it is made indistinct. The basic idea of this type of optimization is a variation of the volumetric efficiency (ratio of volume of massive material to the total volume of the microcells in the structural areas). Here, one tries to approach the volumetric efficiency "0" (no material) and "1" (full material). This method represents a further step towards finding topological layouts of mechanical structures.

– *Geometrical or Macro-approaches*

In this class, the topology of a solid body can be changed, among others, by growing or degenerating material or by inserting holes. The first method recognizes that an optimized shape is defined not only through the boundaries of the original shape, but through the entire domain of a design envelope. Consequently, the optimal design is simply a subset of the design envelope and can be obtained by appropriately removing material from the design domain or envelope, see Atrek and Kodali (1989), Atrek (1991), Atrek and Agarval (1992), Rodrigues and Seireg (1988). The secondly mentioned method consists of an iterative positioning of new holes ("bubbles") at specific points in the topology domain. In each iteration, the holes and the existing variable boundaries of the continuous body are simultaneously subjected to a shape optimization procedure, see Eschenauer and Schumacher (1993a,b), Eschenauer et al. (1994).

- *Emerging methods as optimization algorithms*

a) Use of special methods for solving MDO-problems

The "Three-Columns-Concept" according to Fig. 1.6 shows two classes of optimization algorithms, the Mathematical Programming (MP)- or Optimality Criteria (OC)-Methods on the one hand, and special methods like Design-of-Experiments (DOE)-Method, Response Surface (RS)-Method, etc.

The diverse aims of solving MDO-problems can be achieved by employing methods like Design-of-Experiments (DOE)-Method and the Response Surface (RS)-Method for an efficient evaluation of the (expected) behaviour of the basic behavioural, response or state variables in the decision process, followed by Stochastic Optimization (SO)-Procedures for the formulation and numerical solution of an appropriate deterministic substitute problem incorporating random parameter variations, and eventually by Robust Design (RD)-Methods for the final selection of optimal designs under stochastic uncertainty based on cost/quality-oriented substitute problems. Originally, these procedures have been developed for treating problems from different disciplines like pure mathematical statistics, physics, operations research, economics, etc., but they can easily be transferred to optimal design problems.

The Design-of-Experiments (DOE)-Method, a method for planning, designing, conducting and analysing statistically designed experiments, is the application of geometrical principles to statistical sampling to obtain desired results such as minimizing the number of experiments necessary to obtain the answer to a problem, or minimizing the variance of estimated coefficients obtained through regression. The first of the above-mentioned results affects quantities of efforts, and hence (expected) costs. The second affects quality of results. (Expected) loss and quality are the basic elements of value. Thus, DOE is a powerful tool in designing for value (Schoofs, 1987).

The Response Surface Method (RSM) is a collection of mathematical and statistical techniques that are useful for the modeling and analysis of problems in which a response of interest is influenced by several input variables. By DOE-based regression techniques, for the approximation of the functional relationship between input variables and a real structural response variable the coefficients of an approximate multivariate polynomial model of first, second (or higher) order are estimated; here, the estimation effort for the approximative polynomial model of the expected response function may be controlled by using DOE-techniques.

By means of Stochastic Optimization (SO)-Procedures the structural optimization problem under stochastic uncertainty may be converted into an appropriate deterministic substitute problem involving failure, survival probabilities and/or expected costs, losses (e.g., loss of quality). Moreover, numerous procedures are available for the numerical solution of the substitute problems involving multiple integrals, (probabilities, expectations), as, e.g., RSM-based stochastic approximation methods.

Considering in Robust Design (RD), also termed Quality Engineering (Taguchi, 1986), especially a substitute problem involving the minimization of expected loss of a product and/or process, at the same time guaranteeing a high level of quality in time and cost efficient manner, we obtain a cost-effective product the quality features of which are insensitive towards uncontrollable noise factors.

### b) Design-of-Experiments and Response Surface Methodology

As mentioned before, DOE is a statistical approach for solving problems from engineering to social science (Box et al., 1978, Montgomery, 1991). Statistical experimentation allows to estimate factor effects and to develop response surface approximations from the generated data which are frequently used to replace expensive function evaluations in structural optimization, see Barthelemy and Haftka (1993), Roux et al. (1996), Schoofs et al. (1992). An experimental design describes a sequence of experiments, which are mathematically defined by a matrix $\boldsymbol{X}$.

The rows of X denote experimental runs and the columns the specified factor settings for each run. A large variety of experimental designs exists, e.g., Full and Fractional Factorial Designs, Orthogonal Arrays (OA), or Central Composite Designs (CCD), see Box et al. (1978), Montgomery (1991). In a full factorial design, all possible combinations $m^k$ of factor levels are investigated, where k factors are selected at m levels. It is possible to estimate all main and all interaction effects. The number of experiments increases exponentially with the number of factors. To reduce the number of experiments to a more practical level, fractional factorial designs of the form $m^{k-p}$ are introduced, where the generator p describes the fraction of the full design. It is assumed that certain higher order interactions are negligible. OAs are the smallest fractional factorial design to study main factor effects. It is assumed that factor interactions do not exist, therefore main effects can be confounded with two-factor interactions.

CCDs are probably the most widely used experimental designs to fit response surface models. A CCD generally consists of a two-level full or fractional factorial design portion, where the two levels are coded by $-1$ and $+1$, augmented by two star points on the axis of each factor at a distance of $-\alpha$ and $+\alpha$ from the center, and one (or more) center point(s). Response surface models are created to replace the original structural analysis with an efficient, usually second-order polynomial approximation of the form

$$\hat{y} = b_0 + \sum_{i=1}^{k} b_i x_i + \sum_{i=1}^{k} b_{ii} x_i^2 + \sum_{i=1}^{k-1} \sum_{j=i+1}^{k} b_{ij} x_i x_j \ . \tag{1.24}$$

These models include main effects, two-factor interactions and quadratic terms and the unknown coefficients $\boldsymbol{b}$ are estimated by means of regression analysis based on a least-square method, according to

$$\boldsymbol{b} = [X'X]X'y \ . \tag{1.25}$$

c) Robust Design Techniques

Robust design methods, originally developed by Taguchi in the 1980ies, have been widely used in industry to implement parameter design and tolerance design, while seeking solutions that minimize performance variations $(min\,\sigma^2)$ and achieving specified performance targets $(min(\mu - \mu_0))$. The fundamental principle for achieving robustness is to improve the quality of a product/process by minimizing the effects of the causes (noise factors) of variation without eliminating the causes. Robustness is the product's ability to function according to the specifications besides environmental changes and noise. Instead of eliminating or controlling noise, the nonlinear relationship between a quality (performance) characteristic and the control and noise factors are exploited to find control factor settings that minimize the effects of noise factors (Taguchi, 1986).

To implement the robust design approach, the effects of control and noise factors on mean ($\mu_y$) and variance ($\sigma^2$) of a response must be made available. Response Surface Models for each response are created to model performance and performance variation. Two approaches for creating these models are used in this work: (i) the so-called Dual Response Surface Approach (Combined Array), see Meyers et al. (1995), where a single experimental design that includes control and noise factors is used, and (ii) Taguchi's Crossed Array Approach (Phadke, 1989), where the inner control factor array is crossed with an outer noise factor array for each experimental run. For the dual response surface approach, the design is augmented to include noise factors z and a formal regression model $\hat{y} = f(x, z)$ is postulated. Eq. (1.24) is extended to include noise factors (underlined quantities)

$$\hat{y} = b_0 + \sum_{i=1}^{k} b_i x_i + \sum_{i=1}^{k} b_{ii} x_i^2 + \sum_{i=1}^{k-1} \sum_{j=1+1}^{k} b_{ij} x_i x_j + \underline{\sum_{j=1}^{m} c_i z_i + \sum_{i=1}^{k} \sum_{j=1}^{m} c_{ij} x_i z_j} \ , \tag{1.26}$$

where no quadratic terms are required for noise factors. The exploration of the interactions $x_i z_i$ forms the core of robust design. This model allows to analytically derive individual mean $(\mu_{\hat{y}})$ and variance $(\sigma_{\hat{y}}^2)$ models (expected values, $E$) by means of first order Taylor series expansion according to

$$E[\hat{y}(x,z)] = \mu_{\hat{y}} \approx f(x,\mu_z), E[\sigma^2[\hat{y}(x,z)]] = \sigma_{\hat{y}}^2 \approx \sum_{i=1}^{m} \left( \frac{\partial \hat{y}}{\partial z_i} \right)^2_{|\mu_i} \sigma_{z_i}^2 \, \mathrm{n}, \tag{1.27}$$

where $\sigma_{z_i}$ presents the standard deviation of a noise factor. It is assumed for this approach that all noise factors vary approximately normally. In the cross-array approach, mean and variance for each response are calculated for the outer array experiments of each run of the inner array using standard statistics. Response surface models are then derived, where the responses $\hat{y} = f(x)$ only depend on $x$.

For the solution of optimization problems in structural design, the "Three-Columns-Concept" has been applied successfully in many cases. The robust design approach is implemented into the optimization loop (Fig. 1.10), see Lautenschlager et al. (1999). Experiments are performed for model screening, i.e., using low-order experiments to identify important factors, and for creating approximation models. Mean and variance models are developed for robust design applications, which then enter the optimization loop.

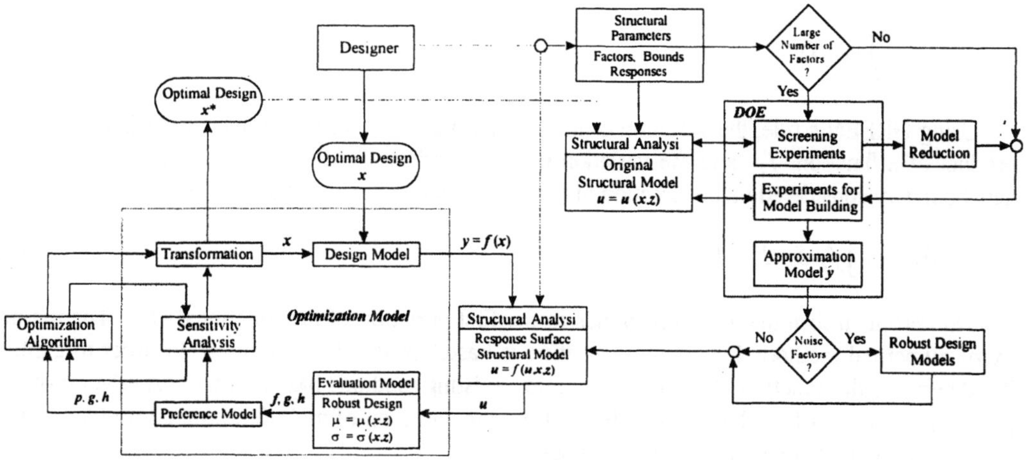


**Figure 1.10** Optimization loop of robust MDO-techniques.

## 1.4 Use of Emerging Procedures for Solving MDO-Problems

### 1.4.1 Application of the Bubble-Method for Finding the Appropriate Topology of an Initial Design

- *Global and local domain variation*

The *Bubble Method* combines global with local domain variation. The global domain variation varies the boundary of the domain, complying with the homeomorphical mapping rules. The interrelations between the elements constituting the domain are maintained, i.e., the topology class of the domain is unchanged (shape optimization). By means of the local domain variation the topology class is modified by inserting small holes (topology optimization). The single steps are briefly described in the following:

a) Global domain variation

The solutions of the global domain variation (Fig. 1.11) are based on the theory of the variation with variable domain, see Courant and Hilbert (1968), Banichuk (1993), where a domain functional is considered that can be expressed by an arbitrary function $f_\Gamma$ :

$$J_\Gamma = \int_\Omega f_\Gamma \, d\Omega \; . \tag{1.28}$$

Using GREEN's rule, the variation of this domain functional can be written as an integral of the product of a function $f_\Gamma$ and the variation of the boundary $\delta s$ (distance between the old and the new boundary considered in the direction of the normal) along the boundary of the domain $\Gamma$ :

$$\delta J_\Gamma = \int_\Gamma f_\Gamma \, \delta s \, d\Gamma = 0 \; . \tag{1.29}$$

This problem formulation corresponds to the first variation of the functional $J_\Gamma$. The majority of numerical methods for domain variation is based on this first variation. In order to explicitly determine the function $f_\Gamma$, one can employ existing methods based on the formulation of the problem by means of the LAGRANGE-function, the variation of multidimensional problems, and on the fundamental equations of structural mechanics.

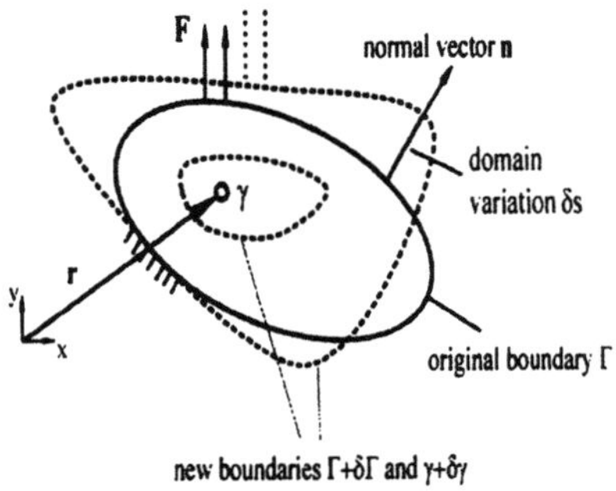


**Figure 1.11** Global domain variation.

b) Local domain variation by hole positioning

The aim of a local domain variation is to determine the optimal position of a new hole (bubble) within the structure. The bubble is then inserted, and by that the topology class is increased. Thus, the confinement to the existing topology class is no longer given.

The insertion of a hole changes the states of stresses and deformations in the elastic body. If an infinitesimally small hole is inserted, it can be treated as a singular disturbance. Two elastic bodies described by the domains $\Omega^{p}$ and $\Omega^{p+1}$, respectively, (Fig. 1.11) possess identical characteristics, i.e., the domain $\Omega^{p+1}$ has an infinitesimal hole with the coordinate vector $\boldsymbol{r}$ and the radius $r_B$ in the case of a circular hole. The difference of an optimization functional $G$ for the two bodies then reads as follows:

$$\Delta G_v = G_v(\Omega^{p+1}) - G_v(\Omega^{p}) \ , \tag{1.30}$$

where it is assumed that $\Delta G_v$ vanishes if $r_B \to 0$. This assumption is to be checked for each $\Delta G_v$ by means of a convergence test. If $\Delta G_v$ does not vanish for $r_B \to 0$, a minor disturbance gains strong influence on the domain functional. In this case (which will not be treated in this paper), we are dealing with an ill-posed problem that has to be solved by special methods (Reinhardt et al., 1993, Wolfersdorf, 1994). For a well-posed problem (i.e., $\Delta G_v$ vanishes for $r_B \to 0$), the radius of the circular hole $r_B$ around $r_B = 0$ can be expanded in a TAYLOR-series:

$$\Delta G_v = \left(\frac{\partial \Delta G_v}{\partial r_B}\right)_{r_B=0} r_B + \left(\frac{\partial^2 \Delta G_v}{\partial r_B^2}\right)_{r_B=0} \frac{r_B^2}{2!} + \cdots + \left(\frac{\partial^n \Delta G_v}{\partial r_B^n}\right)_{r_B=0} \frac{r_B^n}{n!} \ . \tag{1.31}$$

The first term, which does not equal zero, describes the sensitivity of the functional with respect to the positioning of the infinitesimal hole. The solution can be simplified by introducing an evaluation domain $\Omega_A$ (Fig. 1.12), where the radius of the hole shall be substantially smaller than the dimensions of the evaluation domain, and these dimensions shall be much smaller than the component dimensions ($r_B << l << L$). The new hole causes a stress peak in the vicinity of the boundary of the hole, but it decays rapidly according to the Principle of DE SAINT VENANT.

Based on the above assumptions we define that

- the global stress field of the component remains practically unchanged by the infinitesimal bubble (the global stress field depends on the outer shape of the component and on the external loads),
- a local stress concentration is generated in the vicinity of the bubble so that the local stress field in the small evaluation domain $\Omega_A$ depends on the shape of the bubble and on the mean value of the global stress in the domain.

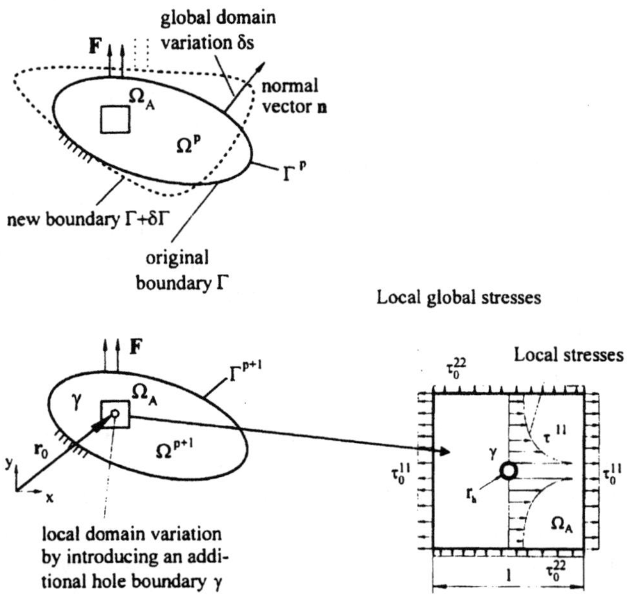


**Figure 1.12** Local variation of the domain by inserting a hole (bubble).

Thus, it is sufficient to evaluate $\Omega_A$ in order to determine the optimal position. By means of GREEN's rule, the variation of the domain integral is to be reduced to a boundary integral, and the local variational problem can be written as an integral of a function $f_\gamma$ over the boundary of the inserted, infinitesimal bubble:

$$\delta J_\gamma = \int_\gamma f_\gamma \delta s_b \, d\gamma = 0 \quad , \tag{1.32}$$

where the function $f_\gamma$ depends on the considered objective and constraint functions of the optimization problem, and on the mechanical conditions. It is required in order to determine the optimal position of the hole.

c) Variation of the total problem

The variation of the total problem can be written as the sum of the global domain variation (1.29) and the local domain variation (1.31):

$$\delta J = \int_\Gamma f_\Gamma \delta s \, d\Gamma + \int_\gamma f_\gamma \delta s_B \, d\gamma = 0 \ . \tag{1.33}$$

In order to solve the simultaneous shape and topology optimization problem, this task is processed by special algorithms. The concept of the simultaneous shape and topology optimization is realized for practical use of the Bubble-method. Fig. 1.13 presents the corresponding optimization loops. The inner loop describes the principle of shape optimization, and the outer loop the principle of topology optimization (Schumacher, 1995).

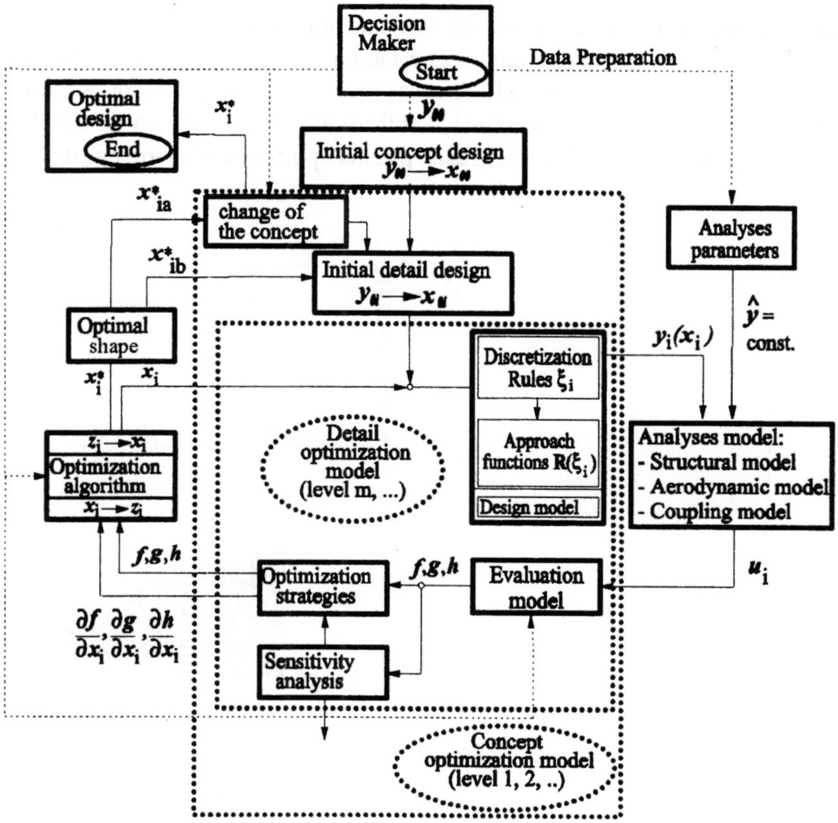


**Figure 1.13** Augmented optimization loop of the "Bubble"-method.

Example 1: Cantilever disk

The task in this example is to find a best-possible initial design for a prescribed topology domain (Fig. 1.14). The disk is clamped at the left-hand side and is subjected to a load on the right-hand side. The optimization problem consists of a minimization of the complementary energy of the component, i.e., minimization of the mean compliance, while considering a volume constraint as equality condition. A triangular disk is chosen as initial design in the half-occupied topology domain. In this case, the problem reads as follows:

$$\underset{x \in \Re^n}{Min} \left\{ U^*(x) \,\middle|\, V(x) = \frac{1}{2} V_{domain} \;,\; \boldsymbol{h}(x) = \boldsymbol{0} \right\} \tag{1.34}$$

with $U^*(x)$ complementary energy,

$V(x)$ volume of the structure,

$\boldsymbol{h}(x)$ vector of $q$ equality constraints.

If a shape optimization is carried out on this structure, a shape as presented in genus 1 is achieved (Fig. 1.14). If a bubble is positioned at the point of the minimum of the characteristic function (in the middle of the supported edge) and a new shape optimization is carried out, we obtain genus 2. By positioning the next bubble at the point of the minimum of the characteristic function, genus 3 is achieved. In the present example, the optimization is terminated at this point.

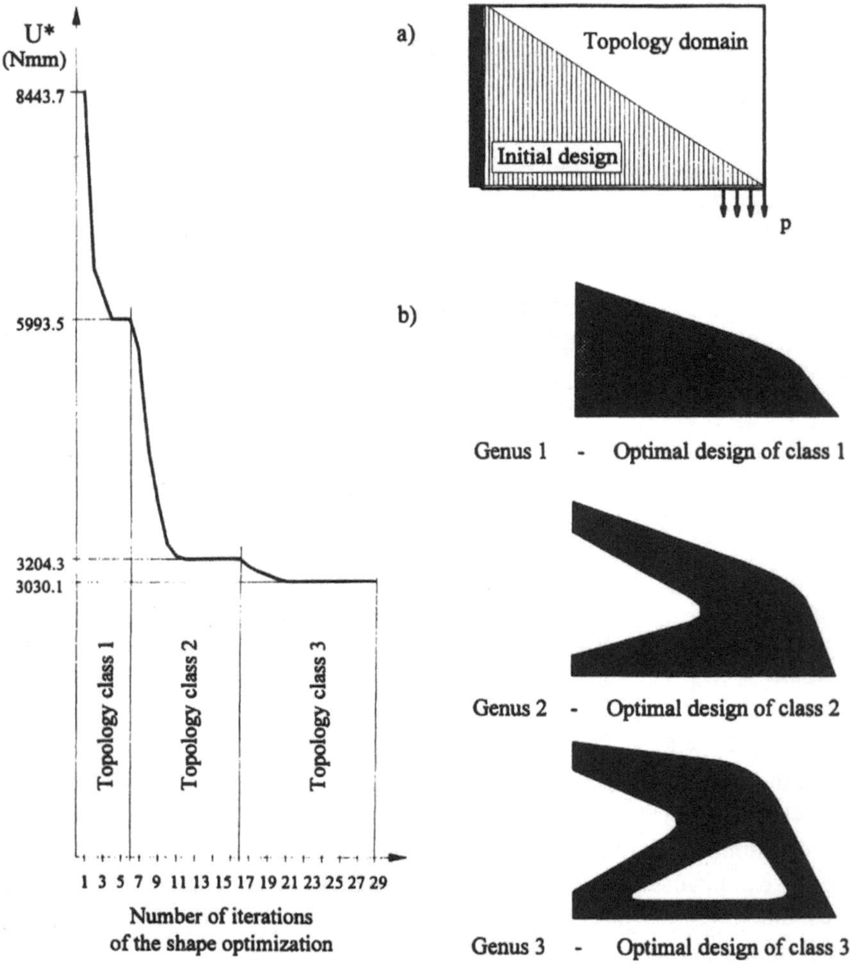


**Figure 1.14** Optimization of a cantilever disk within three topology classes.

Example 2: Wing rib of an aircraft

Wing ribs stabilize the profile of the wing of an aeroplane (see Fig. 1.15a), and they connect the spars and the plating of the supporting structure. The ribs are clamped at the spars at 15% and 75% of the wing depth. The plating is mounted on their outer sides. For reasons of assemblage, the wing ribs must have recesses in the inner part of a wing. These notches shall be optimized by means of a shape and a topology optimization.

The following load cases shall be dealt with in this example:

1. Air-induced forces along the rib length → Pull-out after gliding flight (Fig. 1.15b)
2. Tank pressure (2 bar) (Fig. 1.15c)

The loads are defined on a parameterized geometry model that is independent of the finite element mesh. The outer cover (wall-thickness 5mm) and the spars (wall-thickness 5mm) are simulated by means of beam elements. As an example, calculation is carried out for a stiffener thickness of 5mm and a stiffener distance of 1000mm.

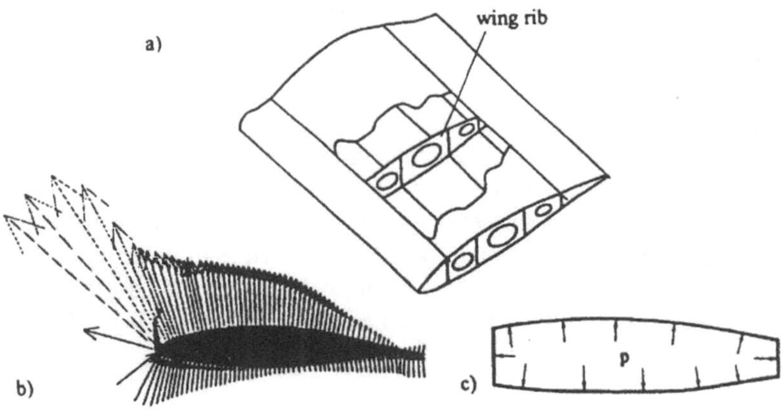


**Figure 1.15** a) Principle sketch of a wing rib of an aeroplane
b) Load case 1: Pull-out after gliding flight
c) Load case 2: Tank pressure (2 bar)

The geometric centre of gravity of the inserted triangular bubble is paced at a point determined by the positioning criterion. The side of the triangle closest to the boundary shall be parallel to the boundary. The coordinates of the corner points of the bubble are guided along the curvilinear coordinate systems of the stiffener boundaries. The sizes of the respective bubbles are kept variable by means of scaling factors, and they are used as design variables in the optimization process.

For the first optimization we consider LC 1 (pull-out after gliding flight). The optimization steps are presented in Fig. 1.16. With unchanged volume, the complementary energy (as a measure of mean compliance) could be reduced from 61818.5 Nmm (class 1) to 60697.6 Nmm (class 3).

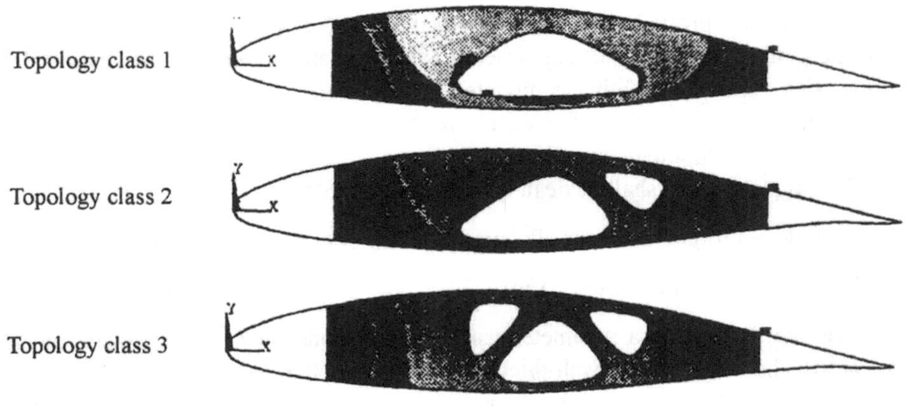


**Figure 1.16** Step-wise positioning and shape optimization of a wing rib in three topology classes.
(Load case 1: Pull-out after gliding flight)

LC 2 (tank pressure) is treated in a second optimization process. The single steps are presented in Fig. 1.17. With unchanged volume, the complementary energy could be reduced from 57725.6 Nmm (class 1) to 17264.6 Nmm (class 2).

Future investigations deal with further single and multiple load cases. In the present example, superposition of the pull-out load and of the tank pressure load leads to the results for the pull-out load, as the sensitivities of the hole positions and their shapes are very low in the load case tank pressure, see Eschenauer et al. (1994), Eschenauer and Schumacher (1993a,b, 1997).

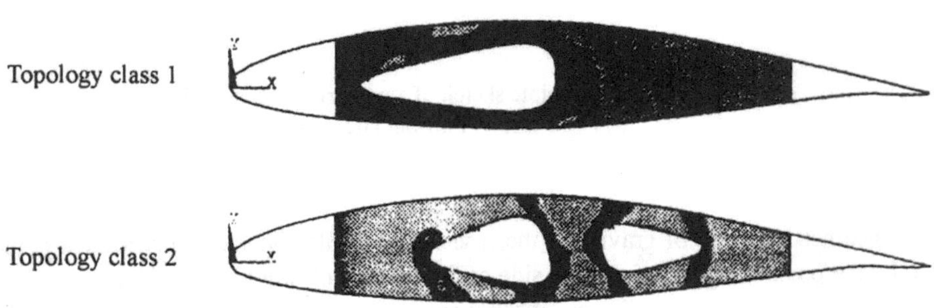


**Figure 1.17** Step-wise positioning and shape optimization of a wing rib in two topology classes
(Load case 2: tank pressure)

### 1.4.2 Application of Simultaneous Product and Process Simulation

- *Optimal layouts of cast rollers regarding process characteristics*

*a) Problem definition of process and product*

Cylindrical components made of cast iron or chilled cast iron are frequently used as thermal rollers in paper manufacturing industries, where the roller forces are exploited for smoothing rough paper surfaces (Fig. 1.18). In addition, the transfer of heat flows also serves the drying of paper strips. During operation, the rollers are generally subjected to "external" thermo-mechanical loads as well as "internal" loads due to the high residual stresses as a result of the casting process. A constructive layout therefore has to consider the resulting total loads in a very early product stage already. In order to find optimal layouts, a multidisciplinary analysis must be carried out, see Beer (1996), Eschenauer and Beer (1998). In order to analyze the structural behaviour of the rollers, a process and product simulation according to Fig. 1.19 is performed that demands both a description of the casting process and of the conditions during operation of cylindrical cast components.

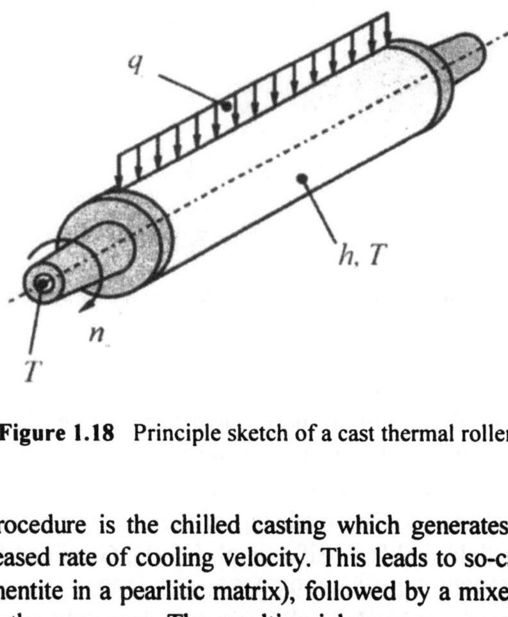


**Figure 1.18** Principle sketch of a cast thermal roller.

A special casting procedure is the chilled casting which generates an inhomogeneous cast structure due to an increased rate of cooling velocity. This leads to so-called "white cast iron" in the boundary areas (cementite in a pearlitic matrix), followed by a mixed intermediate layer, and finally grey cast iron in the core area. The resulting inhomogeneous structure is called "chilled cast iron" which possesses a very hard, brittle boundary layer and a comparatively ductile core, see Beer (1996), Erler (1937).

In the casting process, the temperature curves that are variable with respect to position and duration are the most important process parameters for the resulting cast structure. The grey-white solidification in the cast iron is determined in dependence on the cooling velocity. In order to pre-determine the white solidification, a physical model according to Upadhya et al. (1990) is used which is based on the consideration of the critical cooling velocities. The instationary thermal conduction is determined by thermal transfer conditions within and out of the cast system (thermal conduction in the melt or the unfinished cast object, respectively, and in the ingot mould, as well as the heat transfer conditions in the surrounding of the ingot mould). Immediately after the cooling an air gap is formed that substantially influences the further course of the cooling. The air gap forming can be controlled by a so-called separating layer between melt and ingot mould (Fig. 1.20).

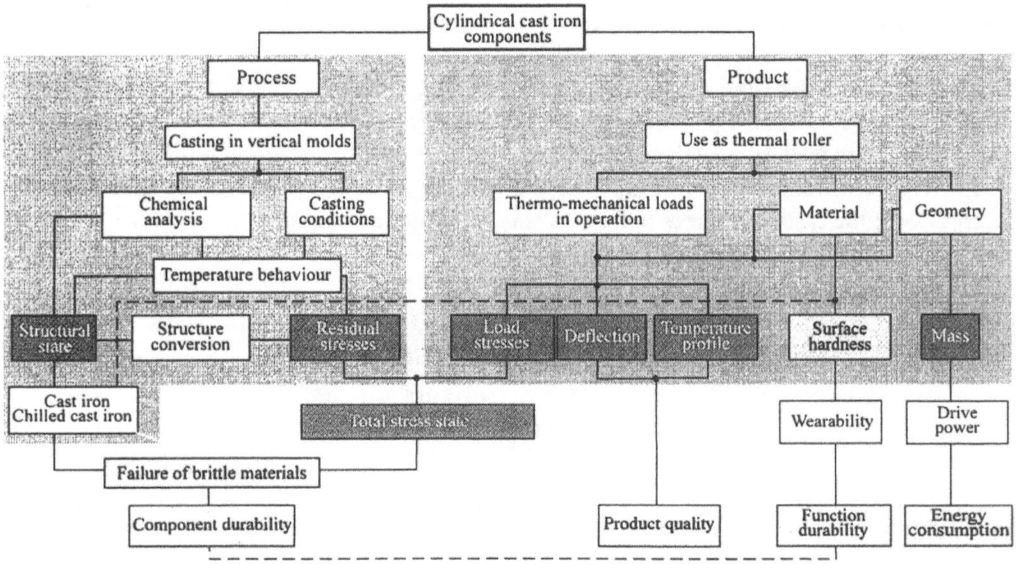


**Figure 1.19** Design and product simulation of a cast iron component.

With the thermal cooling process, cast residual stresses occur. Owing to the local and temporary exceeding of the thermal yield limit due to large temperature differences, thermal residual stresses are generated. One speaks of conversion residual stresses if stresses are produced due to volume changes in transformation processes. Taken together, these stresses are called residual stresses caused by thermal effects. The modeling of the casting process is a coupled thermal-mechanical, elasto-plastic field problem (Eifler et al. (1991), Guan and Sahm (1992), Mack and Gamer (1988), Megahed (1990), Sjöström (1983), Sluzalek (1992), Sahm and Hansen (1984).

Owing to some engineering-related simplifications, a de-coupling of the thermal-mechanical field problem can be carried out in the mathematical-physical modeling of thermal-mechanical processes. Thus, one may neglect the influence of the speed of deformation on the temperature curves, and the influence of the state of stress on the structural conversions (Schimöller (1990), Sjöström (1983), Upadhya et al. (1990)).

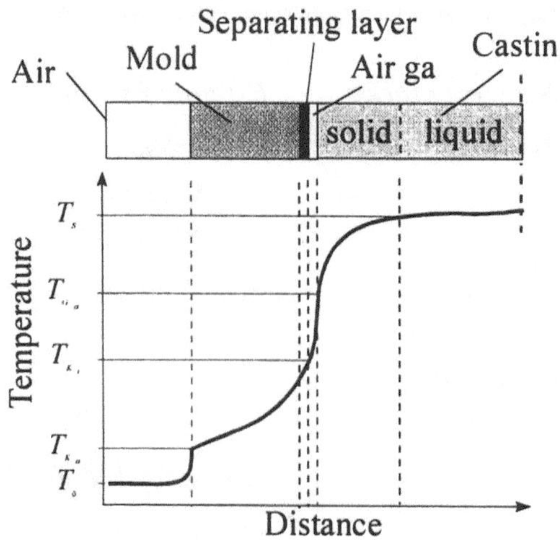


**Figure 1.20** Thermal transfer in the cast roller from the outer to the inner part.

The uncoupled field problem allows to determine the temperature field while only considering those physical material characteristics that depend on temperature and structure. The first equation of the thermal conductivity reads as follows:

$$\rho c \dot{T} + q^i\big|_i = Q^i \tag{1.35}$$

with $T$ temperature field, $q^i\big|_i$ heat flow quantity, $Q^i$ dissipated heat capacity, and $\dot{()}$ time derivative.

Stresses and deformations are calculated on the basis of the instationary temperature field and mechanical material characteristics dependent on temperature and structure as well. The second equation is the condition of equilibrium

$$\left(C_{ep}^{ijkl} \dot{\gamma}_{kl}\right) \dot{\gamma}_{ij} = \dot{f}^i \dot{u}_i + \dot{p}^i \dot{u}_i \tag{1.36}$$

with $C_{ep}^{ijkl}$ elastic-plastic tensor, $\gamma_{ij}$ strain tensor components, $f^i$ volume forces, $p^i$ components of surface forces, $u_i$ components of the displacement vector.

In completion to Fig. 1.19, Table 1.1 presents the single steps for modeling the internal state due to the manufacturing process, and the external state due to operational loads.

**Table 1.1** Comparison of internal and external state.

***Internal state due to manufacturing process***

*Thermo-mechanical cooling process*

- temperature distribution and structural arrangement
- transient temperature calculation
- consideration of latent thermal influences
- white-grey-solidification
- heat transfer and generation of air-gaps
- thermal influence on ingot mould

***Residual stresses***

- elasto-plastic material behaviour
- consideration of creep effects
- residual stresses due to thermal effects (temperature curve, transformations)
- reduction of residual stresses

Static and dynamic failure criteria

- normal stress hypothesis
- latigue limit according to Smith
- linear-elastic fracture mechanics with small area flow
- critical fracture size
- tolerance ratio of the residual load till fracture

***External state due to operational loads***

*Temperature distribution*

- stationary temperature distribution
- heat transfer to paper

***Stresses by operational loads***

- temperature profile
- dead weight
- linear load
- centrifugal force
- state of load stresses
- deformation behaviour

The structural analyses of the cooling process and of the residual stresses due to casting are carried out by means of the finite element method; the stresses induced by external loads are determined analytically and numerically. The total loading of the component consists of residual stresses and load stresses, where the operational load is basically a result of a temperature load and a bending load. For failure evaluation, the fracture of cast components is of particular interest owing to their brittle material structure. The existence of structural inhomogeneities like shrink holes, inclusions or graphite lamellae lead to small cracks prior to the actual operation. In addition to a global consideration using equal stress criteria, fracture mechanics are therefore applied in the local failure evaluation, where stress intensity factors are calculated analytically for different fracture configurations. For more details see Wiese and Dantzig (1988), Weck et al. (1992), Beer (1996).

*b) Comparison of optimization results for two thermal rollers*

First we consider a thermal roller as applied in paper industry. In this example the objectives are low weight and high stiffness. A comparison is carried out between the use of homogeneous cast iron rollers and rollers made of chilled cast iron. Furthermore, we reflect on the temperature dependence of the material properties. Four different optimization models are formulated for the sake of comparison. In general, design variables are the inner radius for both models, and additionally the white iron layer for the inhomogeneous chilled cast iron. Constraints concerning the stresses in various areas of the structure as well as concerning the geometry have to be fulfilled. In order to process this Multicriteria Optimization problem, it is transformed into a suitable scalar optimization problem. The obtained results show that treating the above task as a multicriteria optimization problem is a suitable tool for decision-making. In our case, the design engineer has the possibility to choose between suitable weight and deflection of the rollers. In this example the residual stresses are not considered for reasons of simplification. In order to deal with the important influence of residual stresses, the model must be augmented, see Beer (1996), Eschenauer and Beer (1998).

Subsequently, we consider the optimization of a heterogeneous chilled cast iron roller. Since in paper manufacturing industry the drying temperatures are a measure of the paper quality, the maximum surface temperature of the thermal rollers is defined as the objective function. The design variables are the inner diameter of the thermal roller (constructive), the thickness of the insulating separating layer (casting process), the temperature of the heating medium (operational load), the initial temperature of the ingot mould (casting process), and the ambient temperature of the ingot mould within a certain cooling period (casting process). As constraints we define the deflection as an additional measure for the paper quality, as well as admissible stress values $\sigma_r$ (strength), and a minimum thickness of the chilled cast-iron layer (wear resistance). Owing to the brittle material behaviour of cast iron and chilled cast iron, the Normal Stress Hypothesis is used for the reference stress evaluation.

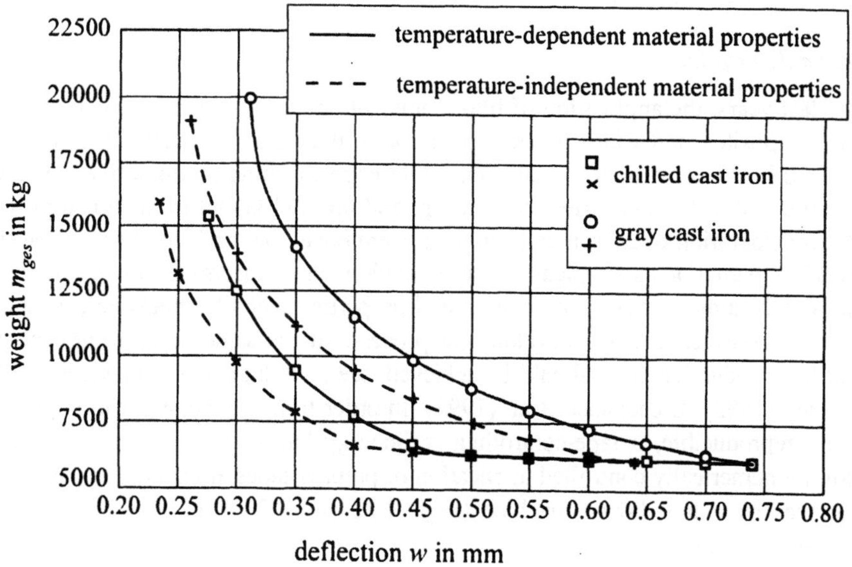


**Figure 1.21** Functional-efficient boundaries for the objectives weight and deflection of cast iron rollers.

By means of the optimization the surface temperature could be increased by almost 40%. The inner radius as well as the thickness of the white iron layer could be increased in such a way that the deflection almost reaches its upper bound. Thus, the temperature of the heating source can be raised substantially, since the temperature gradient within the roller wall has to act on a smaller wall thickness. This fact naturally leads to interrelations between deflection and temperature since the temperature-dependence of the material characteristics has been considered. The thickness of the white cast layer is at its lower bound as, on the one hand, the increased stiffness gains a positive influence and, on the other hand, the low thermal conductivity does not allow for a higher thickness of this layer. The design variables initial temperature and initial ambient temperature of the ingot mould are moved to higher temperatures in order to reduce the level of residual stresses to an admissible range.

Fig. 1.21 shows that the models with homogeneous material structure of cast iron (temperature-dependent as well as temperature-independent) have higher weights at the same deflections as the inhomogeneous chilled cast iron because of the higher stiffness of the white cast layer. If the temperature-dependence of the material properties is taken into consideration within the models, the results shift to less positive values because of the material behaviour at higher temperatures.

- *Optimal layout of composite shell structures regarding manufacturing constraints for a fully automated tape application*

*a) Problem description*

During the last years, the application of fibre composite materials has been continuously growing. Thus, composites find wide-spread application in the fields of aircraft and space technology, as well as in the construction of antennas, telescopes, turbines, automotive parts, rollers in printing, audio and video industries, etc. Design and optimal layout of such complex systems require the consideration of *all* demands on the system behaviour and structural properties.

In aircraft construction as a typical example, requirements concerning failure safety, vibration behaviour, aerodynamic and aeroelastic structural properties, stability (buckling), manufacturing, etc. have to be considered in the development process. At the same time, a minimal structural weight and low production costs should be achieved, see Dollyhigh and Sobieszczanski-Sobieski (1990), Ashley (1992), Eschenauer et al. (1992). In order to make the application process more efficient and reproducible, a tape-controlling technology has been developed in recent years, which allows a numerically controlled application of prepreg-tapes along curved paths on doubly curved surfaces by a tape-laying machine (Figs. 1.22 and 1.23a), cf. Berchtold and Klenner (1992), Geißendörfer (1992).

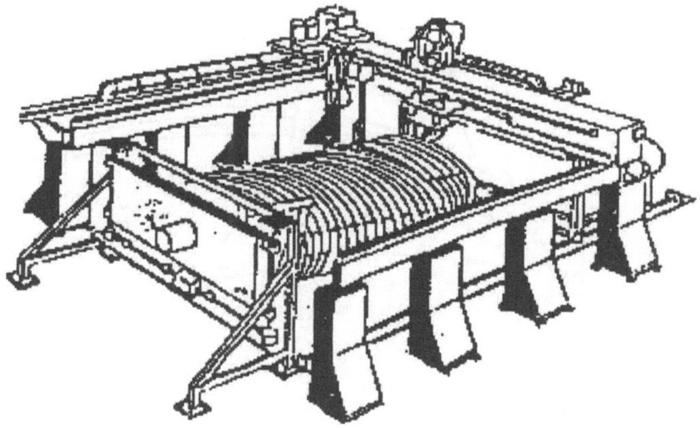

**Figure 1.22** Tape-laying machine for the application of prepreg-tapes along curved courses of doubly curved surfaces.

The additional requirements must be fulfilled for manufacturing reasons:

If one merely bends the tapes when applying them along curved paths, wrinkles would occur because of the different laying length of the fibres across the tape (Fig. 1.23c). In order to prevent this, and to equalise the different laying lengths, the fibres are shifted in the tape-laying process (Fig. 1.23b). The fibre shifting only causes shear strains in the prepreg matrix, while the fibre strains are negligible. The fibre shifting $\gamma_{adm}$ is only possible up to a certain limit because otherwise the resin matrix would become torn. After the application, the shear stresses in the unhardened viscoplastic resin matrix relax completely (Berchtold and Klenner, 1992). Furthermore, the curvature $\kappa_{adm}$ of the paths is limited by machine constraints. In addition to other technological demands, it is especially these curvature and shifting constraints that are the most important manufacturing constraints which have to be considered in the design process. A third constraint comes from the admissible drop-off angle $\alpha_D$ (Fig. 1.23d). If the changes in the thickness distribution are too large, and therefore the drop-off angles exceed a certain value, the shear stresses necessary for force transfer in the overlapping region become so strong that the layers are separated (delamination).

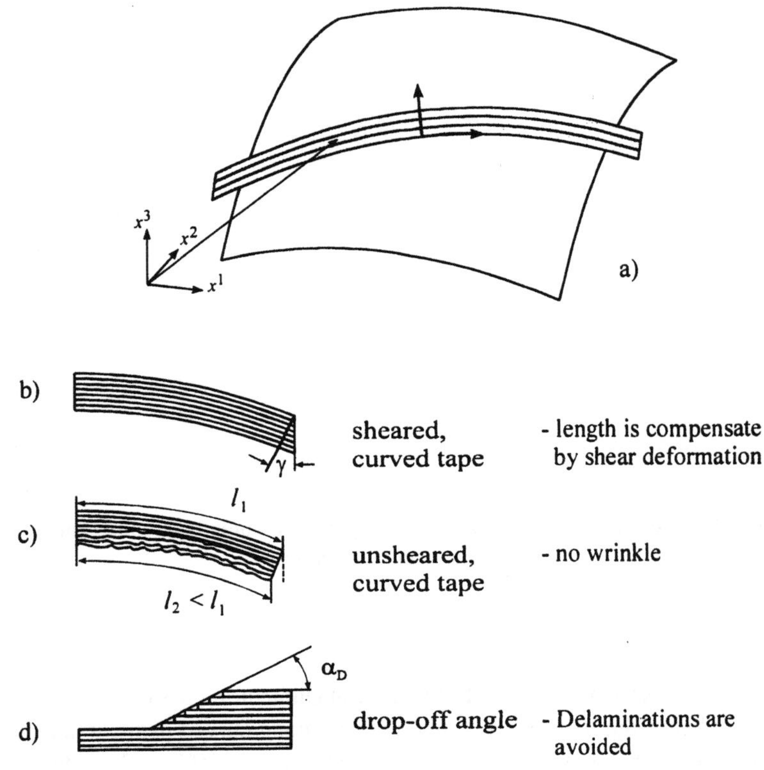


**Figure 1.23** Manufacturing requirements for a fully automated tape application.

Thus, the following manufacturing constraints must be fulfilled in addition to all other constraints:

drop-off angle $$|\alpha_D| \leq \alpha_{D\,adm} \; ; \tag{1.37a}$$

radius of curvature $$|\kappa_B| \leq \kappa_{adm} \; ; \tag{1.37b}$$

tape shear angle $$|\gamma| \leq \gamma_{adm} \; . \tag{1.37c}$$

*b) Results of an aircraft fin optimization*

As a suitable basis for a multidisciplinary optimization using various analysis models, so-called constructive design models have been developed (for more details see Schuhmacher (1995)). The optimization model for the aircraft fin was formulated as follows:

| | | |
|---|---|---|
| *Objective*: | $f(x) = W$ | |
| *Constraints*: | 119 | stresses / load case |
| | 252 | failure / load case |
| | 2 | aeroelastic efficiency |
| | 5 | flutter constraints |
| | 3 | manufacturing constraints |
| *Load cases*: | 5 | for several flight positions |
| *Design variables*: | different number, depending on the model | |
| | sizing | |
| | layer orientations | |

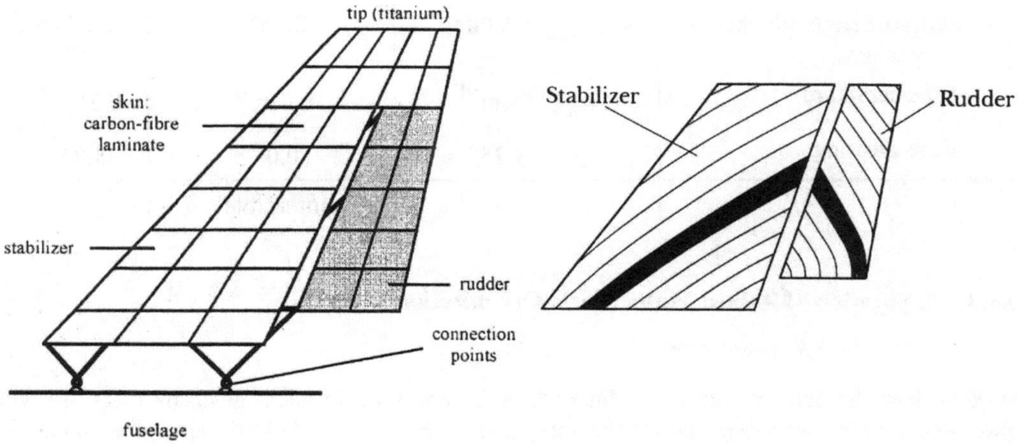


*a)* *b)*

**Figure 1.24** Aircraft fin
(a) Structural model
(b) Curved tape paths for the optimal layout

Table 1.2 lists the demands and compares the computational results for the initial layout (Fig. 1.24a) and the optimal layout. The course of the tape paths for the optimal layout are presented in Fig. 1.24b.

**Table 1.2** Results of the fin optimization.

| ***Criteria*** | ***Requirements*** | ***Initial design*** | ***Optimal design*** |
|---|---|---|---|
| • Objective function: Weight of the top layer | Min | 53.98 kg (100%) | 28.39 kg (52.6%) |
| • Constraints: | | | |
| - Strength constraints | $R_{VS} \geq 1.0$ | > 2.0 | 1.0 |
| - FV-failure safety | $R_{FV} \geq 1.0$ | 0.7695 | 1.0 |
| - Aeroelastic efficiencies (MA 1.8 / 385.8 m/s) | $\eta_{Fin} \geq 0.8$<br>$\eta_{Rudder} \geq 0.5$ | 0.753<br>0.441 | 1.0<br>0.5 |
| - Flutter speed | $v_{Flut} \geq 530$ m/s | 495 m/s | 530 m/s |
| - Drop-off angle (thickness gradient) | $\alpha_{Dadm} \leq 5.00\%$ | 2.05% | 2.239% |
| - Tape curvature | $\kappa_{adm} \leq 4\text{m}^{-1}$ | $0.0\,\text{m}^{-1}$ | $0.25\,\text{m}^{-1}$ |
| - Tape shearing | $\gamma_{adm} \leq 25^\circ$ | $0.0^\circ$ | $8.94^\circ$ |
| | | Infeasible! | |

### 1.4.3 Application of Robust Multicriteria Optimization (RMO)

- *Robust design formulation*

Starting from the general nonlinear Multicriteria Optimization Problem given by (1.2), we will now consider a robust design formulation (see Lautenschlager et al. (1999)). Here, the minimization of the objective function leads to a value that should be achieved with a certain probability. In the formulation of the objective function, *n* standard deviations are added to the mean to express this situation. A deterministic objective function within (1.38) is now formulated as the new stochastic objective function

$$F_i(x,z) = \mu_{f_i}(x) + n\sigma_{f_i}(x,z) \ . \tag{1.38}$$

Similarly, a stochastic formulation has to be considered for the constraints. Otherwise, design points with mean values within the feasible design space may violate a constraint and are actually infeasible with a certain probability. The new stochastic constraint is formulated as

$$G_j(x,z) = 1 - \frac{\mu_{g_j}(x) + n\sigma_{g_j}(x,z)}{g_{j_{adm}}} \geq 0 \ . \tag{1.39}$$

The number *n* is selected on the basis of the desired reliability, a value of $n = 3$ or higher is typical. The larger n is, the smaller is the remaining, feasible design space.

- *Problem definition - Structural modeling of a flywheel*

The classical design objective for a flywheel due to Fig. 1.25 is to maximize the capacity of the stored kinetic energy while satisfying constraints on the mass and stress limits. This problem has been studied extensively in the design literature. In this example the flywheel profile is presented by a BEZIER-curve.

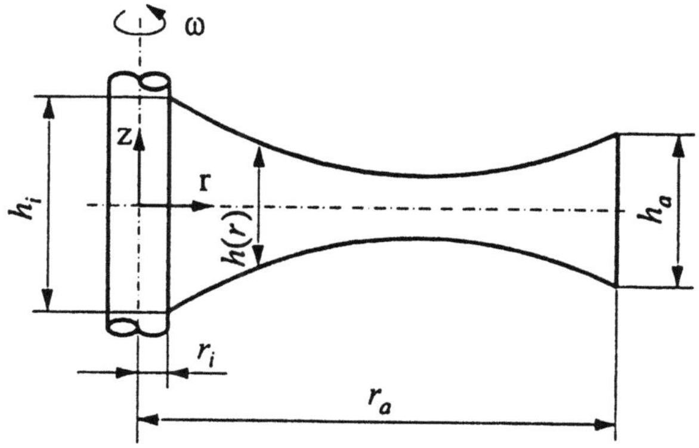


**Figure 1.25** Sketch of a flywheel profile.

Structural analysis of the flywheel comprises both analysis of the shaft and analysis of the flywheel. For reasons of simplicity the following investigations are confined to wheel. Radial and tangential stresses $\tau_{rr}$ and $\tau_{\theta\theta}$, respectively, are assumed to be the principal stresses evenly distributed over the thickness; shearing stresses are neglected. When considering the equilibrium at a small disk element and introducing a stress function $\Phi$ and by applying the material law and considering the radial symmetry, one obtains the following differential equation:

$$r^2 \frac{d^2\Phi}{dr^2} + r\frac{d\Phi}{dr} - \Phi + (3+\nu)\rho\omega^2 hr^3 - \frac{r}{h}\frac{dh}{dr}\left(r\frac{d\Phi}{dr} - \nu\Phi\right) = 0 \ . \tag{1.40}$$

Owing to the non-constant thickness $h(r)$ the above second order differential equation cannot be solved analytically. Instead, a fourth order RUNGE-KUTTA procedure is used to solve (1.40). All integration points ( $h(r), dh/dr$ ) necessary for the calculation are derived from the definition of the BEZIER-curve. If there exists a numerical solution of the stress function $\Phi$, the radial and tangential stresses $\tau_{rr}$ and $\tau_{\theta\theta}$ are calculated by means of the finite differences procedure according to

$$\tau_{rr} = \frac{\Phi}{hr} \ , \tag{1.41a}$$

$$\tau_{\theta\theta} = \frac{1}{h}\frac{d\Phi}{dr} + \rho\omega^2 r^2 \tag{1.41b}$$

As failure criterion the reference stress $\tau_r$ according to VON MISES is introduced which is calculated by

$$\tau_r = \left(\frac{1}{2}\left((\tau_{rr} - \tau_{\theta\theta})^2 + \tau_{rr}^2 + \tau_{\theta\theta}^2\right)\right)^{\frac{1}{2}} . \tag{1.42}$$

If the stresses in the disk exceed the admissible value, failure occurs. The mass $m$ and the kinetic energy $T$ of the disk are calculated numerically with the SIMPSON-rule by solving the integrals

$$m = 2\pi\rho \int_{r_i}^{r_o} hr\,dr \ , \tag{1.43a}$$

$$T = \pi\rho\omega \int_{r_i}^{r_o} hr^3\,dr \ . \tag{1.43b}$$

- *Optimization Modeling – Design Space Investigation*

Optimizing the flywheel design, the principal objective is to maximize the kinetic energy. As secondary objective we define minimization of the maximum reference stress. The afore-described vector optimization problem is realized by transforming the stresses into a constraint, and by defining different admissible values. The design variables are defined for the coordinates of the control points $P_1 - P_4$ of a BEZIER-curve (Fig. 1.26). The four control points yield eight design variables that define the thickness over the entire area of the disk. The end points of the control polygon are identical with the geometric dimensions of the disk. One obtains as design variable vector

$$\boldsymbol{x}^T = (x_1, x_2, x_3, x_4, x_5, x_6, x_7, x_8) \hat{=} (r_i, r_a, h_i, h_a, z_2, r_2, z_3, r_3) \ . \tag{1.44}$$

In the vector of the state quantities

$$\boldsymbol{u}^T = (T, m, \tau_r, h_u, h_o) \tag{1.45}$$

the kinetic energy $T$, the mass $m$, the reference stress $\tau_r$ as well as the smallest and the largest disk thickness $h_u$ and $h_o$, respectively, are considered.

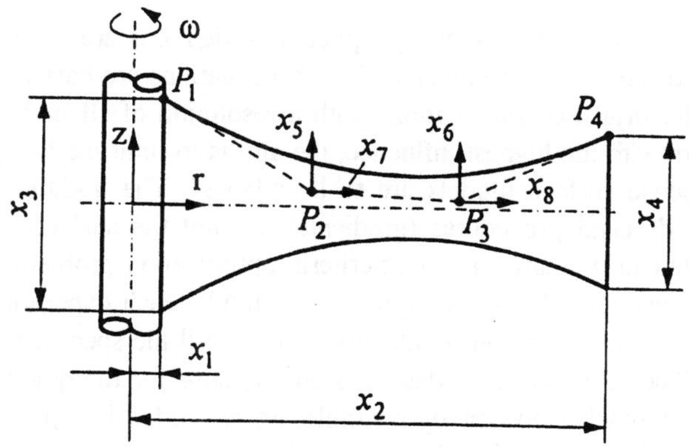


**Figure 1.26** Definition of design variables.

In Table 1.3 the essential features of the optimization model are compiled. In the following a model with six design variables and different upper and lower limits is considered.

**Table 1.3** Optimization model of the flywheel.

| | | |
|---|---|---|
| **Objective function:** $\max\limits_{x} T$ | | |
| **Constraints:** | | |
| Reference stress: | $g_1 = \dfrac{\tau_r}{\tau_{r_{adm}}} - 1 \le 0$ ; | $100\ \text{N/mm}^2 \le \tau_{r_{adm}} \le 250\ \text{MPa}$ |
| Mass: | $g_2 = \dfrac{m}{m_{adm}} - 1 \le 0$ | $m_{adm} = 100$ kg |
| Minimum disk thickness: | $g_3 = \dfrac{\min\limits_{r_i \le r \le r_a}(h)}{h_{min}} - 1 \le 0$ | $h_{min} = 20$ mm |
| Maximum disk thickness: | $g_4 = \dfrac{\max\limits_{r_i \le r \le r_a}(h)}{h_{max}} - 1 \le 0$ | $h_{max} = 100$ mm |
| **Design variable vector:** | $x^T = (r_i, r_a, h_i, h_a, z_2, z_3)$ | |
| **Design variable bounds:** | 20 mm $\le r_i \le$ 50 mm, 20 mm $\le h_i, h_a \le$ 100 mm | |
| | 100 mm $\le r_a \le$ 300 mm, 0 mm $\le z_2, z_3 \le$ 40 mm | |

Within a sequential design strategy design space investigations are carried out to screen the design variables and to reduce the model size. The investigations are based on experiments with two-level, fractional factorial experimentations with a resolution of III or higher. In addition to determining the factors with the highest influence, the aim is to limit the target surfaces in which the higher order response surface models are to be adapted. The traditional response surface method uses steepest descend procedures (gradients) to limit the surfaces. This task becomes much more complicated in the case of a multicriteria optimization problem with several objectives and secondary constraints. Using the optimization model each experiment can be individually evaluated; factor combinations can be identified that fulfill the specified objectives or which come close to them. The reliability of a design is an indicator for the quality of the design domain. An alternative approach is to employ a simple, linear model; if applied in an optimization procedure, it yields valuable information about the design domain.

- *Optimization results*

a) Comparison of original with response surface model

The developed response surface models are integrated into the optimization loop. Optimization calculations with the original structural model are used as reference. The investigations for the model presented here focus primarily on a comparison of the response surface models based on 27 or 43 experiments, respectively. In the scope of this multicriteria optimization task, one objective is minimized, the other one maximized. As solution strategy, the kinetic energy is maximized for different constraint values of the reference stress. The admissible stress values vary in correspondence with the optimization model. All achieved results are compromise solutions towards fulfilling the objective conflict; Fig. 1.27 presents the solutions are functional-efficient or PARETO-efficient curves. A better illustration is achieved by the reciprocal value $1/T$ of the kinetic energy. Here, the boundary is convex; within a wide stress range ($\tau_r$ = 150 - 250 MPa) one obtains high values for the kinetic energy ($T >$ 1000 kJ) while lower values occur for a range between $\tau_r$ = 100 MPa and 150 MPa. The kinetic energy decreases substantially if more mass is transferred from the outer to the inner areas to obtain lower stress values. For all solutions an upper bound of 300 mm is maintained for $r_a$; the design variables $z_2$ and $z_3$ are close to the lower bound.

Comparison between the results of the response surface models and the original structural model shows overestimation in the case of high stress values and underestimation for low values. The response surface model with 43 experiments provides a much better approximation of the original behaviour than the one based on only 27 experiments. For $\tau_r \geq 150$ MPa both models are equally good. In spite of the quantitative differences in the functional-efficient boundary curve an engineer can still derive valuable information on the model behaviour for a wide variety of optimization scenarios.

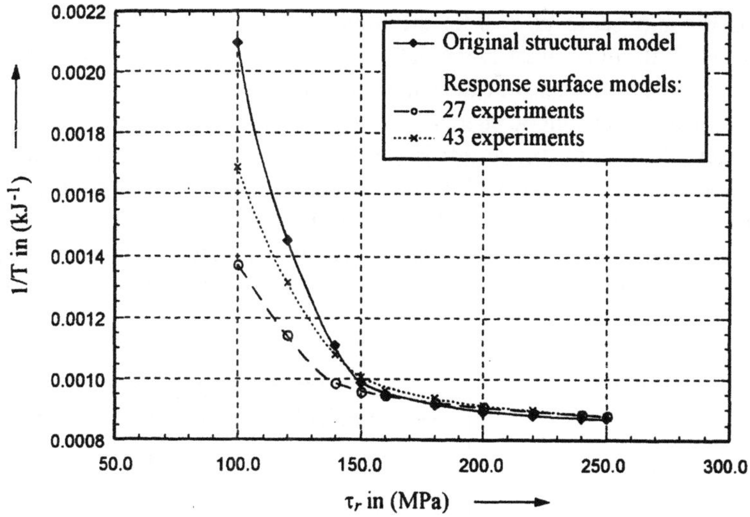


**Figure 1.27** Functional-efficient boundary: stresses vs. reciprocal kinetic energy.

b) Development of Robust Design Models

The formulation of suitable response surface models is the basis also for effectively considering robust design aspects. In this section the optimization of the flywheel is extended to robust design procedures. For this purpose, two robust design strategies are compared and evaluated. As geometrical, stochastic quantities we assume $r_a, h_i$ and $h_a$. A further disturbance quantity is given by the angular velocity $\omega$; any changes of $\omega$ lead to variations of the kinetic energy and of the stress, i.e., both increase with increasing $\omega$. The variations are defined by values of the standard deviations that possess the same dimensions as the considered quantities themselves. The standard deviations are chosen such that they attain 1% of the maximum values in the case of geometrical variables, and 3% for the load (rotation). The admissible stress is also covered by a stochastic consideration. The standard deviations of the five disturbance quantities are constant values that do not change with the mean of the corresponding quantity. They are defined as follows:

- $\sigma_{r_a} = 3$ mm, $\sigma_{h_i} = 1$ mm, $\sigma_{h_a} = 1$ mm, $\sigma_{\omega} = 18.9\ s^{-1}$
- $\sigma_{\tau_{r,adm}} = 7.5$ MPa

Models for the crossed-array and for the combined-array procedure are developed and compared.

c) Results of stochastic optimization calculations

Optimization calculations have been carried out with different objectives and formulations. In a first step, the results obtained by the crossed array and the combined-array procedure, respectively, are compared, where a reliability index of $\beta_G = 3$ has been chosen for the stochastic calculations. In order to provide a possibility of comparison with the deterministic calculations, the value is additionally set to $\beta_G = 0$ for the combined-array model. The results are illustrated graphically in Fig. 1.28. In contrast to the deterministic solutions of the response surface model the functional-efficient boundaries of the robust solutions are shifted upwards, which indicates a lower maximization of the kinetic energy.

The above-described results prove that in the case of a robust consideration fulfillment of the objective function can be expected with certain limitations only. This is mainly due to the constraint formulation of the reference stress, but also to the mass constraint the mean of which leads to a lower weight. However, the differences between the two strategies for realizing the robust design are actually insignificant. Only smaller stress values yield slightly more substantial differences.

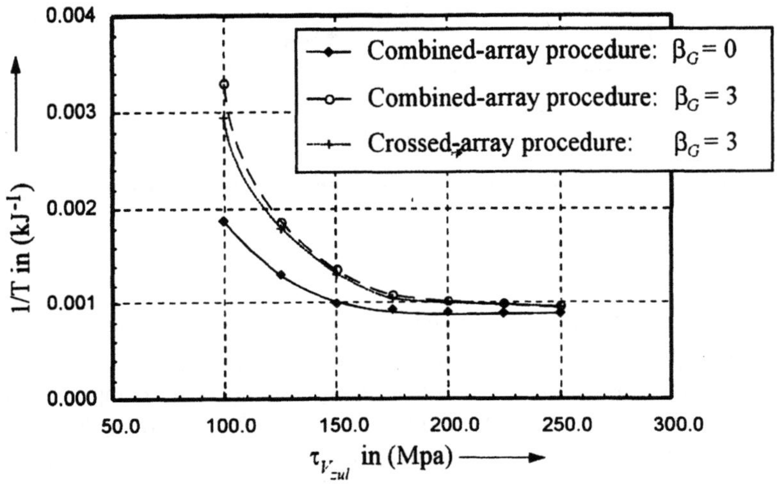


**Figure 1.28** Comparison of different robust design strategies.

## 1.5 Conclusion

The traditional design process comprises a sequential procedure starting from the project definition phase over conceptual and detailed design to production planning, maintenance and recycling. Here, the freedom of decisions of the involved disciplines in the single phases depends on the initial decisions. Current research aims at combining the different disciplines during product development; this design process that is also termed *Concurrent Engineering* (CE) intends to ease inter- and multidisciplinary cooperation. A characteristic of this process is the existence of multiple objectives and demands at the beginning of the design phase. This fact has led to *Multidisciplinary Design Optimization* (MDO) emerging as supporting-tool for CE. MDO comprises, as principal components, design-oriented analysis methods, approximation concepts, mathematical system modeling, decomposition, search algorithms, optimization procedures, and the user as human interface (Sobieszczanski-Sobieski (1995)). One of the aims of CE is to enhance the participation of the single disciplines in the conceptual phase in order to increase the knowledge pool and to provide sufficient freedom of design. Hence, more individuals work in parallel, and the subsequent decisions are based on better and more reliable information.

*Structural Optimization* has been addressing the problem of multiple objectives for a long time. The multidisciplinary character of this field of engineering is clearly indicated by the consideration of objectives and demands from the areas of construction, manufacturing, assembly, or on the material (Eschenauer (1990)). A structure can only be optimal if it fulfils the given specifications with respect to the cooperation of all components/disciplines in a best-possible manner. Thus, structural optimization has laid the foundation for the development of MDO-methods. It is based on mathematical algorithms, and it intends to enable the engineer to achieve a well-aimed improvement of the quality and reliability of structures, and, if at all possible, to even find optimal designs. The layout of commercial, competitive products demands from a CE-based procedure that it be able to predict the behaviour and functionality of a systems in view of the selected parameters as precisely as possible. Therefore, complex, numerical simulation procedures for products and processes must be applied in an early stage already; nowadays, it is no longer acceptable to conduct optimization calculations with a large number of prescribed and hence invariable details. On the contrary, possibilities must be generated to use the existing resources sensibly to apply structural optimization efficiently and effectively at an early design stage. To achieve this aim, primarily topology optimization as an emergent strategy has been developed over the previous years.

In order to be able to integrate these highly efficient analysis procedures into the design process as early as possible and, thus, to converge with the objectives of Concurrent Engineering, it has proved sensible to use statistical procedures, approximation methods, and to consider the stochastic properties of some of the parameters. By this, the existing resources are exploited optimally, and important information about the product behaviour can be obtained by investigations with simple substitute models. This leads to an increased flexibility, and freedoms of design can be maintained as long as possible. This is achieved by partly reducing the model accuracy while increasing the efficiency – especially in the case of multidisciplinary tasks – in a substantial manner.

An important step towards the solution of above-described problems is given by procedures like Design-of-Experiments (DOE)-methods, the Response Surface-method and the Robust Design-method. Originally, these procedures have been developed for treating problems from the field of physics, but they can easily be augmented to numerical tasks. DOE allows to gain a quantitative overview of the effects of single parameters, where the procedure comprises the following steps: planning of experiments, analysis of the experimental results, and mathematical modeling. A well-aimed design of the experiments allows to derive a maximum amount of information with the lowest possible experimental expense. The Response Surface Method (RSM) is an approximation procedure comprising both mathematical and statistical techniques. It is applied for the analysis of problems, where several independent variables influence one dependent variable or system response. In a first step, RSM establishes a functional relationship between a number of independent variables and the investigated system response, using tests of the statistical design of experiments with the aim to optimize the response (Box et al. (1978)). If the design is influenced by stochastic quantities, the Robust Design Method, also termed Quality Engineering according to Taguchi (1986), can be used to improve the quality of a product or process. The principal aim of Robust Design is to minimize the quality loss of a product/process, while being able to achieve a high quality in a time- and cost-efficient manner. The method yields a robust product that is insensitive to any type of uncontrollable noise factors. The use of topology optimization via the Bubble Method as well as of the Robust Design Method is illustrated by means of two application examples.

## 1.6 References


Abadie, J., and Carpentier, J., (1969). Generalization of the Wolfe Reduced Gradient Method to be Case of Nonlinear Constraints. In Fletcher, R.: Optimization, Academic Press, New York

Adelman, H.M. and Haftka, T., (1984). Sensitivity Analysis for Discrete Structural Systems – A Survey, NASA-TM-86333

Allaire, G. and Kohn, R.V., (1993). Optimal design for minimum weight and compliance in plane stress using extremal microstructures, *Eur. J. Mech., A/Solids*, 12, No. 6: 839-878

Ashley, S., (1992). DARPA initiative in concurrent engineering, *Mechanical Engineering*, 54: 45-57

Atrek, E., (1989). SHAPE: A Program for Shape Optimization of Continuum Structures, Proc. First Int. Conf.: Opti'89, Comp. Mechanics Publications, Springer, Berlin, 135-144

Atrek, E., (1993). SHAPE: A Structural Shape Optimization Program. In Hörnlein, H. and Schittkowski, K. (eds.), Software Systems for Structural Optimization, Vol. 110 International Series of Numerical Mathematics, Birkhäuser, Basel, 229-249

Atrek, E. and Agarwal, B., (1992). Shape Optimization of Structural Design. In Billingsley, K.R., Brown III., H.U. and Derohanes, E. (eds.), Scientific Excellence in Supercomputing – The IBM 1990 Contest Prize Papers, University of Georgia, Athens, GA, Baldwin Press

Atrek, E. and Kodali, R., (1989). Optimum Design of Continuum Structures with SHAPE. In Prasad, B. (ed.), CAD/CAM Robotics and Factories of the Future, Vol. 2, Proc. of the 3rd International Conference CARS and FOF '88, Springer, Berlin, 11-15

Banichuk, N.V., (1993). Shape Design Sensitivity Analysis for Optimization Problems with Local and Global Functionals, *Mech. Struct. and Mach.*, 21(3): 375-397

Barthelemy, J.F.M. and Haftka, R.T., (1993), Approximation concepts for optimum structural design – a review, *J. Structural Optimization*, Vol. 5: 129-144

Beer, R., (1996). Multidisziplinäre Optimierung von zylinderförmigen Gußbauteilen mit mehreren Zielvorstellungen, Dissertation Universität-GH Siegen, TIM-Forschungsbericht Nr. T11-08.96

Beer, R., Eschenauer, H.A., Lautenschlager, U. and Hillmer, P., (1994). On the Modeling and Optimization of a Pressure Gas Insulation Component - A Multidisciplinary Optimization Task, 5th AIAA/USAF/NASA/ISSMO Symposium on Multidisciplinary Analysis and Optimization. Panama City, Sept. 7-9, 1994, A Collection of Technical Papers, Part 2, American Institute of Aeronautics and Astronautics AIAA, ISBN 1-56347-097-7, 1994, 423-433

Bendsøe, M.P., (1994). Methods for the Optimization of Structural Topology, Shape and Material, Springer, Berlin

Bendsøe, M.P. and Kikuchi, N., (1988). Generating optimal topologies in structural design using a homogenization method, *Com. Meth. Appl. Mech. Eng.* 71: 197-224

Bendsøe, M.P., Diaz, A. and Kikuchi, N., (1993). Topology and Generalized Layout Optimization of Elastic Structures. In Bendsøe, M.P. and Mota Soraes, C.A. (eds.), Topology Design of Structures, Kluwer Academic Publishers, Dordrecht, 159-205

Berchtold, G. and Klenner, J., (1992). The integrated design and manufacturing of composite structures for aircraft using an advanced tape laying technology, DGLR-Jahrestagung, Bremen / Germany, MBB-Bericht S-PUB-491

Bloebaum, C.L., Hajela, P. and Sobieszczanski-Sobieski, J., (1990). Nonhierarchic system decomposition in structural optimization, Proc. Third Air Force/NASA Symposium on Recent Advances in Multidisciplinary Analysis and Optimization, San Francisco, CA, Sept. 24-26, 1990

Bourgat, J.F., (1973). Numerical Experiments of the Homogenisation Method for Operators with Periodic Coefficients, Lecture Notes in Mathematics 704, Springer, Berlin, 330-356

Box, G., Hunter, W. and Hunter, J., (1978). Statistics for Experimenters, Wiley Inc., New York

Bremicker, M., (1989). Dekompositionsstrategie in Anwendung auf Probleme der Gestaltsoptimierung, Dissertation Universität-GH Siegen. VDI-Fortschrittsbericht, Reihe 1, Nr. 173, VDI, Düsseldorf

Bremicker, M. and Eschenauer, H.A., (1989). Applications of a decomposition technique for treating a shape optimization problem. In Advances in Design Automation, Vol. II, ASME Publ. DE-Vol. 19-2, 1-6

Courant, R. and Hilbert, D., (1968). Methods of Mathematical Physics I, Interscience Publishers Inc., New York

Dollyhigh, S.M. and Sobieszczanski-Sobieski J., (1990). Recent experience with multidisciplinary analysis and optimization in advanced aircraft design, Proceedings of the NASA Symposium on Multidisciplinary Analysis and Optimization, 404-411

Eifler, D., Löhe, D. and Scholtes, B., (1991). Residual stresses and fatigue of metallic materials. In Hauk, V., Hougardy, H.P. and Macherauch, E. (eds.), Residual stresses - measurement, calculation, evaluation, DGM Informationsgesellschaft, Oberursel

Erler, J., (1937). Studies of casting stresses in chilled iron rolls, The Iron Age. 45-51, 81

Eschenauer, H., Koski, J. and Osyczka, A. (1990). Multicriteria Design Optimization. Procedures and Applications, Springer, Berlin, Heidelberg

Eschenauer, H.A., (1989), The 'Three Columns' for Treating Problems in Optimum Structural Design. In Bergmann, H.W. (ed.). Optimization: Methods and Applications, Possibilities and Limitations, Springer, Berlin, Heidelberg, New York, 1-21

Eschenauer, H.A., (1992). Multidisciplinary Modeling and Optimization in Design Processes, *ZAMM - Z. angew. Math. Mech.* 72, (6): T428-T447

Eschenauer, H.A., (1998). Development of Highly Precise Radio Telescopes – a typical Multidisciplinary Problem. In Belegundu, A.D. and Mistree, F. (eds.), Optimization in Industry – 1997, ASME, New York, 13-30

Eschenauer, H.A. and Weinert, M., (1993). Structural techniques as a mathematical tool for finding optimal shapes of complex shell structures. In Gianessi, F. (ed.), Non-Smooth Optimization Methods and Applications. Gordon and Breach Science Publishers, Switzerland

Eschenauer, H.A. and Schumacher, A., (1993a). Bubble Method: A special strategy for finding best possible initial designs, Proc. of the 1993 ASME Design Technical Conferences, 19th Design Automation Conference, Vol 65-2, 437-443

Eschenauer, H.A. and Schumacher, A., (1993b). Possibilities of Applying Various Procedures of Topology Optimization to Components subject to Mechanical Loads, *ZAMM – Z. angew. Math. Mech.*, 73: T392-T394

Eschenauer, H.A. and Schumacher, A., (1997). Topology and shape optimization procedures using hole positioning criteria – theory and applications. In Rozvany, G.I.N. (ed.), Topology Optimization in Structural Mechanics, Volume 374 of CISM Courses and Lectures, Springer, Wien

Eschenauer, H.A. and Weinert, M., (1994). Optimal layouts of complex shell structures by means of decomposition techniques, Proc. of 4th AIAA/USAF/NASA/OAI Symp. on Multidisciplinary Analysis and Optimization, Cleveland, OH. AIAA 1994, 999-1007

Eschenauer, H.A., Kobelev, V.V. and Schumacher, A., (1994). Bubble method for topology and shape optimization of structures, *J. Struct. Opt.* 8: 42-51

Eschenauer, H.A., Geilen, J. and Wahl, H.J., (1993). SAPOP - An Optimization Procedure for Multicriteria Structural Design. In Schittkowski, K. and Hörnlein, H. (eds.), Software Systems for Optimization, Birkhäuser, Basel, ISNM-Volume, 207-228

Eschenauer, H.A. and Beer, R., (1998). Multidisciplinary optimization of cast components regarding process characteristics, *J. Structural Optimization,* 16: 212-225

Eschenauer, H.A., Mattheck, C. and Olhoff, N., (eds.) (1991). Engineering Optimization in Design Processes, Lecture Notes in Engineering, Springer, Berlin, Heidelberg, New York

Etman, P., (1997). Optimization of Multibody Systems using Approximation Concepts, Ph.D.-Thesis, Universiteitsdrukkerij TU Eindhoven

Fleury, C. and Schmit, L.A., (1980). Dual Methods and Approximation Concepts in Structural Synthesis, NASA Contractor Report 3226

Geißendörfer, K., (1992). CALTA-GII development for an integrated design and manufacturing system in aircraft industry, ECU European CATIA Users Association, Annual Conference, Mont Pellier

Guan, J. and Sahm, P.R., (1992). Numerische Untersuchung der thermischen Spannungen in realen 3D-Gußbauteilen, *Giesserei* 79: 318-322

Haug, E.J. and Arora, J.S., (1978). Design Sensitivity Analysis of Elastic Mechanical Systems, *Com. Meth. Appl. Mech. Eng.* 15: 35-62

Haug, E.J., Choi, K.K. and Komkov, V., (1986). Design Sensitivity Analysis of Structural Systems, Academic Press Inc., Orlando

Kneppe, G., (1986). Direkte Lösungsstrategien zur Gestaltsoptimierung von Flächentragwerken, Dissertation, Universität-GH Siegen, VDI-Fortschrittsbericht, Reihe 1, Nr. 135, VDI, Düsseldorf

Lautenschlager, U., Eschenauer, H.A. and Mistree, F., (1999). Design-of-Experiments Methods and their Application to Robust Multicriteria Optimization Problems, *ZAMM Journal Applied Mathematics and Mechanics,* Vol. 79, Supplement 1: GAMM 98 Annual Meeting, Minisymposia

Lawson, M. and Karandikar, H.M., (1994). A Survey of Concurrent Engineering, *J. Concurrent Engineering – Research and Applications*, Vol 2(1): 1-6

Mack, W. and Gamer, U., (1988). Zur Berechnung der Wärmespannungen beim Abkühlen thermisch beanspruchter elastisch-plastischer Bauteile, *Forschung im Ingenieurwesen* 54: 48-52

Megahed, M.M., (1990), Elastic-plastic behaviour of a thick-walled tube with general nonlinear hardening properties, *Int. J. Mech. Sci.,* 32: 551-563

Montgomery, D., (1991). Design and Analysis of Experiments, Third Ed., John Wiley and Sons, New York

Papalambros, P.Y. and Chirehdast, M., (1990), An integrated environment for structural configuration designs, *J. Eng. Design,* 1 (1): 73-96

Parkinson, A. and Wilson, M., (1986). Development of a Hybrid SQP-GRG Algorithm for Constrained Nonlinear Programming, Proceedings of the ASME Design Engineering Technical Conference, Columbus, Ohio, Oct. 5-8, 1986

Parsaei, H.R. and Sullivan, W.G., (eds.) (1993). Concurrent Engineering – Contemporary Issues and Modern Design Tools, Chapman and Hall, London, Glasgow, New York

Powell, M.J.D., (1982). VMCWD: A FORTRAN Subroutine for Constrained Optimization, University of Cambridge, Report DANTP 1982/NA4

Reinhardt, H.J., Seiffarth, F. and Hào, D.N., (1993). Approximate Solutions of Ill-posed Cauchy Problems for Parabolic Differential Equations. In Anger, G. (ed.), Inverse Problems: Principles and Applications in Geophysics, Technology and Medicine, 284-298

Roux, W.J., Stander, N. and Haftka, R.T., (1996). Response Surface Approximations for Structural Optimization, Proc. Sixth AIAA/USAF/NASA/ISSMO Symposium on Multidisciplinary Analysis and Optimization, Bellevue, WA, 565-578

Sahm, P.R. and Hansen, P.N., (1984). Numerical simulation and modeling of casting and solidification processes for foundry and cast-house, CIATF, Zurich / Switzerland

Schimöller, H., (1990). Analytische Behandlung von Eigenspannungszuständen auf der Grundlage der Elastizitätstheorie, VDI-Fortschrittsberichte, Reihe 18: Mechanik / Bruchmechanik, 88

Schoofs, A.J.G., (1987). Experimental Design and Structural Optimization, Ph.D.-thesis, Febodruk, Enschede / The Netherlands

Schoofs, A.J.G., Klink, M.B.M. and van Campen, D.H.. (1992). Approximation of structural optimization problems by means of designed numerical experiments, *J. Structural Optimization*, Vol. 4: 206-212

Schumacher, A., (1995). Topologieoptimierung von Bauteilstrukturen unter Verwendung von Lochpositionierungskriterien, Ph.D.-Thesis, University of Siegen, TIM-Report T09-11.95

Schuhmacher, G., (1995). Multidisziplinäre, fertigungsgerechte Optimierung von Faserverbund-Flächentragwerken, Uni-GH Siegen: Dissertation, TIM-Bericht Nr. T07-03.95

Sjöström , S., (1983). Berechnung der Abschreckeigenspannungen in Stahl. In Macherauch, E. and Hauk, V. (eds.), Eigenspannungen. Entstehung - Messung - Bewertung, Vol. 1, DGM Informationsgesellschaft Oberursel, 155-189

Sluzalek, A., (1992). Introduction to nonlinear thermodynamics, theory and finite element solutions, Springer, Berlin, Heidelberg, New York

Sobieszczanski-Sobieski, J., (1988). Optimization by decomposition: a step from hierarchic to non-hierarchic systems. In Proc. Second NASA/Air Force Symposium on Recent Advances in Multidisciplinary Analysis and Optimization, Hampton, VA, Sept. 28-30, 1988

Sobieszczanski-Sobieski, J., (1989). Multidisciplinary optimization for engineering systems. In Bergmann, H.W. (ed.), Optimization: Methods and Applications, Possibilities and Limitations, Springer, Berlin, Heidelberg, New York, 42-62

Stadler, W., (1988). Multicriteria Optimization in Engineering and in the Sciences, Plenum Press, New York, London

Svanberg, K., (1987). The method of moving asymptotes – A new method for structural optimization, *International Journal for Numerical Methods in Engineering* 24: 259-373

Taguchi, G., (1986). Introduction to Quality Engineering: Designing Quality into Products and Processes, Kraus International Publications, White Plains, N.Y

Upadhya, G., Banerjee, D.K., Stefanescu, D.M. and Hill, J.L., (1990). Heat transfer-solidification of structural transitions: chill formation in grey iron, *AFS Trans.* 90: 699-707

Vanderplaats, G.N., Yang, Y.J. and Kim, D.S., (1990). Sequential linearization method for multilevel optimization, *AIAA J.* 28: 290-295

van Houten, M.H., (1988). Function Approximation Concepts for Multidisciplinary Design Optimization, Ph.D.-Thesis, Universiteitsdrukkerij TU Eindhoven

VDI-Guideline 2235, (1987). Wirtschaftliche Entscheidungen beim Konstruieren (Economic Decisions in Design Processes). VDI, Düsseldorf

Weck, M., Du Maire, E. and Vonderhagen, H., (1992). Optimierung von Gussbauteilen im Werkzeugmaschinenbau, *Giesserei* 79: 11-20

Wiese, J.W. and Dantzig, J.A., (1988). Modeling stress development in grey iron castings. In Giamei, A.F., Abbaschian, G.J. and Bayuzick, R.J. (eds.), Solidification processing of eutectic alloys. The Metallurgical Society, 163-174

Wolfersdorf, (1994). Inverse und schlecht gestellte Probleme, Akademie-Verlag, Berlin

Yoshimura, M. and Takeuchi, A., (1994), Concurrent Optimization of Product Design and Manufacturing on Information of User's Needs, *J. Concurrent Engineering – Research and Applications*, Vol 2(1): 33-44

# CHAPTER 2

# Old and New Non-Gradient Methods in Engineering Optimization

J. Blachut

The University of Liverpool, Mechanical Engineering
Liverpool, UK

**Abstract.** Current interest in non-gradient optimization can be attributed to two, distinct developments. The first stems from studying natural processes and the second from fast progress in the computing environment. The first has resulted in new optimization techniques like simulated annealing, tabu search and genetic algorithms whilst the second has made many of the existing zero order methods computationally affordable. This chapter discusses simulated annealing, tabu search, dynamic programming, random methods and other heuristics. A variety of illustrative examples is provided together with practical examples drawn from structural mechanics.

## 2.1 A Common Platform for Heuristics

### 2.1.1 Introduction

Typical optimization problems encountered in engineering frequently require the satisfaction of conflicting design aims, whilst aiming at an attainment of optimal performance. Despite the availability of a number of optimization techniques, it appears that none of them could be regarded as satisfactory across the entire spectrum of problems encountered in engineering. Various techniques, with variable degrees of sophistication, would usually be needed in practice and they may be classified in a variety of ways. A useful approach could be based on the type of information that must be provided in order to find the optimum. Calculus based methods, utilising gradient information to guide the search, work well in their intended class, i.e. for smooth, unimodal and convex functions, but they may converge to local optimal solutions elsewhere. Additionally, all traditional methods assume deterministic quantities and they have problems dealing with any stochastic effects. Finally, these methods are largely designed for optimization problems with continuous variables and are inadequate for problems with discrete variables.

A class of methods requiring only function values is collectively known as zero-order, non-gradient, or direct search methods. These methods have been in use for many years and they have produced some excellent results. Their strength is in easy programming effort, their ability to handle non-convex and discontinuous functions as well as their ability to work with discrete values of the decision variables (see, e.g., Arora et al., 1995, and Hajela, 1999). The price paid for flexibility and generality of these methods is that they require a large number of function evaluations, leading to an exponentially increasing number of iterations as the problem becomes larger.

Hence, the emphasis has recently been directed toward heuristic techniques despite the fact that they lack any 'a priori' guarantee in terms of solution quality or running time. The term heuristic comes from the Greek 'heuriskein' meaning to find or discover. A loose definition of the term heuristic, as it is currently accepted, is as follows (Reeves, 1995): "A heuristic is a technique which seeks good, i.e. near-optimal solutions, at a reasonable computational cost without being able to guarantee either feasibility or optimality, or even in many cases to state how close to optimality a particular feasible solution is". Three methods, particularly popular recently, have all arisen, at least in part, from a study of nature. They are genetic algorithms, simulated annealing and tabu search. These three methods have a high probability of locating the global optimum in a multimodal search landscape. A brief outline of them is given later in this section whilst the latter two, i.e. Simulated Annealing, SA, and Tabu Search, TS, are addressed in greater detail in sections 2.2 and 2.3, respectively.

Advances in computing hardware, i.e. speed of computing, size of memory and disc space makes the use of some of the 'older' zero-order methods feasible and applicable to a range of new problems. Several of these methods have been selected here to illustrate the feasibility of their use in the new environment. This includes dynamic programming, random techniques and complex method of Box.

Apart from the use of individual methods, it appears that hybridisation can bring further benefits as briefly outlined later.

### 2.1.2 General Optimization - Problem Statement

Optimization can be generally defined as getting the best, i.e. the minimum or maximum of something under given constraints. This would normally require mapping of a physical problem into a mathematical model which is a symbolic representation of relations between factors affecting the problem. Standard terminology is usually associated with description of such model. Clear identification of variables, parameters, the objective function and constraints would normally be necessary. In any decision problem the factors that can be manipulated to achieve the desired result are known as variables. The factors that can affect the objective, but are not controllable, are known as parameters. The value, utility or return associated with particular values of variables and parameters is known as the objective function or the cost function. The limitations placed on variables or parameters, either explicit or implicit, are known as constraints. From a mathematical point of view, it is enough to consider minimization. In the general constrained optimization problem, one aims at finding a design variable vector **x** which minimizes a cost function F(**x**) subject to a set of constraints. This can be formally stated, in a standard form, as follows:

$$\text{Minimize } F(\mathbf{x}) \qquad \text{(cost function)} \tag{2.1}$$

subject to

$$g_i(\mathbf{x}) = 0, \qquad i = 1, \ldots, p \qquad \text{(equality constraints)} \tag{2.2}$$

$$g_i(\mathbf{x}) \leq 0, \qquad i = p+1, \ldots, m \qquad \text{(inequality constraints)} \tag{2.3}$$

$$x_{iL} \leq x_i \leq x_{iU}, \qquad i = 1, \ldots, n \qquad \text{(side constraints)} \tag{2.4}$$

where F and g are cost and constraint functions, respectively; $x_{iL}$ and $x_{iU}$ are lower and upper bounds for the variables $x_i$; p, m and n are the numbers of equality constraints, total constraints and design variables, respectively. The objective function $F(\mathbf{x})$ and the constraint functions $g(\mathbf{x})$ may be linear or non-linear functions of the design vector $\mathbf{x}$. These functions may be explicit or implicit in $\mathbf{x}$ and they may be evaluated by any analytical or numerical technique. Generally it is important that these functions are continuous and that they have continuous first and second derivatives in $\mathbf{x}$. There are a number of classical text books on optimization, where further details can be found, e.g., Gill et al. (1999), Rao (1984), or Vanderplaats (1984).

### 2.1.3 General Optimization - Solution Strategy

**Unconstrained problems.** The unconstrained minimization of $F(\mathbf{x})$ requires finding the vector $\mathbf{x}^*$ such that the gradient of $F(\mathbf{x}^*)$ is to be zero, i.e.:

$$\nabla F(\mathbf{x}^*) = 0 \tag{2.5}$$

This is obviously only a necessary condition. A further check must be made to ensure the sufficiency. This translates into the requirement that the Hessian matrix be positive definite, where the Hessian matrix is the matrix of second partial derivatives of the objective with respect to the design variables. In another words, a given design $\mathbf{x}^*$ is at least a local minimum, when in addition to Eq. (2.5) the Hessian matrix is positive definite, i.e. all of its eigenvalues are positive. There is no convenient method to demonstrate that this local minimum is also a global one unless the Hessian matrix is positive definite for all possible values of the design vector $\mathbf{x}$. This is a recognised deficiency of gradient based methods. It is also rarely possible to solve the above problem directly because the components $\nabla F$ may be highly non-linear implicit functions of $\mathbf{x}$. Instead one resorts to numerical search procedures whereby minimization of $F(\mathbf{x})$ is carried out in an iterative fashion, as briefly outlined below.

**Iterative methods.** Let us assume that the starting point $\mathbf{x}^0$ is known. The gradient $\nabla F(\mathbf{x}^0)$ can be then calculated and the search in the negative, i.e. in the direction of steepest descent, can be carried out by choosing a new vector $\mathbf{x}$ according to the following expression:

$$\mathbf{x} = \mathbf{x}^0 - \alpha \, \nabla F(\mathbf{x}^0) \tag{2.6}$$

If one tries several values for the scalar $\alpha$ and for each resulting vector $\mathbf{x}$, one evaluates $F(\mathbf{x})$ then for some value, say $\alpha^*$, a minimum of $F(\mathbf{x})$ is obtained for this search direction. It is instantly obvious that by using the gradient of $F(\mathbf{x})$ one is able to limit the search to a specific direction rather than searching the whole design space randomly. In principle this method is dependent on the availability of the gradient and it is generally referred to as gradient, or first-order, search method.

It may be more appealing to think about the optimization algorithm adopting the following iterative process:

$$\mathbf{x}^q = \mathbf{x}^{q-1} + \alpha^*(q)\mathbf{S}^q \tag{2.7}$$

where q is the iteration number and $\mathbf{S}^q$ is a vector search direction in the design space. The scalar quantity $\alpha^*(q)$ gives the distance that one wishes to move in direction $\mathbf{S}^q$. Two central issues in the above iterative process are the determination of the search direction $\mathbf{S}^q$ and the determination of the step scalar $\alpha^*(q)$. A number of different algorithms have been proposed for choosing the search direction where the gradient information is not required. These include random search, Powell's search, pattern search methods, simplex method, etc. Further discussion of zero order search methods will be given later.

**Constrained problems.** If at least one constraint on the design is active then, as it is intuitively expected, it is no longer necessary for the gradient of the objective to vanish at the optimum. The conditions to be satisfied at a constrained minimum point, $\mathbf{x}^*$, are known as Kuhn-Tucker conditions. They are the necessary conditions to be satisfied at a local minimum of $F(\mathbf{x})$. In general, these conditions are not sufficient to ensure a local minimum. There is however a class of problems, called convex programming problems, for which the Kuhn-Tucker conditions are necessary and sufficient for a global minimum. Again, details can be found in a number of classical texts. Books by Gill et al. (1999), Rao (1984), and Vanderplaats (1984), seem to be good starting references.

### 2.1.4 Local and Global Optima

In the process of minimizing a function one would normally expect to find a unimodal minimum, i.e. the lowest value of the function in the entire search region. If the function is not unimodal then relative minima may exist and there is no guarantee of finding the true minimum. One usually finds the minimum nearest to the starting point and there is no known mechanism, within the class of gradient based methods, for escaping from the local minimum. It is a perennial problem in optimization - how to deal with the existence of relative minima and in essence how to find a global minimum ? In terms of Fig. 2.1 we would refer to internal points 'C' and 'E' with a further local minimum at point 'A' which, happens to be at the boundary of the interval $[x_1, x_2]$, where the first derivative of the function need not vanish.
Three heuristics outlined later on in this Chapter can, in principle, provide us with a global minimum at 'E'. All three of them have appropriate mechanisms for escaping from being trapped in local minima 'C' and 'A'.

### 2.1.5 Good Solution Versus Optimal Solution

It is not uncommon in practice to see problems in which the cost function becomes discontinuous, non-convex or with internal gaps within an interval - as sketched in Fig. 2.1b. In situations like these, there is only a limited scope for gradient based methods. The direct-search technique(s), on the other hand, can be beneficial here. They typically can be used in situations for which differentiation is a problem or the cost function behaves badly. Most zero-order methods would not be affected by such difficulties. It is however important to appreciate other deficiencies which are embedded in the direct search approach. The gradient methods, for example, are generally supported by optimality proofs whilst most of the zero-order methods are not. As a consequence they would normally produce a 'good solution' rather than an 'optimal solution'. Associated with this difficulty is the problem of detecting convergence and deciding on stopping criteria. It is therefore important to recognise that the direct-search methods have an important role to play in optimization but they should be used only where their characteristics make them more appropriate than gradient based methods. As a result of employing the zero-order methods we can produce a good or better design, and not necessarily the optimal one. Within the realistic time and resource constraints this might be the only practically viable way forward - see Vanderplaats (1999).

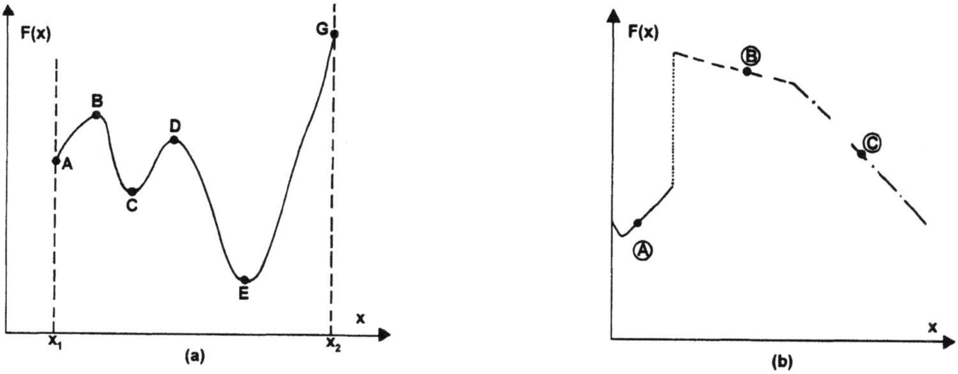


**Figure 2.1** Relative optima and global minimum in an interval $[x_1, x_2]$ – Fig. 2.1a. One-dimensional response function illustrating discontinuity, gap and non-convex character – Fig. 2.1b (various modes of structural response correspond to parts 'A', 'B' and 'C').

### 2.1.6 Combinatorial Optimization - Travelling Salesman Problem

One specific group of problems occurs when the design variables form a discrete set, i.e. where the solution is a set, or a sequence, of integers or other discrete objects. The problem of finding an optimal solution to such problems is known as combinatorial optimization and it is positioned between mathematics, computer science and operations research. Loosely speaking, a combinatorial optimization problem is any optimization problem that has a finite number of feasible solutions and it can be formulated as the one given by Eqns (2.1 - 2.4).

A naive approach to solving a combinatorial problem would be to list all feasible solutions of a given problem, evaluate their objective functions, and identify the best. However, it is immediately obvious that this approach of complete enumeration is likely to be grossly inefficient. Although it is possible in principle to solve any problem in this way, in practice it is not, because of the search space size, which grows exponentially as the number of design variables increases. One further observation here is that since the set is discrete, one is deprived of any feeling of downhill movements in a favourable direction. The most prominent case of the unsolved combinatorial case is that of Travelling Salesman Problem, TSP, despite considerable effort put into devising an efficient algorithm. The TSP problem is formulated as follows: let us consider n cities and an n x n matrix d[p,q], whose elements denote the distance between each pair (p,q) of the n cities - see Fig. 2.2a. A salesman tour is defined as a closed path visiting each city exactly once. The problem is to find a tour of minimal length.

The simplicity of the problem statement and the difficulty involved in obtaining a solution, coupled with wide ranging applications which would benefit from a solution, are the reasons behind the continuous interest in TSP. The methods used to solve TSP can be broadly categorised into two: enumerative and approximation techniques. Cutting planes, branch and bound, nearest neighbour and dynamic programming are examples of enumerative optimization techniques. Genetic algorithms, simulated annealing and tabu search are examples of approximation methods which are frequently adopted in practice. For the case shown in Fig. 2.2a the path of minimum length is shown in Fig. 2.2b. In this case the solution has been obtained using the algorithm published in Press et al. (1986).

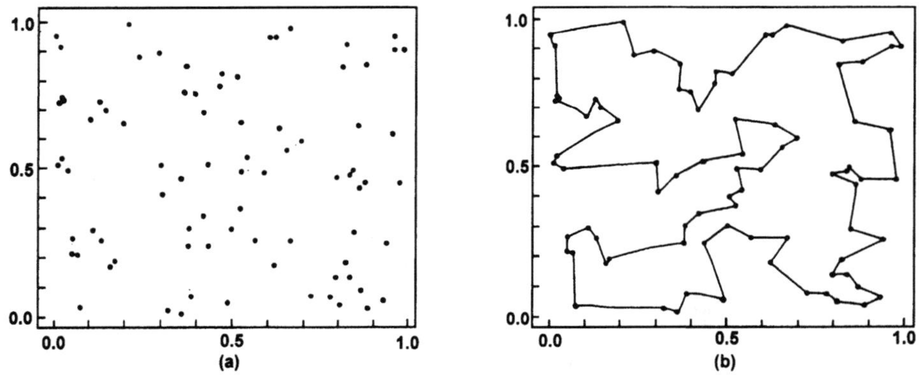


**Figure 2.2** Distribution of n randomly chosen cities to be visited by a salesman (Fig. 2.2a). The minimum path for a TSP-problem has been obtained by employing simulated annealing heuristic (Fig. 2.2b).

### 2.1.7 Neighbourhood and Neighbourhood Search Algorithms

Central to most heuristic search methods is the concept of neighbourhood and that of neighbourhood search. Let us assume that $\mathbf{x}^*$ denotes a solution vector and $\mathbf{X}$ is the set of all feasible solutions. Let us also assume that the cost of solution $\mathbf{x}^*$ is $F(\mathbf{x}^*)$. Then every solution $\mathbf{x}^* \subset \mathbf{X}$ has an associated set of neighbours, $N(\mathbf{x}^*) \subset \mathbf{X}$, called the neighbourhood of $\mathbf{x}^*$. Each solution $\mathbf{x}' \in N(\mathbf{x}^*)$ can be reached directly from $\mathbf{x}^*$ by an operation called a move. Many different types of moves are possible in any particular case. A typical neighbourhood search operates by generating neighbours in an iterative way. A move to a new solution is made whenever certain criteria are met. There are many possibilities in which candidate moves can be chosen for consideration, and in defining criteria for accepting candidate moves, see Aarts and Lestra (1998). Three popular moves are: ascent, steepest ascent and next ascent. The most common case is that of ascent, in which the only moves accepted are to neighbours which lead to improvement of the current solution. Steepest ascent corresponds to the case where all neighbours are evaluated before a move is made, which is the best available move. The next ascent is similar, but the next candidate that improves the current solution is accepted, without necessarily examining the whole neighbourhood. Normally, the search terminates when no moves can be accepted.

For the three methods considered here, i.e. GAs, SA and TS, a common framework of a neighbourhood comprises the following elements:

- *Initialisation.* Choose an initial solution $\mathbf{x}$ to be the current solution and compute its cost $F(\mathbf{x})$.
- *Neighbour generation.* Select a neighbour $\mathbf{x}'$ of the current solution $\mathbf{x}$ and compute $F(\mathbf{x}')$.
- *Acceptance test.* Test whether to accept the move from $\mathbf{x}$ to $\mathbf{x}'$. If the move is accepted, then $\mathbf{x}'$ replaces $\mathbf{x}$ as the current solution. Otherwise $\mathbf{x}$ is retained as the current solution.
- *Termination test.* Test whether the algorithm should terminate. If it terminates, output the best solution generated. Otherwise return to the neighbour generation step.

### 2.1.8 Outline of Genetic Algorithms, GAs, Simulated Annealing, SA, and Tabu Search, TS Methods

**Genetic algorithms, GAs.** Genetic algorithms are stochastic, iterative, evolutionary, general-purpose search techniques rooted in the mechanisms of evolution and natural genetics. The GAs strategies employ the notions of mating and survival of the fittest to evolve a solution to a problem. The concepts of GAs are usually credited to Rechenberg (1973), and Holland (1975). An initial population of candidate configurations is generated by random selection of the individual bits in a binary string, or chromosome, of a given length. The role of the objective function is mapped to the concept of candidate fitness. The artificial recombination among the population of strings is based on the fitness and the accumulated knowledge. In every new generation, a new set of strings is created by using randomised parents selection and crossover from the old set. Small but random changes are also introduced into the algorithm and this operation is called mutation. A number of so called GA-operators have been explored in the past and an elitist operator is just one of them (it ensures that the best chromosome is retained and it always enters the next generation without being altered).
The structure of GAs consists of the following elements:
- *Initialisation*. Chose an initial population P containing M solutions to be current population.
- *Evaluation*. Compute fitness value for each solution M.
- *Reproduction*. Use fitness values to select solutions from P to form a mating pool.
- *Recombination*. Apply crossover, mutation and any other operator to solutions of the mating pool to form a new population which, replaces P as the current population.
- *Termination*. Test whether the algorithm should terminate. If it terminates, output the best solution generated. Otherwise return to the evaluation step.

There are several books devoted to GAs, e.g., Goldberg (1989), Davis (1991), Haupt and Haupt (1998). Chapter six provides more details about GAs.
There have been a large number of papers applying GAs to structural optimization and references by Jenkins (1991), Hajela and Lee (1995), and Nagendra et al. (1996) are just a sample of them.

**Simulated annealing, SA.** Simulated annealing is based on thermodynamics considerations, with annealing interpreted as an optimization procedure. SA uses a probabilistic acceptance test. A move to a neighbour which decreases the objective function value, or leaves it unchanged, is always accepted (we accept $\mathbf{x}'$ as the new current solution if $\Delta \leq 0$, where $\Delta = F(\mathbf{x}') - F(\mathbf{x})$). Moves to $\mathbf{x}'$ which increase the value of the objective function, i.e. if $\Delta > 0$, are accepted with probability $\exp(-\Delta/kT)$. The parameter T is historically called temperature and its value decreases during the search process. Simulated annealing's randomised nature leads to asymptotic convergence to global optimal solutions.

**Tabu search, TS.** Tabu search tries to mimic the human memory process during a neighbourhood search in which a local-optima avoidance mechanism is built. TS uses a tabu list to define prohibited moves that prevent the method of returning a previous current solution after a short sequence of moves. A neighbour is referred as tabu if it corresponds to a prohibited move contained in the tabu list. The tabu list can, however, be overridden if an aspiration criterion is used.

A move from **x** to **x'** is automatically approved if it is non-tabu and F(**x'**) < F(**x**). In the case of the aspiration criterion being used a move is accepted, if **x'** is tabu and F(**x'**) is lower than the objective function value of any previously generated solution.

## 2.2 Simulated Annealing

### 2.2.1 Simulated Annealing in Structural Optimization - Literature Review

Simulated Annealing, SA, has been primarily explored in areas such as Operations Research, Computer Science and Artificial Intelligence with research and applications in the field of combinatorial optimization. A recent comprehensive bibliography, containing 1380 entries, on heuristics can be found in work by Osman and Laporte (1996). In addition to simulated annealing it also covers genetic algorithms and tabu search as well as, other search strategies and hybrids. Apart from referencing past research effort into SA, the above bibliography provides references for applications covering the following problems: assignment, combinatorial, design, function optimization, graph, placement, location and allocation, manufacturing, production planning, load balancing, multi-objective optimization, networks and telecommunications, routing and transportation, scheduling and statistical.
It transpires from the detailed listing and available literature that structural optimization problems have rarely adopted the SA methodology as a solution tool. Balling (1991, 1996) reports on the use of SA in the discrete optimization of three-dimensional steel framed buildings. The total frame weight was minimized subject to design-code specified constraints on stress, buckling and deflection. The design vector consisted of 11 variables. Two approximation strategies were examined as means of reducing the computational effort associated with structural response. Direct comparisons were made between performance of branch-and-bound and SA methods. The latter was found to be more robust and faster (at least in some cases). Further reduction of heavy computing effort usually required by SA is discussed by May and Balling (1992). They introduce a filter through which each design candidate must pass before it is accepted for computationally intensive structural analysis. Once a random candidate configuration is generated, its mass is quickly evaluated together with an approximate estimate of constraints value. A scheme is devised whereby even less feasible candidates can pass through the filter in a probabilistic manner. This methodology is applied to minimization of steel frame's mass with typical constraints on displacements and stresses. Results show that filter size can affect convergence to a global optimum. If, on the other hand, the user has acquired enough information about the problem in-hand, then filter size can speed-up convergence. A large tetrahedral truss has been optimized for minimum surface distortion by Kincaid (1990, 1991). Similarly, Chen et al. (1991) apply the SA to large truss structures in which both passive and active vibration suppression is optimized. In the cooling schedule, the authors modify the standard, canonical form of SA, i.e. the best solution found so far is used as a starting configuration each time the temperature is reduced. Ceranic et al. (2000) apply SA to practical design of reinforced concrete retaining walls. The problem has 7 design variables. Behaviour inequality constraints, which are handled through a probabilistic weight estimates approach, are also considered. In design practice, fresh starts would normally be needed each time a new design specification is made. It would therefore be essential to obtain the global optimum each time the specification is altered. Within this framework authors discuss, in their inter-active implementation of SA process, the effect of different set-ups of generic parameters. Topping et al. (1993) consider simultaneous

optimization of topology and cross sectional design variables. Typical aluminium trusses, for which optimal solutions based on other methods are available, serve as examples. The effects of the generic parameter set-up on the search strategy are discussed (including several approaches to penalty based treatment of constraints). Bennage and Dhingra (1995a) apply the SA algorithm to the design of planar and spatial structures. They comprehensively address the influence of the SA generic parameters on the results. Examples, in which the cost function includes minimization of weight, the minimization of fundamental natural frequency and minimization of deflection of selected nodes, are discussed. Constraints are treated either through rejection of unfeasible configurations or through penalty function approach. The multiobjective optimization problem is solved by combining individual objective functions according to the game bargaining model. Continuous adjustment of the search space, together with gradient assessment of constraint violation, is given in paper by Tzan and Pantelides (1996). Detailed examples are provided for statically and dynamically loaded trusses. Shim and Manoochehri (1997) present a methodology in which shape of structural components is altered by removing or restoring elements in order to minimize the volume of a component, subject to constraints on stress and maintenance of the model connectivity. Examples include design of a cantilevered plate, design of a hook and design of a torque arm. In the SA algorithm itself three different acceptability criteria were considered, i.e. cost acceptability, stress acceptability and connectivity acceptability. In a follow-up work, Shim and Manoochehri (1999) explore features of the enumeration technique and the SA. After determining deficiencies of both methods they propose to combine both approaches whereby the enumeration method is applied to local minimization whilst the simulated annealing is used in the global minimization part. This approach stems from the fact that the enumeration technique had a fast convergence to a local minimum whilst the simulated annealing had the ability to search for the global solution. It is claimed that the proposed hybrid, deterministic/stochastic optimization method, can be almost 10 times faster than the SA and all that without losing globality. The design of a cantilevered plate and a hook serve as illustrative examples.

Szewczyk and Hajela (1993) examine a neural network approximation of planar and spatial truss structures' response through a simulated annealing search strategy for global optimum. Various n-n architectures were set-up with their training based on FE generated input. The authors discuss imperfection limits on the training of n-n and the quality of resulting near-optimal solutions generated by the use of SA.

It is always tempting to compare performance of various methods on a like for like basis. This has been attempted by Manoharan and Shanmuganatham (1999) for SA, TS, GAs and branch-and-bound methods. A test example is a six-bar truss made from hot rolled angle sections. Results suggest that, whilst all four methods arrive at sub-optimal configurations, the tabu search is faster than both SA and GAs. Minimization of weight of a collection of planar and three dimensional bar structures is discussed by Botello et al. (1999). Constraints on stresses and deflections are included in some cases, only. A family of several stochastic algorithms has been used and this includes GAs, SA and hybrids of the latter two. The authors conclude that inclusion of the Metropolis acceptance operator has always led to improved performance of considered methods. But there appears to be no clear cut winner within the class of analysed cases. Other aspects of SA's hybridization have been considered by Lin and Chen (2000).

SA technique typifies a process of sequential decision making. Nevertheless limited speedups have been achieved through introduction of parallel computing. Review of attempts in which parallel modes of operation have been introduced into simulated annealing heuristic is discussed by Balling (1996) and by Leite and Topping (1999).

### 2.2.2 Simulated Annealing - Basic Concepts, the Physics Analogy and SA Parameters

Simulated Annealing, SA, is a probabilistic heuristic based on thermodynamics considerations. In the literature the SA heuristic is usually referred to as 'annealing algorithm', with an implicit meaning of a solution technique implemented in the style of annealing. The technique is motivated by an analogy to the statistical mechanics of annealing in solids. A low energy state usually means a highly ordered state, such as crystal lattice. To accomplish this, the material is first melted, i.e. heated to a temperature that permits many random, disordered, high-energy atomic rearrangements. Then it is cooled slowly and as the temperature is reduced the atoms migrate to a more ordered state with lower energy. The final degree of order depends on the temperature cooling rate. The fast cooling process, known as quenching, is characterised by a monotonic decrease in energy to an intermediate state of semiorder. The slow cooling process, on the other hand, is characterised by a general decrease in the energy level with occasional energy increases. At the final stages of the annealing process, the system's energy reaches a much lower level than in rapid cooling, i.e. in quenching. The atomic arrangement reaches a higher degree of order and the material freezes into a good crystal. Annealing therefore, allows the obtaining of a more global energy minimum than is possible in a quick quenching process, equivalent to a local energy minimum. Numerically realised simulated annealing techniques use an analogous set of 'cooling operations' for non-physical optimization problems, in effect transforming a poor, unordered solution into a highly optimized, desirable solution.

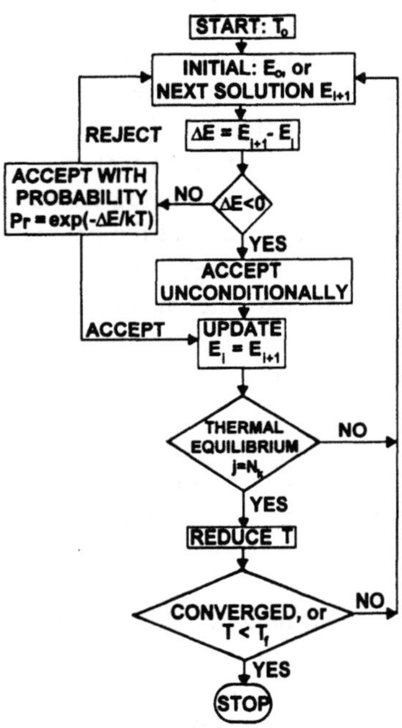


**Figure 2.3** Flowchart of a canonical strategy for simulated annealing.

The probability $P(E_i)$ that at a given temperature T, the system is in state, i, is given by the Boltzman distribution:

$$P(E \equiv E_i) = [1/Z(T)] \exp(-E_i/kT) \tag{2.8}$$

where k is the Boltzman constant, Z(T) is a normalisation factor and $E_i$ is the energy of state, i. Numerical realisation of simulated annealing is based on the theory of Markov chains where it can be shown that as $T \rightarrow 0$, only globally optimum states have a non-zero probability of occurrence and the algorithm converges to these optimum solutions with probability one - Aarts and Korst (1989).

The process of annealing reduces the temperature from an elevated point in a series of controlled steps. At a given temperature, a number of random variations of the system state are considered. If a state results in a lower energy level, it is unconditionally accepted. If, however, a higher-energy state results from the variation, it is only accepted with a probability Pr, defined as follows:

$$Pr = \exp[-(E_2 - E_1)/kT] \tag{2.9}$$

where the state $E_2$ corresponds to a higher energy level than state $E_1$. Fig. 2.3 shows a flowchart of a standard simulated annealing algorithm. The avoidance criterion, of being trapped in a local minimum, centres around occasional acceptance of moves which do not improve the objective. These, probability based, uphill moves are a central feature of any SA algorithm. They mimic the physical processes in which a global energy minimum is obtainable provided that a right cooling process is adopted. Numerical implementations of SA try to retain this central concept as discussed below.

The choices the designer of a simulated annealing heuristic has to make fall into two groups, i.e. problem specific and generic. Problem specific items include: (i) Configuration, (ii) Neighbourhood of a configuration, (iii) Cost of a configuration, and (iv) Initial configuration. There are four generic parameters which must be specified for any SA, i.e.: (i) Initial temperature $T_o$, (ii) Frozen state represented by the final temperature $T_f$, (iii) Cooling ratio, i.e. the rate at which the temperature is lowered between two cooling cycles, and (iv) Epoch length, i.e. the number of random variations at each temperature ($N_k$ in Fig. 2.3). The latter two parameters are collectively referred to as an annealing or cooling schedule. All of them are however problem dependent and devising a good set for a specific problem relies, by-and-large, on empirical studies.

The SA algorithm can be modelled as a process which, given a neighbourhood structure, attempts to transform the current configuration into one of its neighbours. This process, as mentioned earlier, can be modelled as Markov chain. For each temperature value one obtains a single Markov chain of length, $N_k$. The epoch length, $N_k$, should be large enough to achieve a good solution. It should not be overly dependent on the temperature and the fact that at low temperatures the number of moves accepted is small should be taken into account. Otherwise, unnecessarily long Markov chains would exist at low temperatures. The temperature should be decreased in such a way that one does not have to employ very long Markov chains. The last parameter, which must be specified, is that of the stopping criterion which will specify when the system is frozen. A frequent approach is to discontinue the execution once the optimal configuration found so far remains unchanged for a number of temperature reduction stages.

### 2.2.3 Generic Parameters

It is clear from the above considerations that a number of parameters have to be decided upon in an annealing process. Whilst they are generic, there is no single set which would be universally applicable. Merits of various modifications which are available in literature are based on particular, problem dependent, numerical trials. Let us provide a brief outline how the generic parameters might be chosen:

(i) Firstly, the process must start in such a way that virtually all of the moves can be accepted. This means that the initial temperature, $T_o$, must be sufficiently high. Its magnitude is established by monitoring the acceptance rate. The empirical rule here is as follows. Choose a large value for $T_o$ and perform a small number of moves. If the number of accepted moves divided by the number of total moves is smaller than a given acceptance then double the magnitude of the initial temperature $T_o$. The process should continue, until the obtained acceptance ratio exceeds the assumed acceptance ratio. In practice an initial acceptance ratio of between 40% and 90% appears to be a good starting point.

(ii) In theory the annealing process should continue until the final temperature, $T_f$, is zero. In practice however it is sufficient to halt the process when the probability of an uphill move has become small.

(iii) The physical background of SA stipulates that, at a given temperature, the system should be allowed to cool down for a sufficiently long period of time. The number of attempted moves at each temperature remains therefore unknown. Usually the length of the Markov chain is taken as the number of bits, or number of variables, available for random selection. Bennage and Dhingra (1995a), argue that in order for the annealing algorithm to identify a global optimum, it is essential that the system attains thermal equilibrium at each given temperature. The equilibrium configuration would be obtained by sampling enough configurations in the neighbourhood of a current solution. They propose to use an increasing length of the epoch $N_k$ as the temperature T drops, i.e.:

$$N_k(T) = N_f + (N_f - N_o)(T-T_f)/(T_f - T_o) \tag{2.10}$$

where $N_k$, $N_f$ and $N_o$ are the epoch lengths at the current, final and initial temperatures, respectively. They successfully used $N_o = 1$ and $N_f \in [3,6]$ for structural optimization problems.

(iv) The temperature is reduced once the equilibrium is judged to have been reached. A simple way of reducing temperature is based on a geometric schedule - Kirkpatrick et al. (1983), i.e.

$$T_{new} = \alpha T_{current}, \tag{2.11}$$

where $\alpha$ is a constant and it is typically taken from the range 0.9 to 0.99. Glass and Potts (1996) take $N_k$ iterations at each temperature level and then for given values of $T_o$ and $T_f$ they use the constant $\alpha$ given by:

$$\alpha = (T_f/T_o)^{1/(Nk - 1)}\,. \tag{2.12}$$

Lundy and Mees (1986), on the other hand, propose to use a single iteration at each temperature level ($N_k \equiv 1$, nit $\equiv$ number of temperature levels). Temperatures between new and current levels are then given by the following expression:

$$T_{new} = T_{current}/(1 + \beta T_{current}), \quad \text{where} \qquad \beta = (T_o - T_f)/((nit - 1)T_o T_f). \tag{2.13}$$

Under the Lundy-Mees cooling scheme, the temperature decreases faster than in a geometric scheme, and consequently more iterations are performed at lower temperatures. The performance of a geometric scheme is often more sensitive to the choice of initial temperature $T_o$. As an alternative to the above, Balling's (1991) approach to weight optimization of steel frames is as follows. The user inputs three parameters, i.e. $P_s \equiv$ the starting acceptance probability, nit $\equiv$ the number of cooling cycles and $P_f \equiv$ the final acceptance probability. This allows the calculation of the initial temperature from $T_o = -1/\ln(P_s)$. The usual geometric cooling scheme is then used with the coefficient, $\alpha$, given by:

$$\alpha = [\ln(P_s)/\ln(P_f)]^{(1/(nit - 1))} \tag{2.14}$$

The acceptance probability is taken as $Pr = \exp(-\Delta W/CT)$, where $\Delta W$ = Weight of candidate frame - Weight of current frame. The normalisation constant, C, is taken as the running average of $\Delta W$.

It is seen from the above discussion that various implementations of SA use different approaches to the cooling schedule. In conclusion, it remains true that the cooling schedule for simulated annealing can be chosen virtually arbitrarily, as long as the temperature converges to zero in probability - Laarhoven and Aarts (1992).

### 2.2.4 Neighbourhood Configuration for Continuous Domain

The SA technique can also be applied to the optimization of a cost function defined in a continuous domain. The most frequent method of generating a candidate point $\mathbf{x}_{new}$ is based on seeking a random configuration within the neighbourhood of the current configuration $\mathbf{x}_o$. One further possibility based on variable step, is as follows. One starts with the initial, feasible, point $\mathbf{x}_o$. The next point is generated using the following relation:

$$\mathbf{x}_{new} = \mathbf{x}_o + r v_h \mathbf{e}_h , \tag{2.15}$$

where r is a random number generated in the range [-1,1], $v_h$ is the component of the step vector along the $h^{th}$ direction given by the vector $\mathbf{e}_h$. If the $h^{th}$ component of $\mathbf{x}_{new}$ falls outside the bounded domain given by constraints then a fresh random search takes place until a feasible point $\mathbf{x}_{new}$ is obtained. The magnitude of vector **v** is related to the acceptance ratio acc $\equiv n/N_k$, where n is the number of accepted moves and $N_k$ is the total number of evaluated moves ($\equiv$ epoch length). Both n and $N_k$ are at a given temperature. For each direction, h, the new step vector component is taken as:

$$v_h = v^o_h(1 + c(n/N_k - 0.6)/0.4) \qquad \text{if } n > 0.6N_k \tag{2.16}$$

$$v_h = v^o_h/(1 + c(0.4 - n/N_k)/0.4) \qquad \text{if } n < 0.4N_k \tag{2.17}$$

$$v_h = v^o_h \qquad \text{otherwise} \tag{2.18}$$

where c is a constant parameter. The purpose of these variations in step length is to maintain the average percentage of accepted moves at about 50% of the total number of moves, at a given temperature. Papers by Vanderbilt and Louie (1984), Bohachewsky et al. (1986) and Corana et al. (1987) provide a more detailed discussion on choosing a step size for the continuous domain and on adopting other schemes.

### 2.2.5 Localised Simulated Annealing

By the nature of a simulated annealing process, the majority of search time will be spent at low temperature. Spending a lot of time at low temperatures can overlook many good solutions originating from higher temperatures if the corresponding distributions are not close to the stationary ones. Given a limited search time one is faced with two issues in devising a proper annealing schedule: (i) Closely approximating stationary distributions at high temperatures could take most of the available search time, and leave little time for tracing the optimal solution at final stages of SA process; (ii) Greedily focusing on the better solutions in current spaces could sacrifice the approximation of stationary distributions, and this could degrade the probability of finding global optimum. These shortcoming are addressed in an approach known as Localised Simulated Annealing, LSA, see Li et al. (1996). The basic idea is to divide the search space into a set of disjoint subspaces. Different temperatures and annealing speeds could be maintained in different subspaces depending upon adopted evaluation criteria. In the localised simulated annealing method the best solutions in individual subspaces are used to assess these subspaces and the distance between a local best and the global best would measure the goodness of a subspace. The search procedure records the local best solution, $S^*$ for all subspaces and the current global best solution $S^b$. At the end of each temperature level, the procedure measures the distance between $S^b$ and each $S^*$, i.e. $|cost(S^*) - cost(S^b)|$, called the approachability of a subspace. If that distance is greater than a pre-determined parameter, $\varepsilon_T$, i.e. $|cost(S^*) - cost(S^b)| / |cost(S^b)| > \varepsilon_T$, then the local best solution is not reasonably near to the global best solution, and the subspace represented by $S^*$ is declared as 'no-good'. If this happens the temperature in the subspace $S^*$ is left unchanged. As a result, once the search is back to this subspace, it will visit more solutions than SA would normally carry out in this subspace, and these solutions can improve quasi equilibrium of the subspace.

### 2.2.6 Working Mechanisms - a Simple Example

Let us consider the following one-dimensional function:

$$F(x) = x^3 - 50x^2 + 492x + 5000, \tag{2.19}$$

and seek its maximum whilst the variable x is coded as a 5-bit binary integer within $0 \le x \le 31$, i.e.

$$\max_{0 \le x \le 31} F(x) \tag{2.20}$$

This problem has a maximum at (0 0 1 1 0), i.e. at the sampling point x = 6 ($\equiv 0 + 0 + 2^2 + 2^1 + 0$). The maximum value of the above function is $F_{max} = 6368$.
To illustrate the simulated annealing procedure, let us start from the string ( 1 1 1 1 0), binary code for the sampling point x = 30 which gives F = 1760. Let us also assume that all subsequent neighbours are chosen randomly. Table 2.1 shows the history of the first 6 iterations, frequently called cooling cycles. In all of them a new string is obtained by swapping 1 to 0 of the randomly chosen bit. In the case of the first cooling cycle, for example, we change the randomly chosen bit number 3 in the starting string (1 1 1 1 0). This results in the new string being (1 1 0 1 0) for which x = 26. From Eq (2.19) the function value, F(x), at this new configuration is $F^{new} = 1568$. Then the difference between the old and new configurations, $\Delta F$, is evaluated, i.e. $\Delta F \equiv F^{old} - F^{new} = 192$. Since this move does not lead to the improvement of the cost function it should, nominally, be rejected. However the method allows an occasional acceptance of inferior moves, i.e. moves which do not improve the cost function F(x). Hence, the probability associated with the change, $\Delta F$, is calculated as follows:

$$\Pr(\Delta F) = \exp(-\Delta F/T_o), \tag{2.21}$$

**Table 2.1** History of the first 6 iterations for the starting temperature T = 5000 and cooling profile α = 0.1. Note that a local optimum of 6348 is attained at x = 8 in iteration no. 5.

| Cooling | T | Bit | String | x | F(x) | ΔF | Pr(ΔF) | r | Move | New string |
|---|---|---|---|---|---|---|---|---|---|---|
| 1 | 5000 | 3 | 11010 | 26 | 1568 | 192 | 0.96 | 0.69 | Yp | 11010 |
| 2 | 4500 | 1 | 01010 | 10 | 5920 | -4352 | - | - | Y | 01010 |
| 3 | 4050 | 3 | 01110 | 14 | 4832 | 1088 | 0.76 | 0.59 | Yp | 01110 |
| 4 | 3645 | 4 | 01100 | 12 | 5432 | -600 | - | - | Y | 01100 |
| 5 | 3281 | 3 | 01000 | 8 | 6348 | -816 | - | - | Y | 01000 |
| 6 | 2953 | 1 | 11000 | 24 | 1832 | 4416 | 0.22 | 0.20 | Yp | 11000 |

where $T_o$ is the initial temperature and it was arbitrarily taken as $T_o = 5000$ (note that from now on, the Boltzman constant is to be omitted and that the parameter T becomes dimensionless but it is still to be called 'temperature'). Based on the magnitude of Pr(ΔF), a decision is made whether to accept the move or not. It is seen from Table 2.1 that Pr(192) is 0.96 and it is greater than the random number r = 0.69. As a consequence this move is accepted and this is denoted in Table 2.1 by the symbol Yp. The corresponding string, i.e. (1 1 0 1 0) becomes the basis of the next cooling cycle for which the reduced temperature becomes T = 4500. Bit number one is then swapped and the current string becomes (0 1 0 1 0). The resulting difference between the old and new configuration, ΔF, is – 4352. This move leads to the improvement of the cost function and the move is unconditionally accepted with the symbol Y shown in Table 2.1. Details about the further four cooling cycles are also given in Table 2.1. The probability associated with the change, ΔF, decides the Yp-type moves in cycles 1, 3 and 6. Straightforward improvement of the cost function in cooling cycles 2, 4 and 5 leads to Y-type moves. In this example the initial temperature, $T_o$, has been set arbitrarily at

**Table 2.2** History of 22 cooling cycles. Illustration of acceptance and rejection criteria.

| Cooling | T | Bit | String | x | F(x) | ΔF | Pr(ΔF) | r | Move | New string |
|---|---|---|---|---|---|---|---|---|---|---|
| 1 | 15000 | 3 | 11010 | 26 | 1568 | 192 | 0.99 | 0.69 | Yp | 11010 |
| 2 | 13500 | 1 | 01010 | 10 | 5920 | -4352 | - | - | Y | 01010 |
| 3 | 12150 | 3 | 01110 | 14 | 4832 | 1088 | 0.91 | 0.59 | Yp | 01110 |
| ... | ... | ... | ... | ... | ... | ... | ... | ... | ... | ... |
| 16 | 3088 | 2 | 10100 | 20 | 2840 | -1312 | - | - | Y | 10100 |
| 17 | 2780 | 4 | 10101 | 21 | 2543 | 297 | 0.90 | 0.97 | N | 10100 |
| 18 | 2502 | 4 | 10110 | 22 | 2272 | 568 | 0.80 | 0.83 | N | 10100 |
| 19 | 2251 | 4 | 10110 | 22 | 2272 | 568 | 0.78 | 0.86 | N | 10100 |
| 20 | 2026 | 1 | 00100 | 4 | 6232 | -3392 | - | - | Y | 00110 |
| 21 | 1824 | 4 | 00110 | 6 | 6368 | -136 | - | - | Y | 00110 |
| 22 | 1641 | 1 | 10110 | 22 | 2272 | 4096 | 0.08 | 0.19 | N | 00110 |

5000 and in a further 100 cooling cycles the process is unable to escape from the local optimum of 6348. In another trial, the initial temperature, T was set at 15000 and Table 2.2 provides a summary of the first 22 cooling cycles. Again, two types of moves are possible, i.e. "Y" and "Yp" as explained earlier. The optimum was found in the 21st cycle and the search eventually settles at the optimum. In the above example not only the initial temperature was chosen arbitrarily but also the temperature decrease between two cooling cycles was chosen arbitrarily, as $T_{new} = \alpha T_{current}$. The constant, $\alpha$, was assumed to be $\alpha = 0.9$. The next example provides more information on what is the role of generic parameters in this method and how they are chosen in practice.

### 2.2.7 Generic Parameters in Action

Let us illustrate how the generic parameters can be chosen and discuss their influence on the annealing process by examining a single variable function which, has a number of relative minima. Assume that we seek minimum of the following function within $0 \le x \le 10$:

$$F(x) = 6 + \sum_{i=1}^{5} \sin((i+1)x + i), \text{ i.e.} \tag{2.22}$$

$$\min_{0 \le x \le 10} F(x) \tag{2.23}$$

For the illustrative purpose, let us use a five-bit string to cover the range [0,10]. This will result in sampling the function at 32 points with $\Delta x = 10/31$ intervals, as sketched in Fig. 2.4. This discrete sampling leads to the minimum, $F_{min} = 1.091$ at $x_{opt} = 5.161$. The latter value corresponds to the sampling string (1 0 0 0 0), i.e. $x = (16*10)/31$.
Let us assume an initial string to be (1 1 1 1 0) and fix the number of moves at a given temperature to 5 moves. The very first parameter to be selected is the initial temperature.

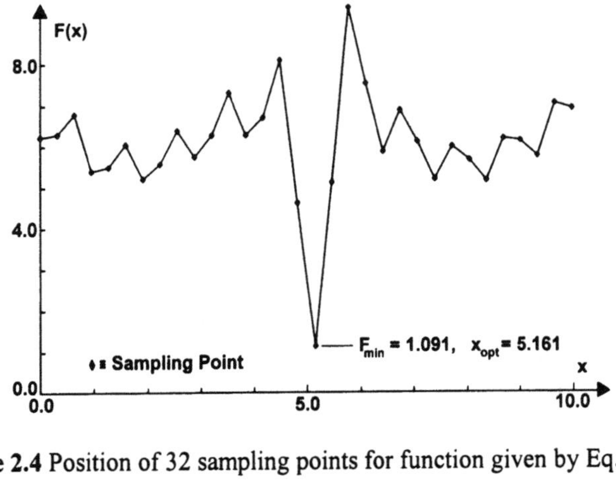


**Figure 2.4** Position of 32 sampling points for function given by Eq. (2.22).

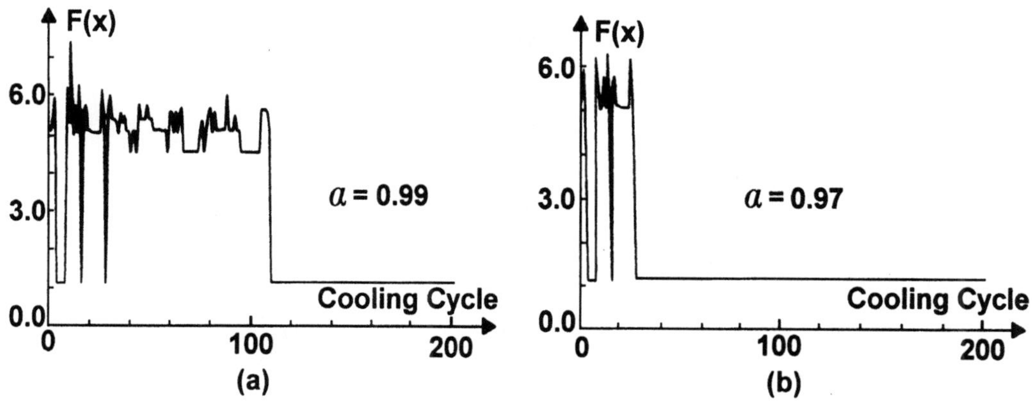


**Figure 2.5** Change of the cost function for two cooling schedules $\alpha = 0.99$ and 0.97.

Several trials narrowed the initial temperature to $T_o = 2$. It gives an acceptance ratio close to unity within the first couple of cooling cycles. Next, the rate of cooling needs to be decided upon. Fig. 2.5 shows the results obtained for two values of constant, $\alpha$, i.e. for $\alpha = 0.99$ and 0.97. In both cases an arbitrary stop criterion was adopted at the end of 200 cooling cycles. It is seen from Fig. 2.5 that adopting the faster temperature decrease, i.e. $\alpha = 0.97$ resulted in attaining the minimum in just 18 cooling cycles. Taking, on the other hand, a slower cooling rate, i.e. $\alpha = 0.99$, required 110 cooling cycles before the search has settled at the optimum $F_{min} = 1.091$, $x_{opt} = 5.161$.

The number of moves at a given temperature and the rate at which temperature is lowered are strongly related through the concept of quasi-equilibrium. It is intuitively clear that large decrement in temperature will make it necessary to carry more moves at the new temperature level to restore quasi-equilibrium. Hence, the larger the decrement in temperature the longer it takes to re-establish the quasi-equilibrium. There is therefore a trade-off between fast decrement of temperature and a small number of moves at fixed temperature. Usually, one opts for small decrements of T, but alternatively one could use a large number of moves in order to restore the equilibrium after large decrements in T. If, on the other hand, the number of moves is fixed then too small temperature decrease can lead to trapment of the search process in a local minima. This is depicted in Fig. 2.6 by vertical bars indicating at which sampling points, and for how long, the search process was trapped. The total length of bars adds up to 100% for $\alpha = 0.99$. For $\alpha = 0.97$ the process escapes the local minima much faster, and as it is seen in Fig. 2.6, it needs less cooling cycles to settle at the optimum.

### 2.2.8 Illustrative Examples

**Example (a): localised annealing.** Let us consider a single variable function given by Eq. (2.22) and let us use a five bit long string to represent the independent variable, x. This representation gives us 32 sampling points within the interval $0 \le x \le 10$. Let us also divide the search space [0, 1, 2, ....., 32] into the following, three disjoint, subspaces: $S_1 \equiv [0, ..., 9]$,

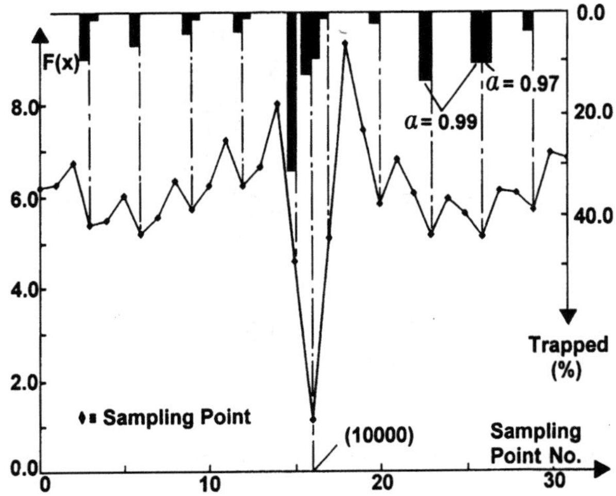


**Figure 2.6** Illustration of how the search process was trapped in local minima for two different cooling schedules. Total length of all bars corresponding to $\alpha = 0.99$ adds up to 100%.

$S_2 \equiv [10, 11, ..., 20]$ and $S_3 \equiv [21, 22, ..., 31]$ - as sketched in Fig. 2.7a. Results of a sample run are shown in Fig. 2.7b. The history of local best values within each sub-space are plotted for the first 50 cooling cycles. The cooling schedule assumed $\alpha = 0.97$ and the parameter $\varepsilon_T = 0.3$. It is seen here that the global

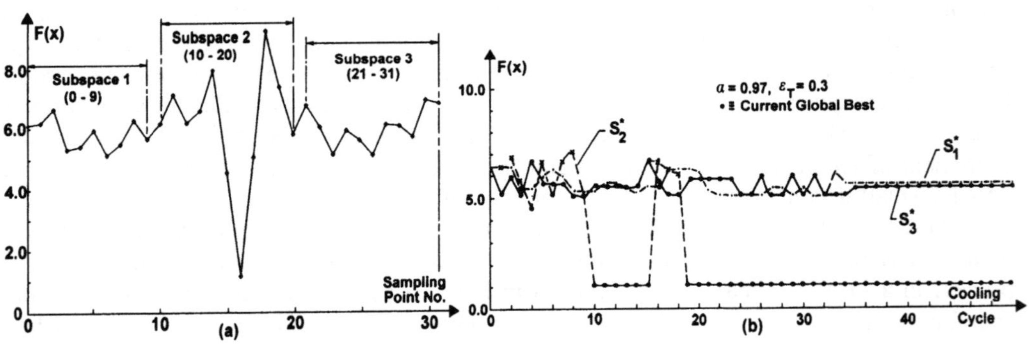


**Figure 2.7** Allocation of sampling points to three different subspaces - Fig 2.7a. History of three local best values obtained in 50 cooling cycles - Fig. 2.7b.

minimum shifts between subspaces and after 19 cycles it settles at the global minimum $F_{min}$ = 1.091. From cooling cycle no. 19 onwards, the separation of local best in $S_1$, $S_3$ and $S_2$ is such that there is no temperature change in these two, $S_1$ and $S_3$, subspaces. The global minimum found in $S_2$ becomes quickly a frozen state.

**Example (b): two-variable cost function in continuous domain.** In this example we consider minimization of a two-variable function $F(x_1, x_2)$ defined by Eq. (2.24) over the domain specified by inequalities (2.25).

$$F(x_1,x_2) = 10 - (5e^{-y_1} + 7e^{-y_2} + 10e^{-y_3} + 6e^{-y_4}) \tag{2.24}$$

where

$$y_1 = 0.03(x_1 - 5)^2 + 0.03(x_2 - 5)^2,\ y_2 = 0.08(x_1 + 4)^2 + 0.4(x_2 - 7)^2,$$

$$y_3 = 0.3(x_1 + 5)^2 + 0.08(x_2 + 5)^2,\ y_4 = 0.08(x_1 - 3)^2 + 0.08(x_2 + 6)^2.$$

$$-10 \le x_1 \le 10 \qquad \text{and} \qquad -10 \le x_2 \le 10 \tag{2.25}$$

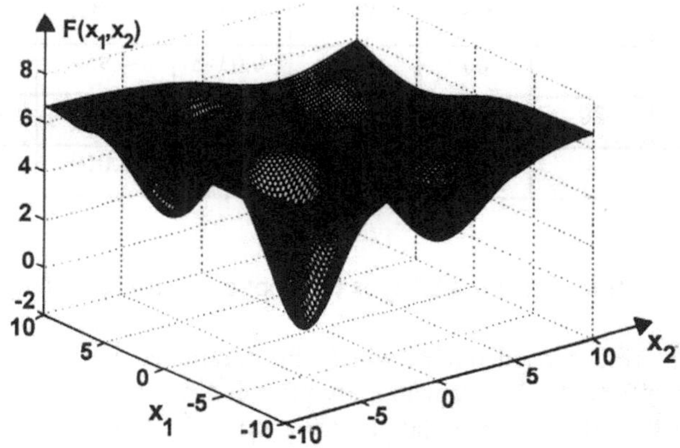


**Figure 2.8** Shape of the test function with its four local minima.

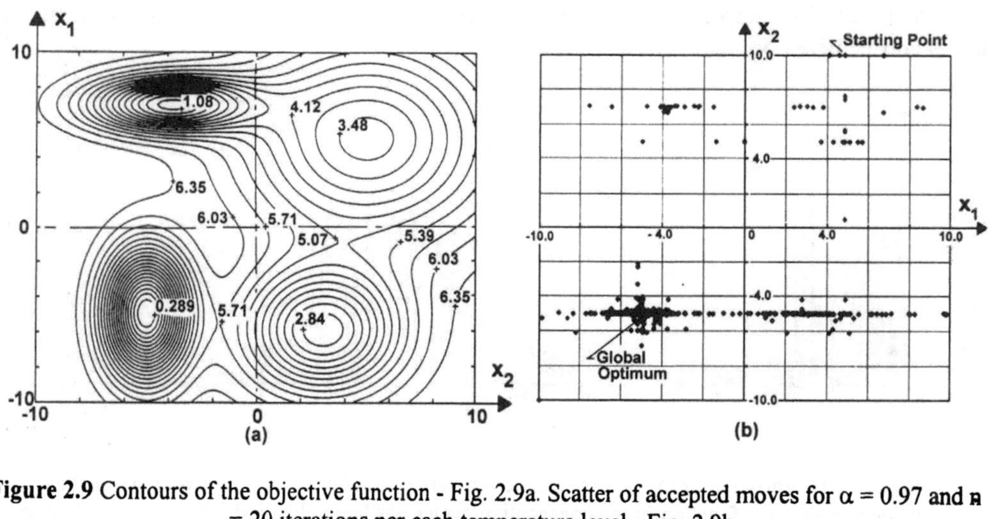


**Figure 2.9** Contours of the objective function - Fig. 2.9a. Scatter of accepted moves for $\alpha = 0.97$ and $n = 20$ iterations per each temperature level - Fig. 2.9b.

This two-variable function has four local minima – as shown in Fig. 2.8. Their contoured profiles are sketched in Fig. 2.9a. Let us use explicit values for variables $x_1$ and $x_2$ instead of

**Table 2.3** Results obtained for minimization of two-variable cost function with multiple local minima. (*) The algorithm will settle at $F_{min}$ = -0.0457 after 180 cycles (≡ 1440 function evaluations) if it is allowed to continue beyond the first encounter of the stop criterion.

| $\alpha$ | $T_{iterations}$ | $n_{cooling}$ | $n_{evaluations}$ | $F_{min}$ | $x_1$ | $x_2$ |
|---|---|---|---|---|---|---|
| 0.97 | 20 | 332 | 13280 | -0.0457 | -4.992 | -5.000 |
| 0.97 | 10 | 394 | 7880 | -0.0457 | -4.991 | -4.999 |
| 0.97 | 4 | 424 | 3372 | -0.0457 | -4.992 | -5.000 |
| 0.90 | 20 | 119 | 4760 | -0.0457 | -4.993 | -5.001 |
| 0.90(*) | 4 | 66 | 488 | -0.0348 | -4.959 | -5.097 |
| 0.80 | 20 | 80 | 3200 | -0.0455 | -4.989 | -4.985 |
| 0.80 | 4 | 42 | 336 | -0.0413 | -5.019 | -4.946 |

their binary representation. One possibility is to adopt a variable step length as given by Eqns (2.16 – 2.18).

A number of runs were carried out in which the parameter c was assumed to equal 0.1. Sample results for different SA parameters are provided in Table 2.3. The magnitude of the initial temperature $T_o$ = 50 and the initial point ($x_1$, $x_2$) ≡ (5, 10) were assumed for all runs. It is seen from Table 2.3 that for all control parameters the method was able to identify position of the global minimum. A number of parameter set-up configurations led to the absolute minimum, $F_{min}$ = -0.0457 although the necessary number of function evaluations varied from 3372 to 13280. The stop criterion was based on the lack of improvement of the cost function in ten consecutive cooling cycles.

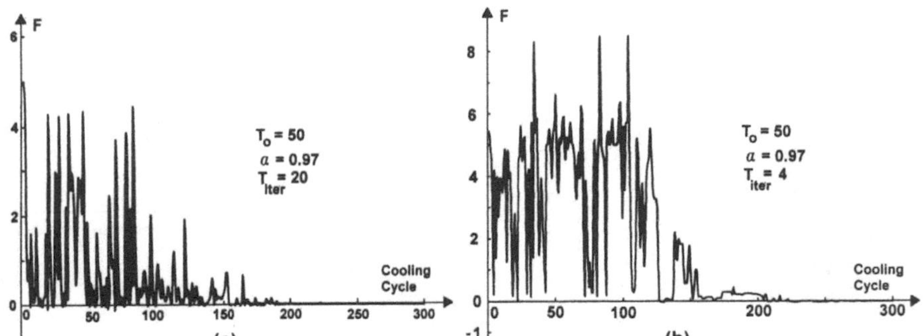


**Figure 2.10** Variation of the objective function versus number of cooling cycle. No further improvement in the cost function in ten consecutive cycles has been obtained for cooling cycle number 332 - Fig. 2.10a. Variation of the cost function versus number of cooling cycle for a different cooling set up - Fig. 2.10b.

Scatter of accepted moves over the search domain is depicted in Fig. 2.9 for the following search parameters: $\alpha$ = 0.97, $T_o$ = 50 and $T_{iterations}$ = 20. It is seen that the algorithm explores all four local minima and it is able to reach the global minimum at $(x_1, x_2) = (-4.992, 5.0)$. The cost function variation for this case is shown in Fig. 2.10. It is seen that the magnitude of the cost function varies very little from the cooling cycle number 200 onwards.

### 2.2.9 Structural Optimization Example - Problem Statement

Let us apply SA to optimal design of bowed out steel shells subjected to static external pressure. The shape of the barrelled out geometry is assumed to be given by the following generalised ellipse – see Fig. 2.11a:

$$\left(\frac{x}{R_o+\Delta}\right)^{n_1}+\left(\frac{y}{0.5L_o+n_3}\right)^{n_2}=1. \tag{2.26}$$

Exponents $n_1$, $n_2$ and $n_3$ are taken as design variables. For a given set of $(n_1, n_2, n_3)$ the wall thickness of the barrelled shell, t, can be evaluated from the condition that the mass of barrelled shell is the same as that of the reference cylinder, i.e. $m_{barrel} = m_{cylinder}$. It is assumed here, that bowed-out mild steel shells have the same radii, $R_o$, at the top and bottom edges as the initial, and mass equivalent, cylindrical shell - see Fig. 2.11b,c.

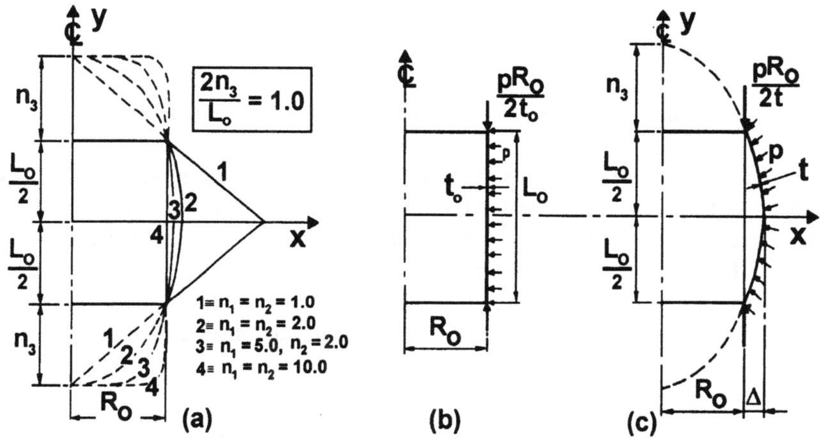


**Figure 2.11** Shape of meridians obtained for different configurations of design vector $(n_1, n_2, n_3)$ - Fig. 2.11a. Geometry of reference cylinder (Fig. 2.11b) and of bowed-out shell (Fig. 2.11c). Both shells have the same length and the same mass. Shells are subjected to static, external hydrostatic pressure, p.

The above shells are subjected to external hydrostatic pressure under which their load carrying capacity could be characterised by yield load, $p_y$, static bifurcation buckling, $p_{bif}$, or axisymmetric collapse, $p_{coll}$, - see Blachut and Wang (2000). The bowed-out configurations capable of supporting the maximum external, hydrostatic pressure are sought. The above problem can be formulated, in a standard form, as follows:

Find $p_{opt}$

$$p_{opt} = \max_{n_1, n_2, n_3} \min\,(p_y, p_{bif}, p_{coll}), \tag{2.27}$$

subject to the following constraints:
- lower and upper bounds on the design variables

$$n_1^{\,L} \le n_1 \le n_1^{\,U} \qquad n_2^{\,L} \le n_2 \le n_2^{\,U} \qquad n_3^{\,L} \le n_3 \le n_3^{\,U} \tag{2.28}$$

- constant mass of barrelled shell

$$m_{barrel} = \text{const} \; (\equiv m_{cylinder}). \tag{2.29}$$

Structural response of the above axisymmetric shells is calculated using BOSOR5 (Bushnell, 1976) and ABAQUS (1998) codes. No axial deflection is allowed at the mid-length of barrels. Also, at the top and bottom edges only, axial displacements are allowed and the magnitude of all other variables is set to zero. Shells are assumed to be made from mild steel with Young's modulus E = 210 GPa, Poisson's ratio $\nu = 0.3$ and the yield point of material $\sigma_{yp} = 300$ MPa. Structural analyses are based on linear elastic, perfectly-plastic modelling of material (with isotropic strain hardening).

**Solution strategy.** It appears that SA is a suitable method for solving the above problem. There is very little a priori knowledge about the design space and the quality of cost function. It is known from past experience however that the cost function may not be continuous at the transition regions between axisymmetric collapse and bifurcation buckling, for example. The following values were assumed for lower and upper bounds on design variables:

$$1.0 \le n_1 \le 10.0 \qquad 1.0 \le n_2 \le 10.0 \qquad 1.0 \le 2n_3/L_o \le 10.0 \tag{2.30}$$

The search process started from a random point $\mathbf{x}_o$ with all subsequent new configurations generated using Eqns (2.15-2.18). Parameter c was taken to be 10. The epoch length, $N_k = 20$, was constant at all temperature levels. A geometric schedule for lowering temperature, as given by Eq. (2.11), was used with the parameter $\alpha = 0.99$ and the initial temperature $T_o$ = 50.0. The number of cooling cycles was not fixed and the annealing process was allowed to continue until there was no change of the cost function in 10 consecutive cycles.

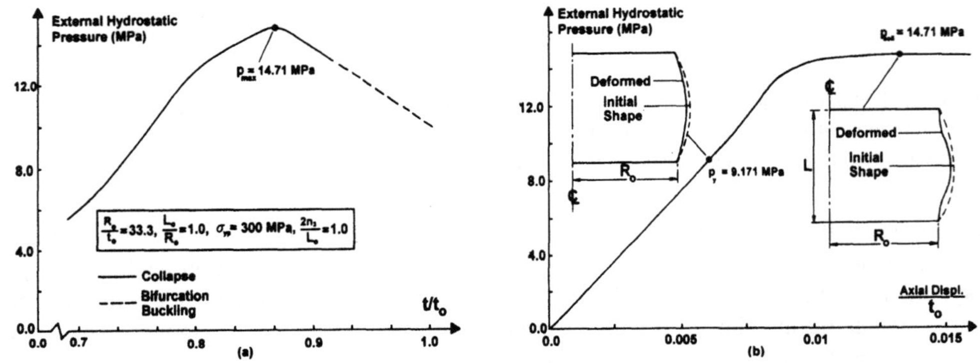


**Figure 2.12** Performance of a barrelled shell within $1.0 \le n_1 = n_2 \le 10.0$ and $2n_3/L_o = 2.0$. Optimal configuration collapses at p = 14.71 MPa - Fig. 2.12a. Deformed shape of optimal barrel at yield pressure, $p_y$, and at collapse pressure, $p_{coll}$ - Fig. 2.12b.

**Results.** Detailed computations were performed for the reference cylindrical geometry given by $L_o/R_o = 1.0$ and $R_o/t_o = 33.33$. The maximum hydrostatic pressure $p_{opt} = 14.71$ MPa, is associated with the axisymmetric collapse as shown in Fig. 2.12a. The optimal design vector $(n_1, n_2, 2n_3/L_o)_{opt}$ equals (2.2, 2.0, 1.0). The plot of external hydrostatic pressure versus end shortening for the optimal barrel is shown in Fig. 2.12b. The initial shape and deformed shape are shown for yield pressure, $p_y = 9.17$ MPa, and for collapse pressure $p_c = 14.71$ MPa. A wider appreciation of the cost function shape can be obtained from a sample of four contour plots shown in Fig. 2.13a, b, c and d. It is seen here that the cost function indeed contains local optima. Yielding of the optimal geometry starts on the inner surface of the barrel at both top and bottom edges and the maximum, effective, plastic strain at the collapse reaches 3.2%. A section through the barrel, at collapse pressure, is depicted in Fig. 2.13e where it is seen that 90% of the cross section has yielded at the

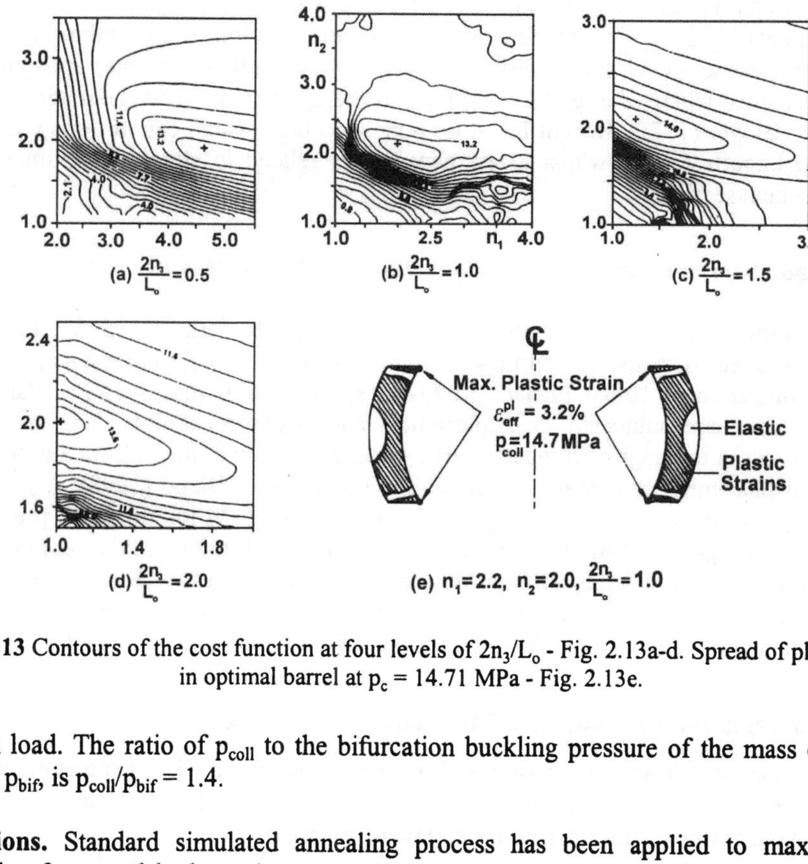


**Figure 2.13** Contours of the cost function at four levels of $2n_3/L_o$ - Fig. 2.13a-d. Spread of plastic zones in optimal barrel at $p_c = 14.71$ MPa - Fig. 2.13e.

optimum load. The ratio of $p_{coll}$ to the bifurcation buckling pressure of the mass equivalent cylinder, $p_{bif}$, is $p_{coll}/p_{bif} = 1.4$.

**Conclusions.** Standard simulated annealing process has been applied to maximize the magnitude of external hydrostatic pressure in a barrel shell in which the shape of meridian was to be found. Non-linear geometry and material properties were taken into account through application of a self contained analysis code. Computational effort of a single re-analysis is such that a predominantly sequential optimization tool, such as SA, is capable to navigate through the design space in an affordable amount of time. Whilst it is not certain whether the obtained solution is a global maximum, it offers some 40% increase over the reference geometry's load bearing capacity.

#### 2.2.10 Closure

It appears that there is not enough practical experience with the method of simulated annealing to say definitively what its future place among structural optimization will be. The body of work in other disciplines has revealed unique and very attractive features of SA when compared with other heuristics. One specific drawback seems to be associated with, the as yet, low success in parallelization of the method.

## 2.3 Tabu Search

### 2.3.1 Introduction

Tabu search is another method inspired by an analogy to natural activities and more specifically to human memory processes. The method is based on neighbourhood search with local optima avoidance policy which centres around the creation of a prohibition list and on using flexible memory cycles of differing time spans. Its usefulness becomes apparent in non-convex problems where the global optimum is being sought. This heuristic is usually attributed to Glover (1977). Recent literature review on tabu search can be found in work by Osman and Laporte (1996) whilst a few new entries related to structural optimization are given in the next section.

### 2.3.2 Tabu Search in Structural Optimization – A Brief Literature Review

Hu (1992) tests the tabu search method on some analytically defined cost functions before applying it to three-bar truss, coil springs and thin-walled channel. Tabu search applied to the weight minimization of 10-bar planar and 25-bar space truss is discussed by Bland (1994, 1995). Structures are subjected to displacement and stress constraints. The same author applies tabu search to weigh minimization of a transmission tower made from aluminium and subject to displacement constraints and reliability constraints - Bland (1998). A tabu search methodology for generation and evaluation of alternate structural topologies is presented by Bennage and Dhingra (1995b). The topological design is explored through a combination of discrete variables for nodal connectivities resulting in a trial geometry. Multiple cost functions are considered, i.e. the minimization of structural weight, the maximization of fundamental natural frequency of vibration and the minimization of deflection of all free nodes.

### 2.3.3 Concepts and Example with 5-bit Long Chromosome

A generic tabu search strategy requires three main operations, i.e. generation of initial solution, generation of neighbours and their selection. Let us provide now more details about the generic tabu search by considering the following function:

$$F(x) = 6 + \sum_{i=1}^{5} \sin((i+1)x + i) \tag{2.31}$$

and seeking its maximum within the range $-10.0 \leq x \leq 10.0$. Figure 2.14a shows that there are several local maxima and that the magnitudes of three of them, $m_1$, $m_2$ and $m_3$ do not differ

much. It is an easy matter to show that the true maximum of the function F(x) given by (2.31) is $F_{max} = 9.3725$ at x = 5.84. As has already been mentioned, the tabu search method requires the continuous problem to be converted into a discrete one. There are a number of approaches here which one can adopt but none appears to be universally applicable. In order to simplify the model let us assume that the variable x is represented by a binary string of 'zeroes' and 'ones'. Let us initially assume that the string's length is five. This gives 32 points where the continuous function F(x) is to be sampled, as shown in Fig. 2.14b. For a 5-bit string the smallest value is 0 and this is equivalent to the string (0 0 0 0 0). The largest value is 31 and it corresponds to the string (1 1 1 1 1). There are therefore 32 points covering the whole range [-10, +10]. A linear mapping will be used to convert the range [-10, +10] into the binary representation. A string (0 1 1 1 1) is equivalent to x = 20*(15/31) - 10 = - 0.322. The binary string (1 1 0 0 1) is equivalent to 20*(25/31) - 10 = 6.129 and the string (1 1 1 1 1) is equivalent to 20*(31/31) - 10 = 10.
There are various prescriptions through which tabu search explores the design space. The first one is associated with short-term memory and it is also known as recency. The first step however is to adopt a procedure for the neighbourhood of a feasible starting point.

**Neighbourhood.** One way of generating candidate configurations is to sequentially flip a bit in the string constituting an initial and feasible configuration. Each flip results in a new string and the number of strings being generated around the feasible point is known as Neighbourhood Size, NS.

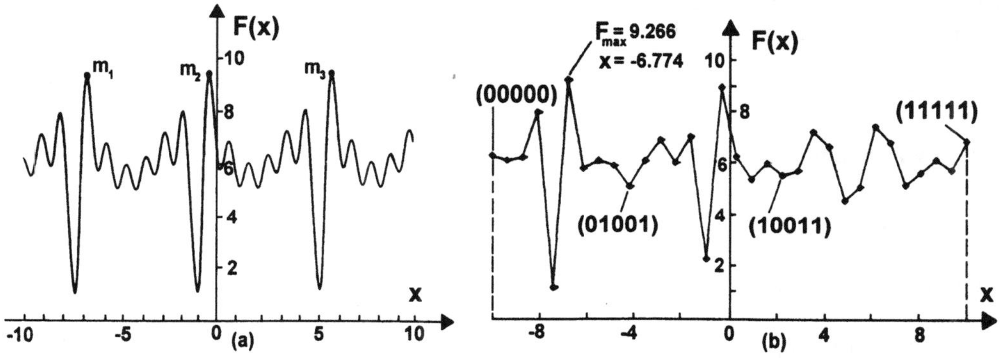


**Figure 2.14** Plot of $F(x) = 6 + \sum_{i=1}^{5} \sin((i+1)x + i)$ - Fig. 214a. Sampling points obtained for a five bit long binary string - Fig. 2.14b.

If a starting string is (1 0 0 1 1) then the neighbourhood corresponding to NS = 4 is shown in Table 2.4. Once the neighbourhood size has been assumed and neighbours have been generated then they have to be evaluated. Firstly, each string's value need to be found - see column 3 in Table 2.4. Next, through the linear mapping to the range [-10, +10] one finds the corresponding value of the argument x - as shown in column 4 of Table 2.4. Finally, values of F(x) are found from Eq. (2.31). They are given in the last column of Table 2.4. Another way of generating neighbours would be to apply some sort of random moves within the design space. An attractive approach based on partitioning of the design space around a current configuration has been explored by Siarry and Berthiau (1997).

**Table 2.4** List of neighbourhood candidates for NS = 4. The associated values of strings, the argument x and cost function F(x), are also given. Starting, 5-bit long, string is (1 0 0 1 1).

| Bit no. to flip | Resulting string | String value | x | F(x) |
|---|---|---|---|---|
| 1 | **0** 0 0 1 1 | 3.0 | -8.065 | 8.028 |
| 2 | 1 **1** 0 1 1 | 27.0 | 7.419 | 5.120 |
| 3 | 1 0 **1** 1 1 | 23.0 | 4.839 | 4.554 |
| 4 | 1 0 0 **0** 1 | 17.0 | 0.968 | 5.348 |

The method assumes that a number of concentric spheres is generated around the current solution point, s, as sketched in Fig. 2.15a for a two-dimensional case. If one assumes that radii of k concentric spheres, or k circles in 2-D, are $r_o$, $r_1$, ...., $r_k$ then partitions around the current solution point s are taken as volumes, $C_i$, between two adjacent and concentric spheres, i and i+1, i.e.

$$C_i(s,r_i,r_{i-1}) = \{s' | \, r_{i-1} \leq \|s' - s\| < r_i\} \tag{2.32}$$

Next, within each partitioned region one generates, in a random manner, a single neighbour as sketched in Fig. 2.15a. The tabu list is updated each time the best configuration is selected. The resulting list contains a set of tabu-patches as shown in Fig. 2.15b. The radius, $r_o$, is a free parameter which needs to be tuned for each case whilst the remaining radii can be selected by using one of the following schemes:

- geometric partitioning

$$r_{k-i+1} = r_k/2^{i-1}, \qquad i = 1, ....., k \tag{2.33}$$

- linear partitioning

$$r_{k-i+1} = r_k * (k - i + 1)/k, \qquad i = 1, ...., k \tag{2.34}$$

- isovolume partitioning

$$h_{i+1} = (h_i^n - h_i^n/k)^{1/n}, \qquad i = 0, ...., k-1 \tag{2.35}$$

Further details about this method and results of numerical tests can be found in work by Siarry and Berthiau (1997).

**Tabu list and tabu tenure.** Out of the neighbourhood sample, NS = 4, the best value of the cost function is found. It is seen in Table 2.4 that in our case this corresponds to F(x) = 8.028. The associated string is taken as a starting configuration for the next iteration. But the position where the bit had been flipped in this configuration, i.e. bf = 1 is now added to a prohibition list known as tabu list. This bit is not allowed to change for a prescribed number of iterations known as Tabu Tenure, TT. Let us assume that TT = 5. The role of tabu tenure will be assessed later. The use of tabu restriction, or the short-term memory, ensures that the search does not return to the local optimum and the algorithm forces the search out in a new direction. The short-term memory, known also as recency, enables the method to leave locally optimal solutions in the search for the global optimum of a cost function.

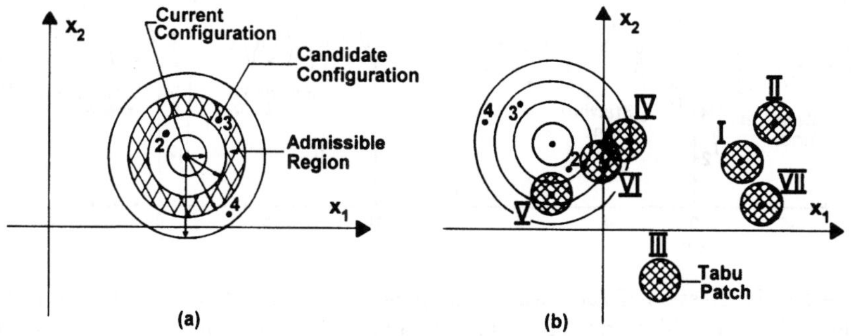


**Figure 2.15** Four partitions with candidate points 2, 3 and 4 which have been chosen randomly (Fig. 2.15a). Current partitions at iteration number VIII. Seven tabu patches which have been created in the previous seven iterations are also shown (Fig. 2.15b).

**Table 2.5** Details about the first 6 iterations. The starting string is (1 0 0 1 1), NS = 4 and TT = 5.

| Iteration no. | Bit to flip | Resulting string | x | F(x) | New string |
|---|---|---|---|---|---|
| 1 | 1 | 0 0 0 1 1 | -8.065 | 8.028** | (bit no. 1 tabu |
| | 2 | 1 1 0 1 1 | 7.419 | 5.120 | until iteration |
| | 3 | 1 0 1 1 1 | 4.839 | 4.554 | number 5) |
| | 4 | 1 0 0 0 1 | 0.968 | 5.348 | 0 0 0 1 1 |
| 2 | 5 | 0 0 0 1 0 | -8.710 | 6.258 | (bit no. 2 tabu |
| | 1 | tabu | - | - | until iteration |
| | 2 | 0 1 0 1 1 | -2.903 | 6.890** | number 6) |
| | 3 | 0 0 1 1 1 | -5.484 | 6.107 | 0 1 0 1 1 |
| 3 | 4 | 0 1 0 0 1 | -4.194 | 5.073 | (bit no. 5 tabu |
| | 5 | 0 1 0 1 0 | -3.548 | 6.090** | until iteration |
| | 1 | tabu | - | - | number 7) |
| | 2 | tabu | - | - | 0 1 0 1 0 |
| 4 | 3 | 0 1 1 1 0 | -0.968 | 2.179 | (bit no. 4 tabu |
| | 4 | 0 1 0 0 0 | -4.839 | 5.890** | until iteration |
| | 5 | tabu | - | - | number 8) |
| | 1 | tabu | - | - | 0 1 0 0 0 |
| 5 | 2 | tabu | - | - | (bit no. 3 tabu |
| | 3 | 0 1 1 0 0 | -2.258 | 5.998** | until iteration |
| | 4 | tabu | - | - | number 9) |
| | 5 | tabu | - | - | 0 1 1 0 0 |
| 6 | 1 | 1 1 1 0 0 | 8.065 | 5.593** | (bit no. 1 tabu |
| | 2 | 0 0 1 0 0 | -7.419 | 1.094 | until iteration |
| | 3 | tabu | - | - | number 10) |
| | 4 | tabu | - | - | 1 1 1 0 0 |

**Iteration history and results.** Details about the first six iterations are given in Table 2.5. The best values of the cost function at each iteration are asterisked. The new string at the end of each iteration is identified as well as entries into the tabu list. It is clear from Table 2.5 that

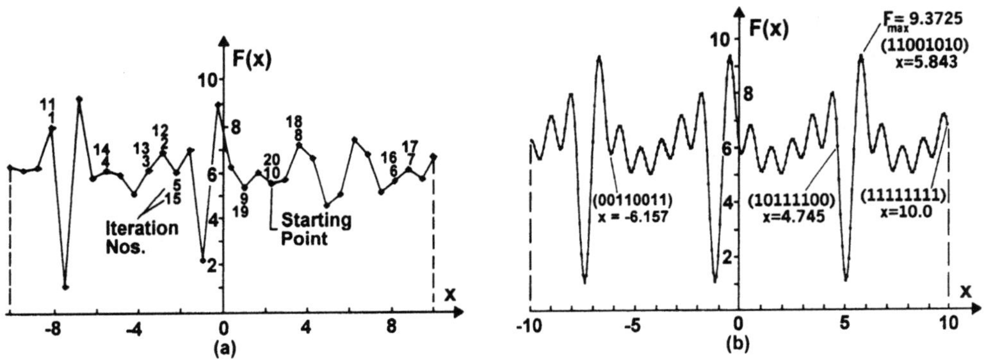


**Figure 2.16** Progression of Tabu Search during the first 20 iterations under short-term memory algorithm, TT = 5 and 5-bit string approximation (Fig. 2.16a). Sampling points for 8-bit long string (Fig. 2.16b).

tabu restrictions can lead to acceptance of inferior values of the cost function. A situation which, one also finds in simulated annealing. The history of the 20 first iterations is provided in Table 2.6 and the best candidates arrived at in each iteration are shown in Fig. 2.16a. It is seen from Table 2.6 that the algorithm starts to return the same values of the cost function from iteration number 11. The maximum value of the cost function is not reached and the method is trapped in a local maximum.

**Table 2.6** History of the first 20 iterations for the starting string (1 0 0 1 1), NS = 4, TT = 5.

| Iteration number | Best configuration per iteration | | Best configuration obtained so far |
|---|---|---|---|
| | x | F(x) | |
| 1 | -8.065 | 8.028 | 8.028 |
| 2 | -2.903 | 6.890 | 8.028 |
| 3 | -3.548 | 6.090 | 8.028 |
| 4 | -4.839 | 5.890 | 8.028 |
| 5 | -2.258 | 5.998 | 8.028 |
| 6 | 8.065 | 5.593 | 8.028 |
| 7 | 8.710 | 6.106 | 8.028 |
| 8 | 3.548 | 7.215 | 8.028 |
| 9 | 0.968 | 5.348 | 8.028 |
| 10 | 2.258 | 5.512 | 8.028 |
| 11 | -8.065 | 8.028 | 8.028 |
| 12 | -2.903 | 6.890 | 8.028 |
| 13 | -3.548 | 6.090 | 8.028 |
| 14 | -4.839 | 5.890 | 8.028 |
| 15 | -2.258 | 5.998 | 8.028 |
| 16 | 8.065 | 5.593 | 8.028 |
| 17 | 8.710 | 6.106 | 8.028 |
| 18 | 3.548 | 7.215 | 8.028 |
| 19 | 0.968 | 5.348 | 8.028 |
| 20 | 2.258 | 5.512 | 8.028 |

The effect of tabu tenure, TT, and the neighbourhood size, NS, on the optimal solution is provided in Table 2.7. It is seen here that the global maximum, $F_{max} = 9.266$, was reached for three out of nine possibilities of (TT, NS) combinations. In all cases the process started to repeat itself, R, and on one occasion all configurations were returned as tabu, T, and the search process has stopped. In the next section we will re-examine the same cost function by adopting an 8-bit long string to sample the continuous function given by Eq. (2.31) over the range $-10 \leq x \leq +10$.

**Table 2.7** Maximum values of the cost function obtained for 5-bit long string and different TT and NS parameters. First number inside brackets gives the iteration number where optimum was reached. The second indicates the iteration number beyond which either the repetitive configurations begun, R, or all configurations were posted as tabu, T.

| Neighbourhood Size, NS | Tabu Tenure, TT | | |
|---|---|---|---|
| | 3 | 4 | 5 |
| 3 | 8.958(15,17R) | 8.028(1,9T) | 8.028(1,11R) |
| 4 | 9.266(9,25R) | 8.028(1,11R) | 8.028(1,11R) |
| 5 | 9.266(5,12R) | 9.266(15,16R) | 8.958(3,11R) |

### 2.3.4 Example with 8-bit Long Chromosome

Let us assume that the initial 8-bit string is (0 0 0 1 0 0 1 1). Its binary value is 19 and the corresponding value of x is obtained by a linear mapping into the range [-10,+10], i.e. x = 20*(19/255) - 10 = - 8.51. Figure 2.16b shows all 256 sampling points and the maximum $F_{max}$ = 9.3725, at x = 5.843. There are two other distinct local maxima, at around - 0.5 and - 6.5, for which there is little difference in the magnitude of F(x). The effect of both the tabu tenure, TT, and neighbourhood size, NS, on the optimal solution was examined by adopting various combinations of these parameters, as shown in Table 2.8. There were 25 various possibilities and the global maximum was reached in 11 of them. It transpires that a short tabu tenure, i.e. TT = 2, under-performs. Equally, a large neighbourhood size may not be the best choice. An aspiration operator is frequently added to the short-term memory operator in the Tabu Search heuristic. Whilst the tabu list prevents the algorithm from returning to the previously visited region of the design space, it may also unnecessarily restrict the search. The aspiration criterion allows to remove the tabu restriction if a specific configuration yields exceptionally good return. In our case, we evaluate each configuration placed on the tabu list. If the value of the cost function, for a nominally prohibited configuration, is better than any other value which has been obtained so far, then one accepts this value. Next, one takes the corresponding chromosome as a starting configuration for the next iteration, and places the bit under consideration back on the tabu list with fresh tabu tenure attached to it. The previous example has been re-run with the aspiration criterion in place in order to assess its effect on the solution. The obtained results are given in Table 2.8. Direct comparison shows that in some cases the aspiration criterion returns the global optimum, in some cases it returns the same result and in one case it returns a marginally worse result. Tables 2.9 and 2.10 provide details about the algorithm for tabu tenure, only and for tabu tenure supplemented with the aspiration criterion. Both tables contain results for TT = 5 and NS = 4. Under the short-term memory

**Table 2.8** Results for various combinations of (NS,TT) parameters and 8-bit representation. The second line in each box refers to the aspiration criterion being used. Numbers in brackets denote iteration numbers for which the maximum was reached.

| TT / NS | 2 | 3 | 4 | 5 | 6 |
|---|---|---|---|---|---|
| 3 | 9.3718(26)<br>9.3718(26) | 9.3725(9)<br>9.3725(9) | 9.3725(9)<br>9.3725(9) | 9.3725(9)<br>9.3725(9) | 9.3725(9)<br>9.3725(9) |
| 4 | 7.7153(3)<br>7.7153(3) | 9.3725(9)<br>9.3725(9) | 9.3725(9)<br>9.3725(9) | 9.1416(15)<br>9.3725(9) | 9.1416(15)<br>9.3725(9) |
| 5 | 9.3718(40)<br>9.3718(40) | 9.3718(8)<br>9.3718(8) | 9.3725(45)<br>9.3725(45) | 9.3718(48)<br>9.3725(45) | 9.222(9)<br>9.3718(37) |
| 6 | 9.3718(26)<br>9.3718(26) | 9.3718(20)<br>9.3718(22) | 9.3718(20)<br>9.3718(18) | 9.3725(27)<br>9.3718(12) | 9.306(28)<br>9.3718(12) |
| 7 | 7.7153(1)<br>7.7153(1) | 7.7153(1)<br>7.7153(1) | 9.3725(21)<br>9.3725(21) | 9.3725(27)<br>9.3725(21) | 9.3725(55)<br>9.3718(14) |

the algorithm reaches the local maximum in 15 iterations. In this case, the tabu list restricts the number of configurations to be re-visited for the period of five consecutive iterations. Once a configuration is prohibited, it is not evaluated. It is seen from Table 2.9 that 24 configurations were prohibited from being assessed under the recency prohibition, out of 60 potential candidates being generated by the algorithm.

The aspiration criterion aims at relaxing the prohibition list. In the current implementation all tabu candidates are being evaluated. In Table 2.10 it applies to candidate number 3, in iteration number 3. The value of the cost function becomes 7.715 and this value is the largest received so far and hence it is immediately accepted. Bit number three for this candidate is put onto the tabu list with the tenure until iteration number 8. The corresponding configuration, i.e. (0 0 0 1 0 1 1 1) is being accepted as a starting configuration for the next iteration. It is seen from all the subsequent iterations that none of the configurations declared tabu is capable of providing a better return of the cost function. The global optimum is reached in fewer iterations, i.e. in 9 iterations requiring 36 evaluations of the cost function of which 11 appeared in the tabu list but had to be examined under the aspiration criterion.

### 2.3.5 Minimization of Two-dimensional Case

Let us now apply the binary coding to the two-dimensional, continuous function given by Eqns (2.24 - 2.25). The previous section on simulated annealing addressed minimization of this function through a variable step sampling of variables $x_1$ and $x_2$. General view of the cost function is shown in Fig. 2.8 and its contours are depicted in Fig. 2.9a. For the purpose of this section each variable is to be represented by a binary string. Both strings are to be connected in a nose-to-tail fashion to form a single chromosome. Let us consider 5-bit strings for each of $x_1$ and $x_2$ variables and assemble them into a 10-bit long string as shown in Table 2.11. The table illustrates how a 10-bit chromosome (1 0 0 1 1 1 0 0 0 0) is divided into two 5-bit strings representing $x_1$ and $x_2$. The corresponding values are obtained from the following mapping: $x_1 = (2^4 + 0 + 0 + 2^1 + 2^0)*(20/31) - 10 = 2.258$ and $x_2 = (2^4 + 0 + 0 + 0 + 0)*(20/31) - 10) = 0.323$. It is easy to establish that the global minimum of the above mentioned function, under 10-bit long chromosome approximation, is $F_{min} = 0.0439$, at $[x_1, x_2] = [-4.839, -4.839]$. It is seen from Table 2.12, for example, that the search process returns, under the aspiration

**Table 2.9** First 15 iterations under short-term memory, only. Starting 8-bit string is (00010011), NS = 4, TT = 5.

| Iteration | Bit to flip | Resulting string | x | F(x) | New string |
|---|---|---|---|---|---|
| 1 | 1 | 1 0 0 1 0 0 1 1 | 1.529 | 6.013 | (bit no. 3 tabu |
| | 2 | 0 1 0 1 0 0 1 1 | -3.490 | 5.939 | until iteration |
| | 3 | 0 0 1 1 0 0 1 1 | -6.0 | 6.093** | number 6) |
| | 4 | 0 0 0 0 0 0 1 1 | -9.764 | 5.916 | 0 0 1 1 0 0 1 1 |
| 2 | 5 | 0 0 1 1 1 0 1 1 | -5.372 | 5.570 | (bit no. 6 tabu |
| | 6 | 0 0 1 1 0 1 1 1 | -5.686 | 6.819** | until iteration |
| | 7 | 0 0 1 1 0 0 0 1 | -6.156 | 5.786 | number 7) |
| | 8 | 0 0 1 1 0 0 1 0 | -6.078 | 5.857 | 0 0 1 1 0 1 1 1 |
| 3 | 1 | 1 0 1 1 0 1 1 1 | 4.353 | 7.623** | (bit no. 1 tabu |
| | 2 | 0 1 1 1 0 1 1 1 | -0.666 | 7.459 | until iteration |
| | 3 | tabu | - | - | number 8) |
| | 4 | 0 0 1 0 0 1 1 1 | -6.941 | 7.595 | 1 0 1 1 0 1 1 1 |
| 4 | 5 | 1 0 1 1 1 1 1 1 | 4.980 | 2.348 | (bit no. 8 tabu |
| | 6 | tabu | - | - | until iteration |
| | 7 | 1 0 1 1 0 1 0 1 | 4.196 | 6.639 | number 9) |
| | 8 | 1 0 1 1 0 1 1 0 | 4.274 | 7.140** | 1 0 1 1 0 1 1 0 |
| 5 | 1 | tabu | - | - | (bit no. 4 tabu |
| | 2 | 1 1 1 1 0 1 1 0 | 9.294 | 5.615 | until iteration |
| | 3 | tabu | - | - | number 10) |
| | 4 | 1 0 1 0 0 1 1 0 | 3.019 | 5.619** | 1 0 1 0 0 1 1 0 |
| 6 | 5 | 1 0 1 0 1 1 1 0 | 3.647 | 7.061** | (bit no. 5 tabu |
| | 6 | tabu | - | - | until iteration |
| | 7 | 1 0 1 0 0 1 0 0 | 2.862 | 5.766 | number 11) |
| | 8 | tabu | - | - | 1 0 1 0 1 1 1 0 |
| 7 | 1 | tabu | - | - | (bit no. 2 tabu |
| | 2 | 1 1 1 0 1 1 1 0 | 8.666 | 5.973** | until iteration |
| | 3 | 1 0 0 0 1 1 1 0 | 1.137 | 5.121 | number 12) |
| | 4 | tabu | - | - | 1 1 1 0 1 1 1 0 |
| 8 | 5 | tabu | - | - | (bit no. 7 tabu |
| | 6 | 1 1 1 0 1 0 1 0 | 8.353 | 5.055 | until iteration |
| | 7 | 1 1 1 0 1 1 0 0 | 8.509 | 5.40** | number 13) |
| | 8 | tabu | - | - | 1 1 1 0 1 1 0 0 |
| 9 | 1 | 0 1 1 0 1 1 0 0 | -1.529 | 5.911 | (bit no. 3 tabu |
| | 2 | tabu | - | - | until iteration |
| | 3 | 1 1 0 0 1 1 0 0 | 6.0 | 8.666** | number 14) |
| | 4 | tabu | - | - | 1 1 0 0 1 1 0 0 |
| 10 | 5 | tabu | - | - | (bit no. 6 tabu |
| | 6 | 1 1 0 0 1 0 0 0 | 5.686 | 8.428** | until iteration |
| | 7 | tabu | - | - | number 15) |
| | 8 | 1 1 0 0 1 1 0 1 | 6.078 | 7.934 | 1 1 0 0 1 0 0 0 |
| 11 | 1 | 0 1 0 0 1 0 0 0 | -4.353 | 5.164 | (bit no. 4 tabu |
| | 2 | tabu | - | - | until iteration |
| | 3 | tabu | - | - | number 16) |
| | 4 | 1 1 0 1 1 0 0 0 | 6.941 | 6.693** | 1 1 0 1 1 0 0 0 |
| 12 | 5 | 1 1 0 1 0 0 0 0 | 6.313 | 6.021 | (bit no. 8 tabu |
| | 6 | tabu | - | - | until iteration |
| | 7 | tabu | - | - | number 17) |
| | 8 | 1 0 0 1 1 0 0 1 | 7.019 | 6.404** | 1 0 0 1 1 0 0 1 |
| 13 | 1 | 0 1 0 1 1 0 0 1 | -3.019 | 6.378** | (bit no. 1 tabu |
| | 2 | 1 0 0 1 1 0 0 1 | 2.0 | 5.059 | until iteration |
| | 3 | tabu | - | - | number 18) |
| | 4 | tabu | - | - | 0 1 0 1 1 0 0 1 |
| 14 | 5 | 0 1 0 1 0 0 0 1 | -3.647 | 6.280 | (bit no. 7 tabu |
| | 6 | tabu | - | - | until iteration |
| | 7 | 0 1 0 1 1 0 1 1 | -2.862 | 7.029** | number 19) |
| | 8 | tabu | - | - | 0 1 0 1 1 0 1 1 |
| 15 | 1 | tabu | - | - | (bit no. 3 tabu |
| | 2 | 0 0 0 1 1 0 1 1 | -7.882 | 6.852 | until iteration |
| | 3 | 0 1 1 1 1 0 1 1 | -0.352 | 9.142** | number 20) |
| | 4 | tabu | - | - | 0 1 1 1 1 0 1 1 |

criterion with NS = 6 and TT = 7, a global minimum in iteration no. 102. The influence of (NS,TT) parameters on the search is also shown in this table. The overall number of allowed iterations was 150 and it is seen from Table 2.12 that in most cases the search process started cycling, e.g. from iteration no. 26 for NS = 5 and TT = 7.

**Table 2.10** History of the first 10 iterations under the aspiration criterion for NS = 4, TT = 5. All shaded candidates, from tabu list, are assessed.

| Iteration | Bit to flip | Resulting string | x | F(x) | New string |
|---|---|---|---|---|---|
| 1 | 1 | 1 0 0 1 0 0 1 1 | 1.529 | 6.013 | (bit no. 3 tabu |
| | 2 | 0 1 0 1 0 0 1 1 | -3.490 | 5.939 | until iteration |
| | 3 | 0 0 1 1 0 0 1 1 | -6.0 | 6.093** | number 6) |
| | 4 | 0 0 0 0 0 0 1 1 | -9.764 | 5.916 | 0 0 1 1 0 0 1 1 |
| 2 | 5 | 0 0 1 1 1 0 1 1 | -5.372 | 5.570 | (bit no. 6 tabu |
| | 6 | 0 0 1 1 0 1 1 1 | -5.686 | 6.819** | until iteration |
| | 7 | 0 0 1 1 0 0 0 1 | -6.156 | 5.786 | number 7) |
| | 8 | 0 0 1 1 0 0 1 0 | -6.078 | 5.857 | 0 0 1 1 0 1 1 1 |
| 3 | 1 | 1 0 1 1 0 1 1 1 | 4.353 | 7.623 | (bit no. 3 tabu |
| | 2 | 0 1 1 1 0 1 1 1 | -0.666 | 7.459 | until iteration |
| | 3 | (T-00010111) | -8.196 | 7.715** | number 8) |
| | 4 | 0 0 1 0 0 1 1 1 | -6.941 | 7.595 | 0 0 0 1 0 1 1 1 |
| 4 | 5 | 0 0 0 1 1 1 1 1 | -7.568 | 2.128 | (bit no. 8 tabu |
| | 6 | tabu | -8.509 | 6.048 | until iteration |
| | 7 | 0 0 0 1 0 1 0 1 | -8.352 | 6.745 | number 9) |
| | 8 | 0 0 0 1 0 1 1 0 | -8.274 | 7.253** | 0 0 0 1 0 1 1 0 |
| 5 | 1 | 1 0 0 1 0 1 1 0 | 1.764 | 5.645** | (bit no. 1 tabu |
| | 2 | 0 1 0 1 0 1 1 0 | -3.254 | 5.627 | until iteration |
| | 3 | tabu | -5.764 | 6.827 | number 10) |
| | 4 | 0 0 0 0 0 1 1 0 | -9.529 | 5.635 | 1 0 0 1 0 1 1 0 |
| 6 | 5 | 1 0 0 1 1 1 1 0 | 2.392 | 6.001** | (bit no. 5 tabu |
| | 6 | tabu | 1.45 | 5.904 | until iteration |
| | 7 | 1 0 0 1 0 1 0 0 | 1.607 | 5.998 | number 11) |
| | 8 | tabu | 1.843 | 5.397 | 1 0 0 1 1 1 1 0 |
| 7 | 1 | tabu | -7.647 | 3.236 | (bit no. 4 tabu |
| | 2 | 1 1 0 1 1 1 1 0 | 7.411 | 5.116 | until iteration |
| | 3 | tabu | 4.902 | 3.512 | number 12) |
| | 4 | 1 0 0 0 1 1 1 0 | 1.137 | 5.121** | 1 0 0 0 1 1 1 0 |
| 8 | 5 | tabu | 0.509 | 6.817 | (bit no. 6 tabu |
| | 6 | 1 0 0 0 1 0 1 0 | 0.823 | 5.986** | until iteration |
| | 7 | 1 0 0 0 1 1 0 0 | 0.980 | 5.307 | number 13) |
| | 8 | tabu | 1.215 | 5.243 | 1 0 0 0 1 0 1 0 |
| 9 | 1 | tabu | -9.215 | 6.77 | (bit no. 2 tabu |
| | 2 | 1 1 0 0 1 0 1 0 | 5.843 | 9.372** | until iteration |
| | 3 | 1 0 1 0 1 0 1 0 | 3.333 | 6.698 | number 14) |
| | 4 | tabu | 2.078 | 5.061 | 1 1 0 0 1 0 1 0 |
| 10 | 5 | tabu | 5.215 | 1.246 | (bit no. 8 tabu |
| | 6 | tabu | 6.156 | 7.164 | until iteration |
| | 7 | 1 1 0 0 1 0 0 0 | 5.686 | 8.428 | number 15) |
| | 8 | 1 1 0 0 1 0 1 1 | 5.921 | 9.185** | 1 1 0 0 1 0 1 1 |

**Table 2.11** Details about coding and decoding a 10-bit string (1 0 0 1 1 1 0 0 0 0) containing the vector $[x_1, x_2]$.

| bit no | 1 2 3 4 5 | 6 7 8 9 10 |
|---|---|---|
| variable | $x_1$ | $x_2$ |
| coded | (1 0 0 1 1 | 1 0 0 0 0) |
| decoded | 2.258 | 0.323 |

**Table 2.12** Results for various combinations of (NS,TT) parameters. Number in brackets denotes iteration number for which minimum was found and the iteration number where cycling begins, R.

| TT / NS | 5 | 6 | 7 | 8 |
|---|---|---|---|---|
| 5 | 0.6705(8,18R) | 0.6705(8,18R) | 0.6705(8,26R) | 0.6705(8,26R) |
| 6 | 0.6705(8,14R) | 0.0439(66) | 0.0439(102) | 0.2087(16,36T) |
| 7 | 0.0439(20) | 0.0439(36) | 0.0439(40) | 0.6705(8,57R) |
| 8 | 0.6705(8,15R) | 0.0439(18) | 0.0439(50) | 0.0439(56,92R) |

Adoption of 20-bit long chromosome led to more focused exploration of four local minima. With a starting string (00100100001100100100), equivalent to $(x_1, x_2) \equiv (-7.185, 5.718)$ one can reach the global minimum $F_{min}$ = - 0.0456, at $(x_1, x_2) = (-4.995, -4.995)$ in 44th iteration (NS = 9, TT = 12).

### 2.3.6 Optimization of Bowed out Composite Shells - Background

The static buckling of axisymmetric shells with positive Gaussian curvature has, in the past, been studied for two separate types of external loading, i.e. for axial compression and for external pressure. The latter type of loading includes external pressure where barrelling becomes a very effective way of increasing the magnitude of static buckling load. A brief overview of work in this area can be found in Blachut and Wang (1999). It seems that there are very few experimental results on barrel-shaped shells subjected to either axial compression, external pressure or combination of both. One particular series of tests was associated with the development of natural gas container ships, Odland (1981). One additional experimental result is due to Schmidt and Krysik (1991). There appears to be no studies into the static stability performance of barrel-shape, composite shells, subjected to either pure axial compression or combined axial and external pressure loading. Structural optimisation of these components has not been addressed, as yet.

The combined stability domain of barrel type shells is expected to outperform that of the plain and equivalent cylindrical geometry. There are various ways in which the domain of combined, static, stability could be enlarged. This includes variable meridional profiling, variable wall thickness, appropriate configuration of internal/external reinforcing and stacking sequence of plies. This example concentrates on composite shells. Filament winding and the use of woven fabric are possible manufacturing techniques.

**Shell geometry and failure mechanisms.** Let us consider a cylindrical shell of radius $R_0$, length $L_0$, and the wall thickness $t_0$ which is subjected to external hydrostatic pressure p (see Fig. 2.17a). It is assumed that a mass equivalent shell having the same top and bottom radii, $R_0$, and constant meridional curvature, r, can be obtained by shortening the initial length $L_0$ to the length L. The wall thickness, $t_0$, is the same for both the cylinder and bowed out shell. The wall itself is made from N-plies, each oriented at the angle $\theta_i$ with respect to the meridional direction. In all the pertaining analysis the boundary conditions are applied at the top and bottom openings and the only non-zero Boundary Condition, (BC), is the axial movement of the shell's edge. All other quantities are set to zero, i.e. radial and circumferential deflections as well as end-rotation. Bosor4 code has been adopted as a numerical tool for evaluation of the bifurcation buckling, Bushnell (1977). Calculations have shown that in addition to bifurcation buckling and collapse mechanisms, a strength criterion also needs to be considered. A polynomial, First Ply Failure (FPF) criterion based on the Tsai-Wu approach written in stress space, has been adopted.

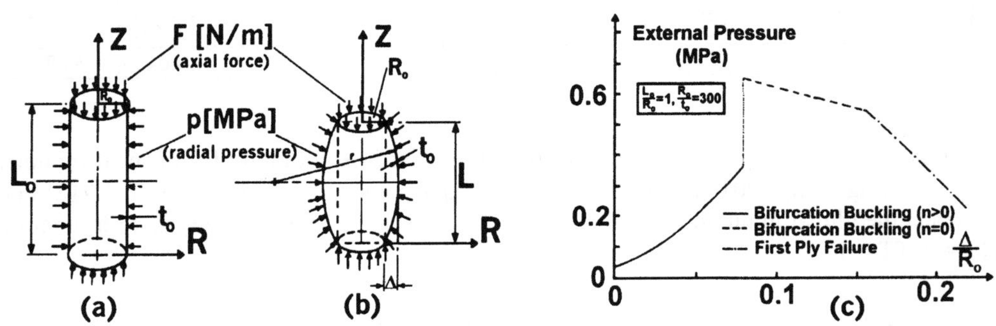


**Figure 2.17** Hydrostatically loaded cylinder (Fig. 2.17a) and barrelled out shell (Fig. 2.17b). The radius $R_0$ and wall thickness $t_0$ of both shells remain the same. Different failure modes obtained for various amount of barrelling Δ. The value $\Delta/R_0 = 0$ corresponds to cylindrical geometry (Fig. 2.17c).

Typical material constants for carbon fibre woven fabric were assumed in calculations and they were: $E_1 = E_2 = 70$ GPa, $G = 5$ GPa, $\nu_{12} = 0.1$, $X_t = Y_t = 600$ MPa, $X_c = Y_c = 570$ MPa and $S = 90$ MPa (constants $X_t$, $X_c$ are longitudinal tensile and compressive strengths; $Y_t$, $Y_c$ are transverse tensile and compressive strengths and S is the in-plane shear strength). Numerical results, which are given below, correspond to 6-ply wall for which quasi-isotropic $[0^\circ/60^\circ/-60^\circ]_s$ stacking sequence was assumed.
Numerical results obtained for $L_0/R_0 = 1$ are shown in Fig. 2.17c. It is seen that, for small amounts of barrelling, the shell loses its load carrying capacity through asymmetric bifurcation buckling with a number of circumferential waves greater than zero. Then at $\Delta/R_0 = 0.08$ there is a sudden change of the buckling mode to bifurcation buckling with n = 0. For larger barrelling, i.e. for $\Delta/R_0 > 0.16$ the shell loses its load carrying capacity through the first ply failure mechanism. Another observation here is that any amount of bowing out leads to a shell which has a higher load carrying capacity than a corresponding mass equivalent cylinder. It is clear from Fig. 2.17c that, for a given lamination stacking sequence, the structural response function is discontinuous (jump), and it can also be non-smooth at the bifurcation-first ply failure transition (cusp).

**Optimization problem and results.** It is seen from the previous considerations that the resulting cost function could therefore be associated with either bifurcation, collapse or First Ply Failure. It has been decided from the outset that the tabu search technique will be adopted in order to seek the maximum of the cost function, $p_{max}$, i.e.

$$p_{max} = \max \{\min (p_{bif}, p_{coll}, p_{FPF})\} \tag{2.36}$$

subject to the following constraints:

*on design variables:*

meridional radius of curvature

$$r_l \leq r \leq \infty \tag{2.37}$$

lamination angle

$$\theta_l \leq \theta_i \leq \theta_u, \qquad i = 1,\ldots\ldots, N; \quad \text{and} \tag{2.38}$$

*on constant mass*:

$$m_{barrel} = m_{cylinder}, \tag{2.39}$$

where N denotes number of plies. The mass of a cylinder is known once $R_o/t_o$ and $L_o/R_o$ are given. Equation (2.39) can then be solved explicitly for the unknown length, L, of the barrelled shell (or value of the meridional curvature, r). This removes the equality constraint (2.39) and leaves the optimization subject to bounds on the design variables, only.
Explicit values of bifurcation buckling and collapse were obtained directly from Bosor4. Some external Fortran programming was required to calculate the First Ply Failure pressure. For illustrative purposes only the effect of barrelling has been assessed. The design variable, $\Delta$, has been discretised using the binary representation with 5-bit and 8-bit strings. Numerical trials have shown that the algorithm had difficulties reaching the area around $\Delta = 0.08$ (large discontinuity). After closer examination it has occurred that a particular bit in the string has never been swapped in the search process. This has made a certain area of the design space non-reachable. As a result, an additional 'offending'- operator has been added according to the following course of action. From a certain iteration onwards, say from 10th iteration, all bits which had not been swapped during the last 10 iterations are identified and one of them, chosen randomly, is swapped. This allows the extension of the search into previously unseen parts of the design space. In our case it was enough to swap one, 'resistant', bit in order to reach the global maximum at $\Delta = 0.08$ ($p_{max} = 0.644$ MPa).

In conclusion, this example shows that the magnitude of the load carrying capacity of a barrel-type composite shell can be a discontinuous and non-smooth function of design variables. It has been demonstrated that a sizeable increase in the load carrying capacity could be achieved. On practical side, the tabu search algorithm has utilised an internal interruption mechanism based on the UNIX 'fork' command.

### 2.3.7 Closure

The major difficulty with the tabu search method is the possibility of cycling. As a 'memoryless' method it can easily be trapped in visiting the same parts of the design space. The stopping criterion appears to be problem dependent and there are no specific rules here. Introduction of an 'offending'-operator shows how a simple measure can greatly contribute to the diversification of the search.

## 2.4 Dynamic Programming

### 2.4.1 Introduction

Most of our decisions are made sequentially, e.g. at different points of time, at different points of space or at different time and space operations. Problems in which decisions are made successively, or sequentially, are known as sequential decision problems or multistage decision problems. These decisions can be made in such a way that they minimize the cost function subject to certain constraints. In cases like that, dynamic programming would be a suitable mathematical technique to be applied to facilitate an optimal approach.
The dynamic programming method, when applicable, decomposes a multistage decision problem into a sequence of single stage decision making. An n-variable problem is represented as a sequence of n single variable problems which are solved successively. These n sub-problems are usually easier to solve than the original problem. It is immaterial then what optimization technique is used for solving the resulting n-single variable problems. Methods ranging from enumerative to differential calculus can be applicable. The latter however would normally require continuous functions, and that they are differentiable, and that the optimum points do not lie on the boundary. Dynamic programming, on the other hand, can tackle discrete variables, non-convex, discontinuous and non-differentiable functions.
The theoretical background to dynamic programming can be found in books by Bellman (1957), Bellman and Dreyfus (1962), or Nemhauser (1966).

### 2.4.2 Dynamic Programming in Structural Optimization

It appears that dynamic programming has not been widely used for structural optimization. The available literature also points towards limited range of applications. Palmer (1968) applies dynamic programming formalism to the optimal design of multispan beams and frames. The same method is also applied to a thin cylindrical shell subjected to a ring loading (Palmer, 1969). For elastic-perfectly plastic material, the paper develops the method for evaluation of the limit ring load. The problem of optimizing the design of pin-jointed structures is studied by Palmer and Sheppard (1970). A cantilever truss is also studied. Distefano and Rath (1975) apply dynamic programming to minimize weight of elastic trusses with constraints imposed on displacements and stresses. Numerical results are provided to illustrate the solution methodology for both statically determinate and statically indeterminate trusses. Distefano (1975) studies optimization of linear structural systems by adopting dynamic programming methodology. The optimization of a multispan, continuous beam, is provided as an example. The individual spans are of the same length and they are assumed to have piecewise constant moment of inertia. Inequality constraints on stress and deflection are considered. The minimum weight profile of an electrical power transmission tower, made from steel members, was searched for in work by Raj and Durrant (1976). Andreev and Dyachenko (1977) propose to apply dynamic programming to study load-deflection curves for non-shallow, elastic shells of revolution. Equilibrium is obtained via global minimum of the potential energy. A spherical segment, loaded through the edge bending moment, is provided as an illustrative example. Gero et al. (1978) extend the serial dynamic programming into a non-serial dynamic programming containing feed-forward loops which do not intersect. The paper provides optimal design of a platform and optimal design of a multi-storey office building as typical examples of serial and non-serial dynamic programming. Gero and Kaneshalingam (1978) discuss benefits of the dynamic programming method when applied to building design. It transpires that small changes to the design specification may not necessarily require repetition of the entire computational effort due to the fact that some sub-optimal paths are already available to the designer. Minimum weight design of tower structures is discussed by Howell and Doyle (1978). Dynamic programming is used for automatic selection of geometry and member sizing. Several examples are provided in which the number of structural members varies from 15 to 336. Hamada et al. (1981) seek mass distribution of an elastic multi-segment, stepped cantilever. Both lengths of individual segments and magnitude of the cross-sections are taken as simultaneous design variables. Casings used in oil wells are usually composed from a string of tubes. Osman and Parent (1982) seek minimum mass of a multi-segment tube subjected to external pressure, working internal pressure and its own weight. Applicability of DP is exemplified by a 4000m long tube, which was divided into 8 segments. Thevendran (1982) studies minimum weight of a spherical cap loaded by a concentrated load applied to a rigid boss at its apex. Continuously varying wall thickness was considered under stress constraints. In a similar study, Thevendran (1984) considers minimum-weight design of circular plates subjected to transverse uniformly distributed pressure. Continuously varying wall thickness with Tresca and Mises yield criteria are discussed. Simply supported or clamped boundaries are considered. Jendo et al. (1985) use dynamic programming approach to illustrate pareto-optimization in which the statically determinate truss serves as an example. Schnack and Sporl (1986) seek an optimal contour along which the maximum magnitude of Mises stress would be minimal. They apply Bellman's dynamic programming to solve the resulting min-max problem. Examples include shaping a hole in plates under bi-axial loading and shaping a connecting rod under tensile load and internal pressure.

### 2.4.3 Bellman's Principle of Optimality

Dynamic programming is a sequential decision making method based on the optimality principle, Bellman (1957). It can be stated as: "An optimal policy has the property that whatever the initial state and initial decisions are, the remaining decisions must constitute an optimal policy with regard to the state resulting from the first decision".

The requirements that a specific problem must satisfy, in order to be soluble using dynamic programming, all relate to the user's ability to decompose the problem into sequential stages. These requirements manifest themselves as separability and monoticity conditions. As it has already been mentioned, the aim of a multistage decision problem is to find the design variables $x_1$, ...., $x_n$ in order to optimize the function of the individual stage returns, e.g. $F(R_1, \ldots, R_n)$, where $R_i$ is the return function for stage number i - see Fig. 2.18a. In order to have separability of the objective function one has to be able to represent the objective as the composition of the individual stage returns. This requirement is satisfied, for example, for objective functions which are additive, i.e.

$$F = \sum_{i=1}^{n} R_i(x_i, s_{i+1}) \tag{2.40}$$

where $x_i$ represents decision variable at stage i and $s_{i+1}$ represents the input state to stage i whilst $s_i$ represents the output state at the stage number i (see Nemhauser, 1966, or Rao, 1984).
Let us now provide, after Rao (1984), the recurrence formula, which forms the basis of sub-optimization procedure. Let us assume that we seek the following minimum:

$$\min_{x_n, \ldots, x_i, x_{i-1}, \ldots, x_1} F \tag{2.41}$$

Where the cost function, F, is given by Eq (2.40) in which the state and decision variables are related by the following design equations:

$$s_i = T_i(s_{i+1}, x_i), \qquad i = 1, \ldots., n \tag{2.42}$$

Consider the first sub-problem by starting from the final stage, i = 1 - as sketched in Fig. 2.18b. If the input to this stage, $s_2$, is specified, then according to the principle of optimality, $x_1$, must be selected such that $R_1(x_1, s_2)$ is an optimum for the input $s_2$. If this optimum is denoted by $F_1^*$, we have:

$$F_1^*(s_2) = \min_{x_1} R_1(x_1, s_2) \tag{2.43}$$

This is called a one-stage policy since once the input state $s_2$ is specified, the optimal values of $R_1$, $x_1$ and $s_1$ will be completely defined. Eq. (2.43) is a parametric equation giving the optimum $F_1^*$ as a function of the input parameter $s_2$.

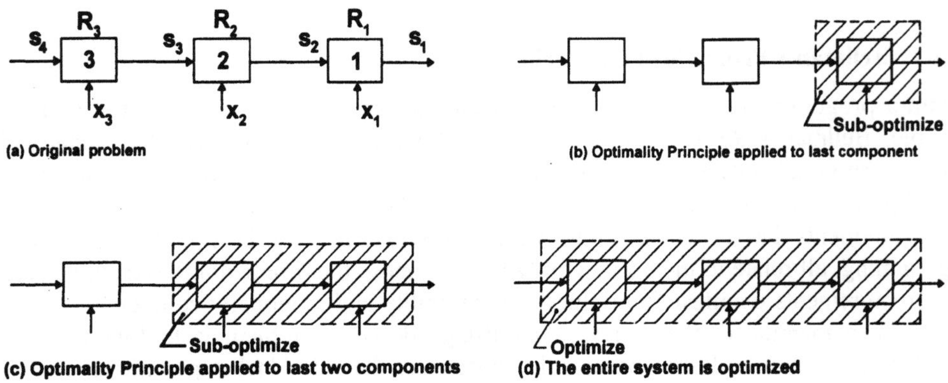


**Figure 2.18** Details about stage-by-stage application of optimality criterion.

Next, consider the second sub-problem by grouping the last two stages together as sketched in Fig. 2.18c. If $F_2^*$ denotes the optimum objective value of the second sub-problem for a specified value of the input $s_3$, we have

$$F_2^*(s_3) = \min_{x_1, x_2} [R_2(x_2, s_3) + R_1(x_1, s_2)] \tag{2.44}$$

The optimality principle requires that $x_1$ be selected in such a way as to optimize $R_1$ for a given $s_2$. Since $s_2$ can be obtained once $x_2$ and $s_3$ are specified, Eq. (2.44) can be written as:

$$F_2^*(s_3) = \min_{x_2} [R_2(x_2, s_3) + F_1^*(s_2)] \tag{2.45}$$

This means that $F_2^*$ represents the optimal policy for the two-stage sub-problem. It is seen that the optimality principle reduced the dimensionality of the problem from two, in Eq. (2.44) to one in Eq. (2.45). It can also be seen that for a specified input, $s_3$, the optimum is determined solely by a suitable choice of the decision variable $x_2$.
This idea can be generalised further and the ith problem defined by:

$$F_i^*(s_{i+1}) = \min_{x_i, x_{i-1}, \ldots, x_1} [R_i(x_i, s_{i+1}) + R_{i-1}(x_{i-1}, s_i) + \ldots\ldots + R_1(x_1, s_2)] \tag{2.46}$$

can be re-written as

$$F_i^*(s_{i+1}) = \min_{x_i} [R_i(x_i, s_{i+1}) + F_{i-1}^*(s_i)] \tag{2.47}$$

where $F_{i-1}^*$ denotes the optimal value of the objective function corresponding to the last i-1 stages, and $s_i$ is the input to the stage i-1. It is worth noting here that problem given by Eq. (2.46) requires the simultaneous variation of i decision variables $x_1, \ldots, x_n$, to determine the optimum value of $F_i = \sum R_k$ for any specified input $s_{i+1}$. By using the optimality principle, this problem has been decomposed into i separate problems, each involving only one decision variable. Equation (2.47) is known as the recurrence relationship valid for i = 1, 2,..., n.
The next paragraphs will provide details about numerical implementation of the above technique.

### 2.4.4 Illustrative Example

Let us consider a routing problem which involves finding a shortest distance between two end cities shown in Fig. 2.19a. Distances between all intermediate cities are given in this figure in some artificial units, and it is our aim to find the shortest path between city number one and city number 10. The problem has already been sketched as a four-stage model. Let us begin our journey from city no. 1, and whilst at stage 2 it is obvious that there are three possible routes, i.e. from cities 2, 3 and 4. After moving to stage three we apply the optimality principle in turn, i.e. whilst being at cities 5, 6 and 7. The best strategy from city no. 5 amounts to the distance 1230 (with the other two values being 1480 and 1280, respectively). The best forward strategies from cities 6 and 7 are 1340 and 1600, respectively. By moving

backwards by one stage we obtain, as best forward strategies, the following distances: 1840 from city number 8 and 1870 from city number 9. Finally, we move to stage number five where we see that the path via town number 9 gives the smallest distance equal to 2870.
The process is complete once the whole path associated with the optimal distance of 2870 is recovered. It is seen from the tabulated data that the optimal route is 10 → 9 → 7 → 4 → 1. It transpires that from city 7 there is an alternative path through city number 2, i.e. 10 → 9 → 7 → 2 → 1.

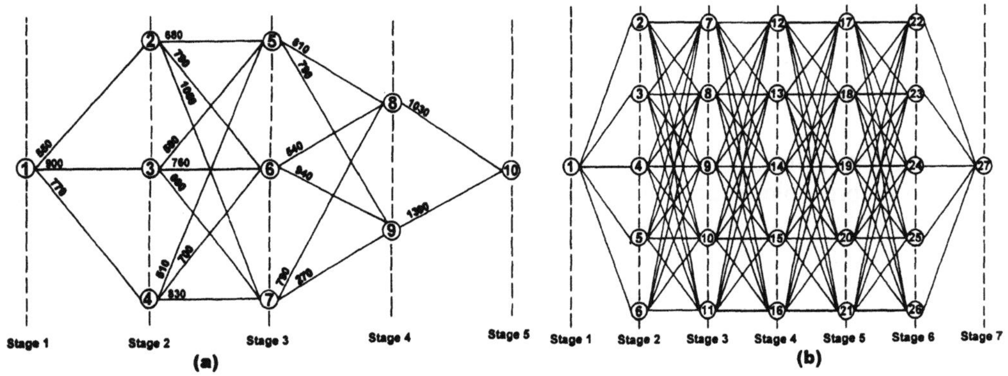


**Figure 2.19** Illustration of a networking problem posed as 5-stage decision problem (Fig. 2.19a). Illustration of dynamic programming efficiency for a 27-node, 7-stage networking case (Fig. 2.19b).

It appears that for the above example it would be an easy matter to find the shortest path from node 10 to node 1 by enumerating all the possible paths. Thus in this problem, the use of dynamic programming did not really show great benefits. For larger networks, however, dynamic programming is much more efficient for determining a shortest path than the explicit enumeration of all paths. To illustrate this, let us consider the network sketched in Fig. 2.19b. It is seen from this figure that it is possible to travel from any node in stage, i, to any node in stage, i+1. Suppose we want to determine the shortest path from node 1 to node 27. In explicit enumeration there are $5^5$ possible paths between node 1 and 27. It takes five additions to determine the length of each path. Explicit enumeration of the length of all paths requires $5^5 \times 5 = 5^6 = 15625$ additions.
The shortest path from node 1 to 27 can be found using dynamic programming, as illustrated in the previous example. The optimal path can be found after 105 additions. Since explicit enumeration requires 15625 additions, one can see that dynamic programming requires only 0.7 % additions, i.e. a very small fraction of what would be needed in direct enumeration. In conclusion it can be noted that the computational savings effected by dynamic programming are substantial.

### 2.4.5 Application to Large Deflection of Struts

Let us consider an elastic bar of length $l$, subjected to an axial force P as sketched in Fig. 2.20a. Let us also assume that the beam's stiffness $\alpha = EI$ is piecewise constant, i.e. $\alpha_1 = EI_1$ for $y \in [0, l_1]$ and $\alpha_2 = EI_2$ for $y \in (l_1, l]$ - see Fig. 2.20b. Let us formulate the optimization problem as follows:

Find the magnitude $\delta$ ($\equiv \alpha_2/\alpha_1$) which minimizes the beam's end deflection, $x_k$, as shown in Fig. 2.20a, i.e.:

$$\min_{\delta \in U_1} x_k \tag{2.48}$$

where $U_1 \equiv \{\delta : 0 < \delta \leq 1.0\}$, subject to the following constraints:
- on the total volume, $V_o$, of the beam:

$$V_o = \text{const} \tag{2.49}$$

- on the beam's equilibrium which amounts to the minimum of the potential energy, E, i.e.:

$$\min_{\varphi \in U_2} E \tag{2.50}$$

where $U_2 \equiv \{ \varphi : 0 \leq \varphi(s) \leq \pi \wedge \varphi(0) = 0\}$, $0 \leq s \leq 1$.
It can be shown that potential energy, E, for the system shown in Fig. 2.20 can be written as:

$$E = \sum_{R=N+1}^{M} [\frac{1}{2}(\frac{\varphi_R - \varphi_{R-1}}{\Delta})^2 + c_1 \cos(\varphi_R)]\Delta + \sum_{R=1}^{N} [\frac{\delta}{2} (\frac{\varphi_R - \varphi_{R-1}}{\Delta})^2 + c_1 \cos(\varphi_R)]\Delta \tag{2.51}$$

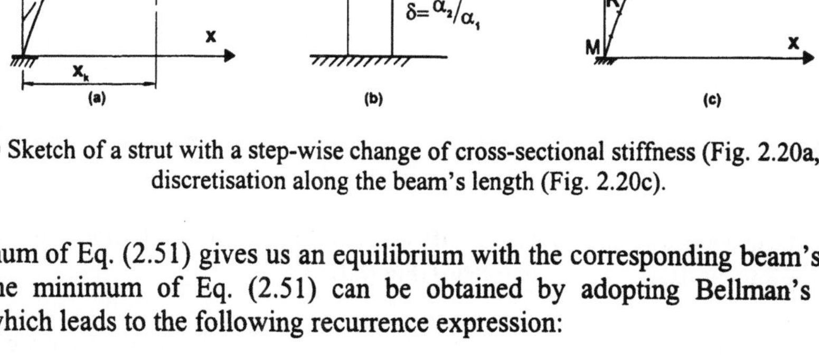


**Figure 2.20** Sketch of a strut with a step-wise change of cross-sectional stiffness (Fig. 2.20a,b). Adopted discretisation along the beam's length (Fig. 2.20c).

The minimum of Eq. (2.51) gives us an equilibrium with the corresponding beam's deflection profile. The minimum of Eq. (2.51) can be obtained by adopting Bellman's optimality principle which leads to the following recurrence expression:

$$F_R(\varphi_R) = \underbrace{\min}_{\varphi_{R-1} \in U_2} \left\{ \left[ \frac{\delta}{2}(\frac{\varphi_R - \varphi_{R-1}}{\Delta})^2 + c_1 \cos(\varphi_R) \right] \Delta + F_{R-1}(\varphi_{R-1}) \right\} \tag{2.52}$$

for $0 \leq R \leq N$.

Numerical evaluation of the above equation can start from a free end and progress towards the clamped end where the boundary condition, $\varphi(0) = 0$, must be satisfied.
Volume of a beam with rectangular cross-section can be written as $V = [A_1 s_1 + A_2(s_2 - s_1)]l$, where $A_1$ and $A_2$ denote cross-sections of the beam. Introducing the parameter $\delta$ one can write the expression for the volume, V, as follows:

$$V = (12Pl^3)[s_1 + \delta(s_2 - s_1)]/Eb^2c_1 \tag{2.53}$$

If one introduces a dimensionless volume, $V_o$, as $V_o = (VEb^2)/(12Pl^3)$ then the equation (2.53) can be written as follows:

$$c_1V_o = s_1 + \delta(s_2 - s_1) \tag{2.54}$$

Figure 2.21a shows minimum end-deflection curve which was obtained for a fixed value of beam's volume, i.e. $V_o = 0.30$. This solution was obtained for the parameter $\delta$ varying between 0.0 and 1.0. A selection of deflected profiles is provided in Fig. 2.21b for arbitrarily chosen points a, b, c and d on the corresponding ($x_k$ versus $\delta$) graph. It is seen from Fig. 2.21a, for example, that there are two possible minima of $x_k$ and they are denoted by points 'a' and 'b', respectively. It is seen from the corresponding Fig. 2.21b that the upper portion of the beam undergoes very large deflection for point 'a'. Hence the acceptable solution corresponds to the profile denoted by point 'b'. Additional calculations have shown that further increase of the volume $V_o$ enlarges the domain where the beam does not lose its stability (gap between 'b' and 'c').

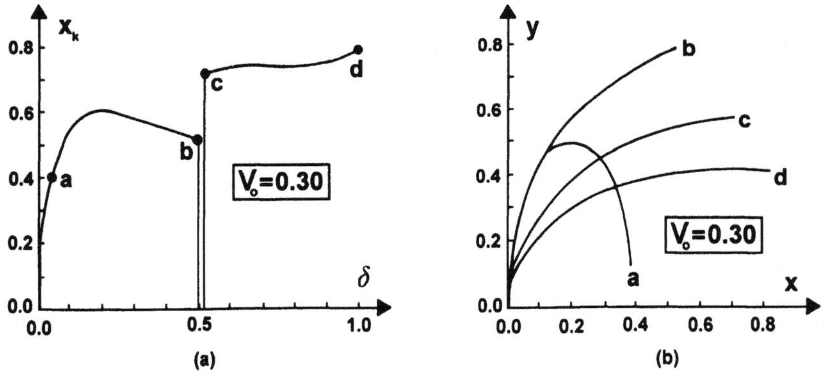


**Figure 2.21** End deflection versus the ratio of stiffness split (Fig. 2.21b). Deflection lines for a sample of δ-ratios (Fig. 2.21b).

Self-weight of the beam can have significant effect on the end behaviour of the stepped beam considered above. Further details and the corresponding results can be found in Blachut (1977).
In the above example sequential decisions were made in order to find static equilibrium of an elastic beam for a given distribution of stiffness between two, upper and lower, segments. The method proved to be efficient and it could easily accommodate further extensions, e.g. self-weight, larger number of segments of constant or variable length.

### 2.4.6 Conclusions

Unlike other programming techniques, dynamic programming has no standard mathematical formulation for the optimization procedures. A number of solution methods, ranging from enumeration to differential calculus and non-linear programming, are available in dynamic programming. However, all these methods are based on a recursive optimization principle known as the optimality principle. Depending on the conditions prescribed for the system, a dynamic programming problem can be solved as an initial-value or as a final-value problem. This technique proves to be a very efficient tool for networking problems. Its relative under-use in structural optimization has primarily been associated with computer memory requirements. Nowadays, for large disc memories with shorter and shorter access time the dynamic programming has an increased potential for application.

## 2.5 Other search algorithms

### 2.5.1 Introduction

Let us consider an unconstrained minimization problem. There are several methods available for solving this class of problems. These methods can conveniently be classified into the following two broad categories:

- Direct search methods, e.g. random search, Monte Carlo, univariate search, pattern search methods (Powell's, Hooke and Jeeves), Rosenbrock's method of rotating co-ordinates, simplex or complex.
- Descent methods, e.g. steepest descent, conjugate gradient (Fletcher-Reeves), Newton's method, variable metric (Davidon-Fletcher-Powell).

The direct search methods require only objective function evaluations and they do not require any derivatives of the cost function in the process of finding the minimum. As a consequence they are frequently called the non-gradient methods or zero-order methods.

The descent methods, on the other hand, require evaluation of both the function and its derivative(s). They are frequently called the gradient methods. These methods are generally more efficient compared to the direct search methods since they rely not only on values of the cost function but also on values of its derivatives. All these methods are iterative in nature. In what follows we will concentrate on some of direct search methods and we will highlight their unique features. This includes random based techniques and the complex method of Box. The latter can handle inequality constraints as well.

### 2.5.2 Random Search Methods

These methods are considered to be computationally inefficient, especially for large number of design variables. On the other hand they are the most easily implementable techniques and one should not ignore them from a practical point of view. Let us now illustrate the random jumping, random walk and random walk with direction exploitation. Classical texts by Rao (1984) or Vanderplaats (1984) provide further details.

**Random jumping.** The simplest way of finding a minimum of $F(\mathbf{x})$ subject to the lower, $\mathbf{x}_L$, an upper, $\mathbf{x}_U$, bounds on the vector $\mathbf{x}$, i.e. $\mathbf{x}_L \leq \mathbf{x} \leq \mathbf{x}_U$, is to randomly generate n candidates in the design space. Let us assume that the set $r \equiv (r_1, r_2, \ldots, r_n)$ represents n random numbers which are uniformly distributed between 0 and 1. Each set, r, of these numbers is used to find a new point $\mathbf{x}^i$ inside n-dimensional cube according to the following equation:

$$\mathbf{x}^i = \mathbf{x}_L + \mathbf{r} * (\mathbf{x}_U - \mathbf{x}_L) \tag{2.55}$$

Next the value of the cost function is evaluated at this point, $\mathbf{x}^i$. By generating a large number of random trial points and evaluating the value of the function at each of them, one takes the smallest value of the F(**x**) as the minimum. A quick illustration of obtained results for minimization of the function given by Eqns. (2.24 - 2.25) is given in Table 2.13. The random jumping process was terminated after 10000 iterations and only those iterations showing improvement are listed in Table 2.13. It is seen that no improvement has been recorded beyond the iteration number 7137 for which, $F_{min} = 0.0115$ is very close to the true minimum. The above algorithm is attractive for its simplicity but its efficiency can be increased, as shown in the next method of random walk.

**Table 2.13** History of random jumping search. Only those configurations are shown which improve the value of the cost function F.

| Iteration | Random numbers | | Design variables | | Value of |
|---|---|---|---|---|---|
| i | $r_1$ | $r_2$ | $x_1$ | $x_2$ | $F(x_1, x_2)$ |
| 1 | 0.6928 | 0.1732 | 3.8565 | -6.5353 | 4.3814 |
| 18 | 0.2033 | 0.1923 | -5.9345 | -6.1530 | 3.0679 |
| 161 | 0.2694 | 0.2742 | -4.6125 | -4.5165 | 0.5482 |
| 643 | 0.2674 | 0.2258 | -5.2168 | -5.4840 | 0.2885 |
| 2187 | 0.2392 | 0.2479 | -5.2111 | -5.0417 | 0.0982 |
| 7137 | 0.2394 | 0.2516 | -5.1295 | -4.9676 | 0.0115 |

**Random walk method.** This method explores the design space along a random direction in n-dimensional cube. The direction is taken as that of a unit vector **u** which is defined as follows. Firstly, a set of random numbers $r_1, \ldots.., r_n$ is generated in the interval [-1,+1] and length of the corresponding vector is computed as $R = (r_1^2 + \ldots..+ r_n^2)^{1/2}$. The generated set of random numbers is accepted only if $R \leq 1.0$ and it is discarded otherwise. Next, the set of $r_1, \ldots.., r_n$ is normalised in order to obtain R = 1 and finally the vector **u** is found from the following relation:

$$\mathbf{u} = 1.0/(r_1^2 + \ldots..+ r_n^2)^{1/2} * [r_1, \ldots.., r_n]^T \tag{2.56}$$

One starts from a feasible point $\mathbf{x}_i$, at iteration number i, and seeks the improved approximation iteratively by creating a new point given by:

$$\mathbf{x}_{i+1} = \mathbf{x}_i + \lambda \mathbf{u}_i \tag{2.57}$$

where $\lambda$ is a given scalar step length, $\mathbf{u}_i$ is a unit random vector generated in iteration number i. For a given value of scalar $\lambda$ the algorithm is run for a prescribed number of times, N, before $\lambda$ is reduced. All moves which, improve the cost function, are accepted unconditionally. If a move does not improve the cost function then a new direction is tried by

computing a fresh vector **u**. After N-trials the step length lambda is reduced and the search repeats itself.
As an example let us consider the same example an in the previous section, i.e. minimization of the function given by Eqns. (2.24 – 2.25). Table 2.14 provides summary of obtained results. They were obtained for N = 100, the initial value of $\lambda = 10$ with a simple geometric reduction, i.e. $\lambda_{new} = 0.5\lambda_{old}$. The starting point was taken as $[x_1, x_2] = [0.0, 0.0]$. All designs which, had fallen outside the bounds given by inequality (2.25) were rejected. It is seen from Table 2.14 that for some values of the step size there was no improvement in the cost function for all N-iterations in a loop (e.g. for $\lambda = 5.0$ or 0.3125). The algorithm was however able to navigate to the global minimum. Further improvements to random search methods include exploitation of a successful search direction until it fails to improve the cost function any further (Random Walk Method with Direction Exploitation – see Rao, 1984). It should be stressed here again that efficiency of random search methods drops significantly once larger problems are attempted.
In addition to random search techniques, a string of other direct search methods has been developed in the past and we will concentrate on one of them which is applicable to both

**Table 2.14** History of random walk.

| No. of F-evaluations | Random numbers $r_1$ | Random numbers $r_2$ | $\lambda$ | Design variables $x_1$ | Design variables $x_2$ | Value of $F(x_1,x_2)$ |
|---|---|---|---|---|---|---|
| 1 | 0.5910 | -0.7282 | 10.0 | 6.3017 | -7.7646 | 8.0088 |
| 19 | 0.0835 | -0.8128 | 10.0 | -5.0585 | -3.2048 | 2.2308 |
| 202 | 0.1067 | -0.6313 | 2.50 | -4.6419 | -5.6699 | 0.6510 |
| 301 | -0.1542 | 0.3908 | 1.25 | -5.1007 | -4.5071 | 0.1803 |
| 401 | 0.0562 | -0.4853 | 0.6250 | -5.0288 | -5.1280 | -0.0282 |
| 606 | 0.3835 | 0.5286 | 0.1562 | -4.9370 | -5.0015 | -0.0368 |
| 704 | -0.9118 | 0.3003 | 0.0781 | -5.0112 | -4.9771 | -0.0442 |
| 813 | 0.0737 | -0.6191 | 0.0391 | -4.9878 | -5.0142 | -0.0455 |
| 905 | -0.0434 | 0.6005 | 0.0195 | -4.9892 | -4.9947 | -0.0457 |
| 1408 | -0.6016 | -0.7919 | 0.006 | -4.9918 | -4.9987 | -0.0457 |

unconstrained and constrained optimization. This method is known as simplex (for unconstrained cases) and complex method of Box (for constrained optimization). Further details on other direct-search techniques applicable to constrained optimization can be found in work by Swann (1974).

### 2.5.3 Background to the Simplex and Complex Methods

The simplex method derives its name from the use of regular geometric figure, called simplex, to explore the design space. A simplex is the structure formed by n + 1 points, not in the same plane, in an n-dimensional space. Once a simplex is created it is moved throughout the design space by a set of three operations called reflection, contraction and expansion. This allows the simplex to adapt itself to the local landscape and shrink to the final minimum. The method is usually attributable to Spendley et al. (1962) and Nelder and Mead (1965). The simplex is formed from a set of points, called corners or vertices. The basic idea of the cost function minimization in the simplex method is to compare values at all vertices and then to replace the worst vertex in the simplex, i.e. the one having the highest value of F(x), by its reflection in the centroid of the others. This leads to a new simplex. Nelder and Mead (1965) improved the performance of the method by allowing the simplex to re-scale itself according to the local geometry of the function by incorporating expansion and shrinkage. The method proved to be a very robust technique for unconstrained optimization, see Parkinson and Hutchinson (1972), for example. Box (1965) has expanded the above idea into constrained optimization. He developed a new constrained simplex method using $q \geq n + 1$ vertices and termed this figure a complex - see Fig. 2.22a for a sketch of 5-corner complex. The additional vertices were introduced to prevent the simplex from losing dimensions, and eventually

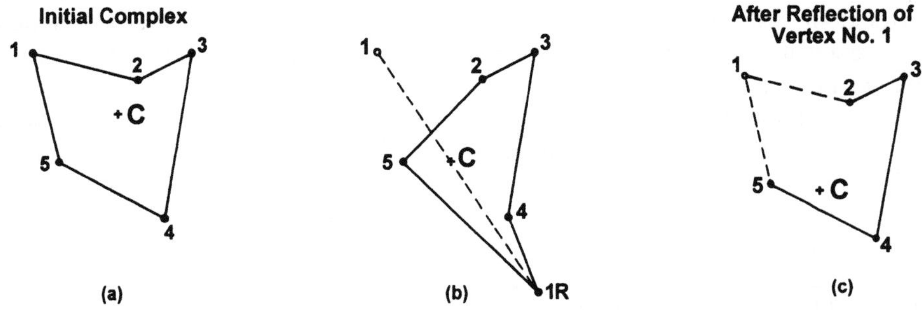


**Figure 2.22** Five corner complex – Fig. 2.22a. Position of a centroid once the vertex no. 1 is discarded – Fig.2.22b. New five-corner complex obtained after reflection through C – Fig. 2.22c.

collapsing, when confronted with constraints. The resulting method is known as complex method of Box. The basic mechanisms through which the simplex is driven into optimum remain the same. As in the unconstrained simplex procedures, the worst vertex is replaced in each iteration, and again the basic move is a reflection in the centroid of other vertices. Fig. 2.22b shows the centroid 'C' of the remaining vertices 2, 3, 4, and 5 whilst the vertex no. 1 has the highest value of the cost function and it is to be removed from the next complex. The resulting new complex, after the reflection, is sketched in Fig. 2.22c. In the next section we will consider the complex method of Box in greater detail, seeing it as a more general technique.

### 2.5.4 The Complex Method

The complex method of Box has been extensively researched and, over the years, a number of modifications have been proposed - see Swann (1974), Kneppe and Sattler (1982). The algorithm has been applied to optimization of space trusses by Lipson and Gwin (1977). The same technique was applied to the minimum weight of a rotating gas turbine disc by Luchi and Poggialini (1980). Blachut (1988, 1991, 1992, 1997) has applied complex method of Box to optimization of metallic and composite pressure vessel closures. The first three papers consider externally pressurised domes whilst the latter one considers internally pressurised steel head. Some of numerical results were subject of experimental verification.

The complex method is to be applied now to structural optimization of pressure vessel closures subjected to external pressure.

**Example – externally pressurised dome.** Let us consider an elastic/elastic-plastic torisphere of constant thickness t and diameter *D* under uniform, static external pressure p. Let us also assume that there is no flange, i.e. $L \equiv 0$, and that the torisphere is fully clamped at the edge (see Fig. 2.23). If $R_s$ and *r* denote the radius of the spherical part and the radius of the knuckle, respectively, then the ratio $W_T/W_H$ indicates the amount of material allocated to the torisphere, with respect to the reference hemispherical end closure weight, $W_H$, and this equals:

$$W_T/W_H = 4(D/t_o)/(D/t)\{(r/D)(0.5 - r/D)(0.5\pi - \Phi) + [(r/D)^2 - (R_s/D)^2]\cos\Phi + (R_s/D)^2\} \quad (2.58)$$

Let us imagine a hemispherical pressure vessel of diameter *D* and thickness $t_o$ weighing $W_H$. The given amount of material $W_T$ (or any arbitrary fraction of it) can be utilised in many

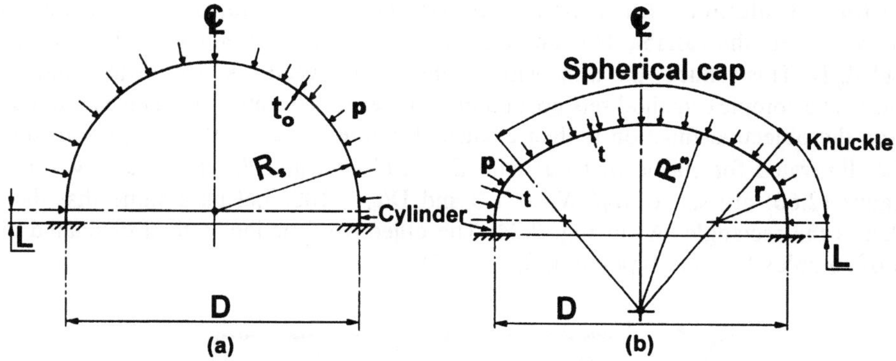


**Figure 2.23** Geometry of hemispherical and torispherical domed closures loaded by uniform external pressure, p.

different ways to form an infinite number of torispherical end closures, all having the same weight, $W_T$, but having a different shape and thickness. In turn, the buckling strength of these closures would vary. The following optimization problem can then be formulated. Find the torispherical dome, i.e. the vector $X_{opt} = [r/D, R_s/D, D/t]$, which:

$$\max(\min p_i) \qquad (2.59)$$
$$X \subset \Omega, i = 1, 2$$

where the index i = 1 means failure by bifurcation and i = 2 corresponds to axisymmetric snap-through;
- with side constraints:

$$0.6 \le W_T/W_H \le 1.0 \qquad 100 \le D/t_o \le 1000 \qquad (2.60)$$

- and limits on design variables:

$$D/t > 0, \qquad 0.05 \le r/D \le 0.45, \qquad 0.55 \le R_s/D \le 1.25 \qquad (2.61)$$

The choice of upper and lower limits in (2.61) is arbitrary, but the following factors were taken into account. The lower limit on the knuckle shallowness (i.e. r/D = 0.05) is close to the BS 5500 specification (r/D = 0.06). Some previous computations showed an unexpected drop in the buckling strength within the range $0.45 \leq r/D \leq 0.5$ and it was felt justifiable to limit the knuckle-to-diameter ratio to $r/D \leq 0.45$. In addition to the optimization posed by (2.59 - 2.61) the following optimization problem is also formulated:

$$\min_{X \subset \Omega,\ i=1,2} (\min p_i) \tag{2.62}$$

subject to constraints (2.60 – 2.61). The results of these two separate optimization processes are intended to indicate the range of the failure pressures one might obtain if the end closures are formed on a non-optimum basis.

**The method of solution – illustration for pressure maximization.** Once the side constraints (2.60) are chosen, the variable D/t can be directly eliminated via equation (2.58) for a given r/D and $R_s/D$. This reduces the dimension of the control vector X from 3 to 2 independent variables. The complex method sets up a number of vertices (arbitrarily taken as 4 in our case) at which the objective function is then evaluated. Some stages of the complex method will now be illustrated for elastic material with $E$ = 207 GPa and Poisson's ratio $\nu = 0.3$. The constraints (2.60) are set to $W_T/W_H = 1.0$ and $D/t_0 = 100$. Let us assume that the initial complex is the rectangle 'A' in Fig. 2.24. The objective function is then evaluated at each vertex of complex 'A' (i.e. at points 1, 2, 3 and 4).

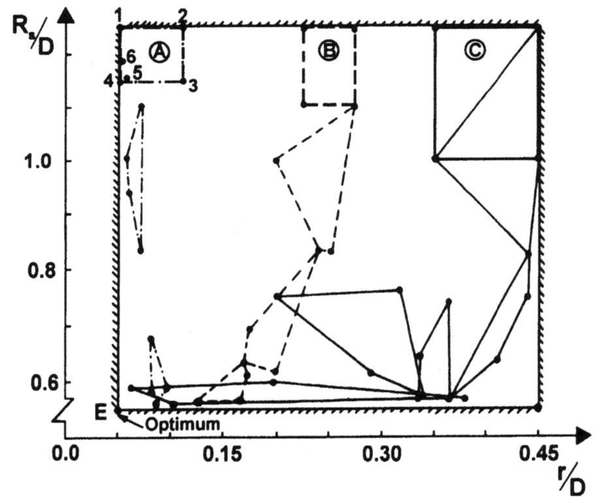


**Figure 2.24** Paths of three different, initial complexes A, B and C from the initial position to the optimum at E.

Next, the vertex with the smallest value of the cost function, $X_l$, is replaced by an operation called reflection. The reflection of $X_l$ is denoted by $X_r$, and its co-ordinates are given by the following relation in which $X_0$ is the centroid of the remaining vertices:

$$X_r = (1+\alpha)X_o - \alpha X_\ell \quad \text{, where} \qquad X_o = \frac{1}{k-1}\sum_{i=1, i\neq \ell}^{k} \|X_i - X_\ell\|. \tag{2.63}$$

In our initial complex 'A' the least value is obtained for the vertex number 2. Using the suggested reflection factor $\alpha = 1.3$ (Box, 1965), one obtains from equation (2.63) a new trial vector $X_r$.
In our case this trial vertex lies outside the feasible region and it is moved halfway towards the centroid of the remaining points using the relation:

$$(X_r)_{new} = 0.5\,(X_o + X_r) \tag{2.64}$$

Two contractions using equation (2.64) produce the permissible point $X_k$, which constitutes the vertex number 5 in Fig. 2.24. Vertices 1, 3, 4 and 5 belong to the next complex, in which BOSOR 5 is used to evaluate the buckling pressure at the vertex number 5. It then becomes obvious that point 3 has to be discarded from future complexes. Its reflection is denoted in Fig. 2.24 as the point 6.
The above process continues until the following criteria of convergence are satisfied simultaneously:

$$\frac{1}{p_{av}}\sum_{j=1}^{4}\sqrt{(p_{av} - p(j))^2} \le \varepsilon_1\,, \qquad \sum_{j=1}^{4}\|X_o - X_j\| \le \varepsilon_2 \tag{2.65}$$

where $p_{av}$ is an averaged value of the objective function over 4 vertices of the complex and $\varepsilon_1 = 0.001$, $\varepsilon_2 = 0.05$ have been assumed as suitable in all calculations. Figure 2.24 shows the search path which starts at the complex 'A' and reaches the optimum at the point 'E'. Some vertices belonging to that path are also shown together with two interim complexes. As the method discussed does not guarantee a global extremum, two additional starting complexes 'B' and 'C', in Fig. 2.24, were tried. In both cases they converged to the same optimum. In all subsequent calculations the complex 'B' served as the starting one. The flow chart showing all stages of the computations can be found in Blachut (1988).
For elastic material the maximum pressures were obtained for $X \equiv [0.05, 0.55]$ and D/t given by Eq. (2.58). The increase in buckling pressure above that of the corresponding reference hemisphere were 53%, 48%, 37% and 19% for $D/t_o = 100$, 250, 500 and 1000, respectively. Solving the problem (2.62) shows that torispheres with an inadequate geometry can sustain very much less pressure than the same torispheres shaped to carry the maximum pressure by a factor of 14 in some cases (for $D/t_o = 100$). Similar set of results were also obtained for mild steel having the yield point $\sigma_{yp} = 414$ MPa. Further details about the method of solution and results can be found in Blachut (1988).

### 2.5.5 Closure

It has been frequently argued that the best use of the most advanced optimization techniques requires the creation of integrated analysis and optimization software from the beginning.

This can then provide an efficient use of analytical gradients, constraint deletion, approximations, etc. It should however be realised that this is not a universal approach and some of the above results, especially those based on complex method of Box, would be difficult to obtain by any method which is gradient based.

## 2.6 References

Aarts, E., and Korst, J. (1989). *Simulated Annealing and Boltzman Machines.* Chichester - New York: John Wiley and Sons. 272p.

Aarts, E., and Lenstra, J.K., eds, (1998). *Local Search in Combinatorial Optimization.* Chichester - New York: John Wiley and Sons. 512p.

ABAQUS: Hibbit, Karlsson and Sorensen Inc., (1998). ABAQUS. *User's and Theory Manual,* Version 5.8, Pawtucket, RI 02860-4847, USA.

Andreev, L.V., and Dyachenko, V.E. (1977). Application of dynamic programming to studying the stability of shells of revolution with finite displacements. *Soviet Aeronautics* 20:1-6.

Arora, J.S., Elwakeil, O.A., and Chahande, A.I. (1995). Global optimization methods for engineering applications: a review. *Structural Optimization* 9:137-159.

Balling, R.J. (1991). Optimal steel frame design by simulated annealing. *Journal of Structural Engineering* 117:1780-1795.

Balling, R.J. (1996). Application of the simulated annealing algorithm to structural design. In Grierson, D.E., and Hajela, P., eds., *Emergent Computing Methods in Engineering Design.* Berlin: Springer. 283-293.

Bellman, R.E. (1957). *Dynamic Programming.* Princeton: Princeton University Press. New Jersey.

Bellman, R.E., and Dreyfuss, S.E. (1962). *Applied Dynamic Programming.* Princeton: Princeton University Press. New Jersey.

Bennage, W.A., and Dhingra, A.K. (1995a). Single and multiobjective structural optimization in discrete-continuous variables using simulated annealing. *International Journal for Numerical Methods in Engineering* 38:2753-2773.

Bennage, W.A., and Dhingra A.K. (1995b). Optimization of truss topology using tabu search. *International Journal for Numerical Methods in Engineering* 38:4035-4052.

Blachut, J. (1977). Optimal design of a compressed rod with large deflections by means of dynamic programming. *Mechanika Teoretyczna i Stosowana* 15:375-385.

Blachut, J. (1988). Search for optimal torispherical end closures under buckling constraints. *International Journal of Mechanical Sciences* 31:623-633.

Blachut, J. (1991). Shape optimization of FRP dome closures under buckling constraints. In Eschenauer, H.A., Matteck, C., and Olhoff, N., eds., *Engineering Optimization in Design Processes.* Berlin: Springer-Verlag. 155-164.

Blachut, J. (1992). Influence of meridional shaping on the collapse strength of FRP domes. *Engineering Optimization* 19:65-80.

Blachut, J. (1997). Minimum weight of internally pressurised domes subject to plastic load failure. *Thin-Walled Structures* 27:127-146.

Blachut, J., and Wang, P. (1999). On the performance of barrel-shaped composite shells subjected to hydrostatic pressure. In Toropov, V.V., ed., *Engineering Design Optimization.* Bradford: MCB University Press. 59-65.

Blachut, J., and Wang, P. (2000). Buckling of barreled shells subjected to external hydrostatic pressure. In Deardorff, A.F., ed., *PVP-Vol. 407 Pressure Vessel and Piping Codes and Standards.* New York: ASME. 107-114.

Bland, J.A. (1994). A tabu search approach to engineering optimisation. In Rzevsky, G., Adey, R.A., and Russell, D.W., eds., *Application of Artificial Intelligence in Engineering IX.* Southampton: Computational Mechanics Publ Limited. 423-430.

Bland, J.A. (1995). Discrete-variable optimal structural design using tabu search. *Structural Optimization* 10:87-93.

Bland, J.A. (1998). Structural design optimization with reliability constraints using tabu search. *Engineering Optimization* 30:55-74.

Bohachewsky, I.O., Johnson, M.E., and Stein, M.L. (1986). Generalized simulated annealing for function optimization. *Technometrics* 28:209-217.

Botello, S., Marroquin, J.L., Onate, E., and van Horebeek, J. (1999). Solving structural optimization problems with genetic algorithms and simulated annealing. *International Journal for Numerical Methods in Engineering* 45:1069-1084.

Box, M.J. (1965). A new method of constrained optimization and a comparison with other methods. *Computer Journal* 8:42-52.

Bushnell, D. (1976). Bosor 5 - program for buckling of elastic-plastic complex shells of revolution including large deflections and creep. *Computers and Structures* 6:221-239.

Bushnell, D. (1977). Bosor4 program for stress, buckling, and vibration of complex shells of revolution. In *Structural Mechanics Software Series*. Charlottesville: University Press of Virginia. VA. 11-143.

Ceranic, B., Fryer, C., and Baines, R.W. (2000). An application of simulated annealing to the optimum design of reinforced concrete retaining structures. *Computers and Structures*, to appear.

Chen, G-S., Bruno, R.J., and Salama, M. (1991). Optimal placement of active/passive members in truss structures using simulated annealing. *American Institute of Aeronautics and Astronautics Journal* 29:1327-1334.

Connor, A.M., Seffen, K.A., Parks, G.T., and Clarkson, P.J. (1999). Efficient optimisation of structures using tabu search. In Toropov, V.V., ed., *Engineering Design Optimization*. Bradford: MCB University Press. 127-133.

Corana, A., Marchesi, M., Martini, C., and Ridella, S. (1987). Minimizing multimodal functions of continuous variables with the simulated annealing algorithm. *ACM Transactions on Mathematical Software* 13: 262-280.

Davis, L., ed., 1991. *Handbook of Genetic Algorithms*. New York: Van Nostrand Reinhold.

Distefano, N. (1975). Dynamic programming and the optimization of two-point boundary value systems. *Journal of Mathematical Analysis and Applications* 52:142-150.

Distefano, N., and Rath, A. (1975). A dynamic programming approach to the optimization of elastic trusses. *Journal of Optimization Theory and Applications* 15:13-26.

Gero, J.S., Sheehan, P.J., and Becker, J.M. (1978). Building design using feedforward non-serial dynamic programming. *Engineering Optimization* 3:183-192.

Gero, J.S., and Kaneshalingam, K. (1978). A method for the optimum design of traditional formwork. *Engineering Optimization* 3:249-251.

Gill, P.E., Murray, W., and Wright, M.H. (1999). *Practical Optimization*. London: Academic Press. 401p.

Glass, C.A., and Potts, C.N. (1996). A comparison of local search methods for flow shop scheduling. *Annals of Operations Research* 63:489-509.

Glover, F. (1977). Heuristic search - an approach to the nonconvex optimization problem. *Decision Sci.* 8:156-166.

Goldberg, D.E. (1989). *Genetic Algorithms in Search, Optimization and Machine Learning*. Reading: Addison-Wesley. Massachussets.

Hajela, P., and Lee, E. (1995). Genetic algorithms in truss topological optimization. *International Journal of Solids and Structures* 32:3341-3357.

Hajela, P. (1999). Nongradient methods in multidisciplinary design optimization - status and potential. *Journal of Aircraft* 36:255-265.

Hamada, M., Seguchi, Y., and Tada, Y. (1981). An optimal design of beam with variable cross section. *Bulletin of the Japanese Society of Mechanical Engineers* 24:621-627.

Haupt, R.L., and Haupt, S.E. (1998). *Practical Genetic Algorithms*. New York: John Wiley. 177p.

Holland, J.H. (1975). *Adaptation in Natural and Artificial Systems*. Ann Arbor: University of Michigan Press. Michigan.

Howell, G.C., and Doyle, W.S. (1978). Dynamic programming and direct iteration for the optimum design of skeletal towers. *Computers and Structures* 9:621-627.

Hu, N. (1992). Tabu search method with random moves for globally optimal design. *International Journal for Numerical Methods in Engineering* 35:1055-1070.

Jendo, S., Marks, W., and Thierauf, G. (1985). Multicriteria optimization in optimum structural design. *Large Scale Systems* 9:141-150.

Jenkins, W.M. (1991). Towards structural optimization via genetic algorithm. *Computers and Structures* 40:1321-1327.

Kincaid, R.K. (1990). Minimizing distortion and internal forces in truss structures by simulated annealing. In Proceedings of 31st AIAA/ASME/ASCE/AHS/ASC Structural Materials and Dynamics Conference, Long Beach: AIAA. USA. 327-333.

Kincaid, R.K. (1991). Minimizing distortion in truss structures: a comparison of simulated annealing and tabu search. In Proceedings of 32nd AIAA/ASME/ASCE/AHS/ASC Structures, Structural Dynamics, and Materials Conference, Baltimore: AIAA. USA. 424-430.

Kirkpatrick, S., Gelatt Jr, C.D., and Vecchi, M.P. (1983). Optimization by simulated annealing. *Science* 220:671-680.

Kneppe, G., and Sattler, H.J. (1982). Modifikation des Boxschen Complex-Verfahrens. *Zeitschrift fuer Angewandte Mathematik und Mechanik* T 371-373.

Laarhoven van, P.J.M., and Aarts, E.H.L. (1992). *Simulated Annealing - Theory and Applications.* Dordrecht: Kluwer Academic Publishers. 187p.

Leite, J.P.B., and Topping, B.H.V. (1999). Parallel simulated annealing for structural optimization. *Computers and Structures* 73:545-564.

Li, Y.H., Richards, E.B., Jiang, Y.J., and Azarmi, N.A.S. (1996). Localised simulated annealing in constraint satisfaction and optimization. In Rayward-Smith, V.J., Osman, I.H., and Reeves, C.R., eds., *Modern Heuristic Search Methods.* Chichester: John Wiley. 27-39.

Lin, C.Y., and Chen, W.T. (2000). Stochastic multistage algorithms for multimodal structural optimization. *Computers and Structures* 74:233-241.

Lipson, S.L., and Gwin, L.B. (1977). The complex method applied to optimal truss configuration. *Computers and Structures* 7:461-468.

Luchi, M.L., and Poggialini, A. (1980). An interactive optimization procedure applied to the design of gas turbine discs. *Computers and Structures* 11:629-637.

Lundy, M., and Mees, A. (1986). Convergence of an annealing algorithm. *Mathematical Programming* 34:111-124.

Manoharan, S., and Shanmuganathan, S. (1999). A comparison of search mechanisms for structural optimization. *Computers and Structures* 73:363-372.

May, S.A., and Balling, R.J. (1992). A filtered simulated annealing strategy for discrete optimization of a 3D steel frameworks. *Structural Optimization* 4:142-148.

Nagendra, S., Jestin, D., Gurdal, Z., Haftka, R.T., and Watson, L.T. (1996). Improved genetic algorithm for the design of stiffened composite panels. *Computers and Structures* 58:543-555.

Nelder, J.A., and Mead, R. (1965). A simplex method for function minimization. *Computer Journal* 7:308-313.

Nemhauser, G.L. (1966). *Introduction to Dynamic Programming.* New York: John Wiley. 256p.

Odland, J. (1981). Theoretical and experimental buckling loads of imperfect spherical shell segments. *Journal of Ship Research* 25:201-218.

Osman, C., and Parent, P. (1982). Piecewise smooth dynamic programming application to a casing design for ultradeep drills. *Engineering Optimization* 5:249-256.

Osman, I.H., and Laporte, G. (1996). Metaheuristics: A bibliography. *Annals of Operations Research* 63:513-623.

Palmer, A.C. (1968). Optimal structure design by dynamic programming. *Proceedings of the American Society of Civil Engineers* 94:1887-1906.

Palmer, A.C. (1969). Limit analysis of cylindrical shells by dynamic programming. *International Journal of Solids and Structures* 5:289-302.

Palmer, A.C., and Sheppard, D.J. (1970). Optimizing the shape of pin-jointed structures. *Proceedings of the Institution of Civil Engineers* 47:363-376.

Parkinson, J.M., and Hutchinson, D. (1972). An investigation into the efficiency of variants on the simplex method. In Lootsma, F.A., ed., *Numerical Methods for Non-linear Optimization*. New York: Academic Press. 115-135.

Press, W.H., Flannery, B.P., Teukolsky, S.A., and Vetterling, W.T. (1986). *Numerical Recipes - The Art of Scientific Computing*. Cambridge: Cambridge University Press. UK. 818p.

Raj, P.P., and Durrant, S.O. (1976). Optimum structural design by dynamic programming. *Journal of the Structural Division, American Society of Civil Engineers* 102:1575-1589.

Rao, S.S. (1984). *Optimization - Theory and Applications*. New Dehli: Wiley Eastern Limited. 747p.

Rayward-Smith, V.J., Osman, I.H., Reeves, C.R., and Smith, G.D., eds., (1996). *Modern Heuristic Search Methods*. Chichester: John Wiley. 294p.

Rechenberg, I. (1973). *Evolution Strategy: Optimization of Technical Systems According to Biological Evolution*. Stuttgart: Frommann-Holzboog.

Reeves, C.R., ed., (1995). *Modern Heuristic Techniques for Combinatorial Problems*. London: McGraw-Hill. 320p.

Salama, M., Bruno, R., Chen, G.S., and Garba, J. (1990). Optimal placement of excitations and sensors by simulated annealing. In *Recent Advances in Multidisciplinary Analysis and Optimization*, NASA CP-3031, 1441-1458.

Schmidt, H., and Krysik, R. (1991). Towards recommendations for shell stability design by means of numerically determined buckling loads. In Jullien, J.F., ed., *Buckling of Shell Structures, on Land, in the Sea and in the Air*. London New York: Elsevier Applied Science. 508-519.

Schnack, E., and Sporl, U. (1986). A mechanical dynamic programming algorithm for structure optimization. *International Jounal for Numerical Methods in Engineering* 23:1985-2004.

Shim, P.Y., and Manoochehri, S. (1997). Generating optimal configurations in structural design using simulated annealing. *International Journal for Numerical Methods in Enginering* 40:1053-1069.

Shim, P.Y., and Manoochehri, S. (1999). A hybrid deterministic/stochastic optimization approach for the shape configuration design of structures. *Structural Optimization* 17:113-129.

Siarry, P., and Berthiau, G. (1997). Fitting of tabu search to optimize functions of continuous variables. *International Journal for Numerical Methods in Engineering* 40:2449-2457.

Spendley, W., Hext, G.R., and Himsworth, F.R. (1962). Sequential application of simplex designs in optimisation and evolutionary operation. *Technometrics* 4:441-461.

Swann, W.H. (1974). Constrained optimization by direct search. In Gill, P.E., and Murray, W., eds., *Numerical Methods for Constrained Optimization*. New York: Academic Press. 191-268.

Szewczyk, Z., and Hajela, P. (1993). Neural network approximations in a simulated annealing based optimal structural design. *Structural Optimization* 5:159-165.

Thevendran, V. (1982). Dynamic programming in minimum weight design of shells. *International Journal for Numerical Methods in Engineering* 18:1349-1360.

Thevendran, V. (1984). Dynamic programming in minimum-weight design of axisymmetric plates. *Journal of Optimization Theory and Applications* 44:689-700.

Topping, B.H.V., Khan, A.I., and de Barros Leite, J.P. (1993). Topological design of truss structures using simulated annealing. In Topping, B.H.V., and Khan,. A.I., eds., *Neural Networks and Combinatorial Optimization in Civil and Structural Engineering*. Edinburgh: Civil-Comp Press. 151-165.

Tzan, S.R., and Pantelides, C.P. (1996). Annealing strategy for optimal structural design. *Journal of Structural Engineering, Transactions of the American Society of Civil Engineers* 122:815-827.

Vanderbilt, D., and Louie, S.G. (1984). A Monte Carlo simulated annealing approach to optimization over continuous variables. *Journal of Computational Physics* 56:259-271.

Vanderplaats, G.N. (1984). *Numerical Optimization Techniques for Engineering Design*. New York: McGraw-Hill Book Company. 333p.

Vanderplaats, G.N. (1999). Structural design optimization status and direction. *Journal of Aircraft* 36:11-20.

# CHAPTER 3

# Optimal Engineering Design by Means of Stochastic Optimization Methods

K. Marti

Federal Armed Forces University Munich, Aero–Space Engineering and Technology
Neubiberg/München, Germany

**Abstract.** The optimization problems arising in engineering design contain many model parameters (material, loading, costs, tolerances, etc.) which are not given, fixed quantities, but must be considered as random variables with a given probability distribution. Evaluating the performance of the system by expected cost functions or taking into account the reliability of designs, the basic optimization problem under stochastic uncertainty is replaced by a certain deterministic substitute problem taking into account random parameter variations. The mathematical properties of the substitute problems are examined and (approximate) solution techniques are provided by applying stochastic optimization methods.

## 3.1 Optimum Structural Design Problems under Stochastic Uncertainty

In the optimum design of structural systems two basic classes of criterions occur:
A weight, volume or more general cost (of design, construction, maintenance, etc.) function

$$G_0 = G_0(a, X) \tag{3.1a}$$

and a certain number of safety margins or limit state functions (e.g. some displacement, stress, force, moment components)

$$M_i = M_i(a, X),\ i = 1, \ldots, m_M \tag{3.1b}$$

Here,

$$X \in D \tag{3.1c}$$

denotes the $r$-vector of design variables, e.g. structural dimensions, sizing variables, such as cross-sectional areas, thickness, lying in a certain convex subset $D$ of $\mathbb{R}^r$; hence, (3.1c) describes e.g. box constraints. Moreover, $a$ is the parameter $(m + \nu)$-vector

$$a = \begin{pmatrix} R \\ p \end{pmatrix} \tag{3.1d}$$

composed of the $m$-vector $R$ of the acting external loads, e.g. wave, wind loads, payload, etc., and the $\nu$-vector $p$ of the remaining model parameters containing the material parameters, like strengths parameters (yield/allowable stress, elastic modulus, etc. of each member of the structure), the manufacturing tolerances and the weight or more general cost coefficients.

For the survival of the structure/structural system one has then the condition

$$M_i(a, X) > 0 \text{ for all } i = 1, \ldots, m_M. \tag{3.2a}$$

Hence, the structural system fails if and only if the structure is in the $i$-th failure mode (failure domain)

$$M_i(a, X) \leq 0 \tag{3.2b}$$

for at least one index $i, 1 \leq i \leq m_M$.

Note that the safety margins or limit state functions $M_i = M_i(a, X), i = 1, \ldots, m_M$, may be determined e.g. in optimal plastic design by considering the extreme points of the admissible domain of the dual pair of static/kinematic LP's related to the equilibrium and linearized convex yield condition.

Solving optimum structural design problems of the type

I)

$$\min G_0(a, X) \tag{3.3a}$$

s.t.

$$M_i(a, X) > 0, i = 1, \ldots, m_m \tag{3.3b}$$

$$X \in D \tag{3.3c}$$

or

II)

$$\text{"}\min\text{"} \Big(M_i(a, X)\Big)_{1 \leq i \leq m_M} \tag{3.4a}$$

s.t.

$$G_0(a, X) \leq G^{\max} \tag{3.4b}$$

$$X \in D, \tag{3.4c}$$

a first basic problem is that the vector $a$ of model parameters $a$ is **not** a given, fixed quantity.

In practice, due to several types of uncertainties, such as ( Thoft-Christensen et al., 1986)

* physical uncertainty (variability of physical quantities, like material, loads, dimensions, etc.)
* statistical uncertainty (e.g. estimation errors of parameters due to limited sample data)
* model uncertainty (model errors),

the $(m + \nu)$-vector $a$ of model parameters must be modelled by a random vector

$$a = \begin{pmatrix} R \\ p \end{pmatrix} = \begin{pmatrix} R(\omega) \\ p(\omega) \end{pmatrix}, \omega \in (\Omega, \mathcal{A}, P), \tag{3.5a}$$

on a certain probability space $(\Omega, \mathcal{A}, P)$ with sample space $\Omega$ having elements $\omega$. For the mathematical representation of the corresponding probability distribution $P_{a(\cdot)}$ of the random vector $a = a(\omega)$, two main models are taken into account in practice:

i) discrete probability distributions
ii) continuous probability distributions.

In the first case there is a finite or countably infinite number $l_0 \in \mathbb{N} \cup \{\infty\}$ of realizations $a^l \in \mathbb{R}^{m+\nu}, l = 1, \dots, l_0,$

$$P\Big(a(\omega) = a^l\Big) = \alpha_l, l = 1, \dots, l_0, \tag{3.5b}$$

taken with probabilities $\alpha_l, l = 1, \dots, l_0$. In the second case, the probability that the realization $a(\omega) = a$ lies in a certain subset $B \subset \mathbb{R}^{m+\nu}$ is described by the integral

$$P\Big(a(\omega) \in B\Big) = \int_B \varphi(a)\, da \tag{3.5c}$$

with a certain probability density function $\varphi = \varphi(a) \geq 0, a \in \mathbb{R}^{m+\nu}, \int \varphi(a)\, da = 1.$

The properties of the probability distribution $\pi = P_{a(\cdot)}$ may be described - fully or in part - by certain numerical characteristics, called parameters of $P_{a(\cdot)}$. These distributional parameters $\theta = \theta_h$ are obtained by considering expections

$$\theta_h := Eh\Big(a(\omega)\Big) \tag{3.6a}$$

of functions

$$(h \circ a)(\omega) := h\Big(a(\omega)\Big) \tag{3.6b}$$

of the random vector $a = a(\omega)$ with certain mappings

$$h : \mathbb{R}^{m+\nu} \longrightarrow \mathbb{R}^{s_h}, s_h \geq 1. \tag{3.6c}$$

According to the type of the probability distribution $\pi = P_{a(\cdot)}$ of $a = a(\omega)$, the expectation $Eh\Big(a(\omega)\Big)$ is defined by

$$Eh\Big(a(\omega)\Big) = \begin{cases} \sum\limits_{l=1}^{l_0} h\Big(a^l\Big)\, \alpha_l, & \text{in the discrete case (3.5b)} \\ \int\limits_{\mathbb{R}^{m+\nu}} h(a)\varphi(a)\, da, & \text{in the continuous case (3.5c).} \end{cases} \tag{3.6d}$$

Further distributional parameters $\theta$ are functions

$$\theta = \Psi(\theta_{h_1}, \dots, \theta_{h_s}) \tag{3.6e}$$

of certain "$h$-moments" $\theta_{h_1}, \dots, \theta_{h_s}$ of the type (3.6a). Important examples of the type (3.6a), (3.6e), resp., are the expection

$$\bar{a} = Ea(\omega) \text{ (for } h(a) := a, a \in \mathbb{R}^{m+\nu}) \tag{3.7a}$$

and the covariance matrix

$$Q := E\Big(a(\omega) - \bar{a}\Big)\Big(a(\omega) - \bar{a}\Big)^T = Ea(\omega)a(\omega)^T - \bar{a}\bar{a}^T \tag{3.7b}$$

of the random vector $a = a(\omega)$.

Due to the stochastic variability of the random vector $a(\cdot)$ of model parameters, and since the realization $a(\omega) = a$ is not available at the decision making stage, the optimal design problem (3.3a-c) or (3.4a-c) under stochastic uncertainty can not be solved directly.

Hence, appropriate deterministic substitute problems must be chosen taking into account the randomness of $a = a(\omega)$.

### 3.1.1 Deterministic Substitute Problems in Structural Design

Reliability based optimization (RBO) requires (Schuëller et al., 1998) a combined use of numerical structural reliability analysis and structural optimization. As a basic substiute problem in (RBO) one has the minimization of the total expected costs including the expected costs of failure

$$\min c_G EG_0\Big(a(\omega), X\Big) + c_f p_f(X) \tag{3.8a}$$

s.t.

$$X \in D \tag{3.8b}$$

where $p_f = p_f(X)$ is the probability of failure defined by

$$\begin{aligned} p_f(X) &:= 1 - P\left(M_i\Big(a(\omega), X\Big) > 0, i = 1, \dots, m_M\right) \\ &= P\left(M_i\Big(a(\omega), X\Big) \leq 0 \text{ for at least one index } i, 1 \leq i \leq m_M\right), \end{aligned} \tag{3.8c}$$

$c_f > 0$ describes the failure costs, and $c_G > 0$ is a certain scale factor.

Obviously, the objective function (3.8a) may be interpreted as the Lagrangian of the following constrained substitute problems:

$$\min EG_0\Big(a(\omega), X\Big) \tag{3.9a}$$

s.t.

$$p_f(X) \leq \alpha^{\max} \tag{3.9b}$$
$$X \in D, \tag{3.9c}$$

where $\alpha^{\max} > 0$ is a prescribed maximum failure probability, e.g. $\alpha^{\max} = 0.01$.

The "dual" version to (3.9a-c) reads

$$\min p_f(X) \tag{3.10a}$$

s.t.

$$EG_0\Big(a(\omega), X\Big) \le G^{\max} \tag{3.10b}$$
$$X \in D \tag{3.10c}$$

with a maximum (upper) cost bound $G^{\max}$.

More general substitute problems are obtained by considering the expected failure costs

$$\Gamma(X) = E\gamma\Big(M\Big(a(\omega), X\Big)\Big). \tag{3.11a}$$

arising from a structural failure. Here,

$$M\Big(a(\omega), X\Big) := \Big(M_1\Big(a(\omega), X\Big), \ldots, M_{m_M}\Big(a(\omega), X\Big)\Big)^T, \tag{3.11b}$$

and

$$\gamma : \mathbb{R}^{m_M} \to R^{m_\gamma} \tag{3.11c}$$

is a nonincreasing scalar or vector valued cost function, i.e. we require here that

$$\gamma(z) \ge \gamma(w), \; if \; z \le w, \tag{3.11d}$$

where inequalities between vectors are defined componentwise.

*Example 3.1.1.* If $\gamma(z) = 1$ for $z \not> 0$ and $\gamma(z) = 0$ for $z > 0$, then $\Gamma(x) = p_f(X)$.

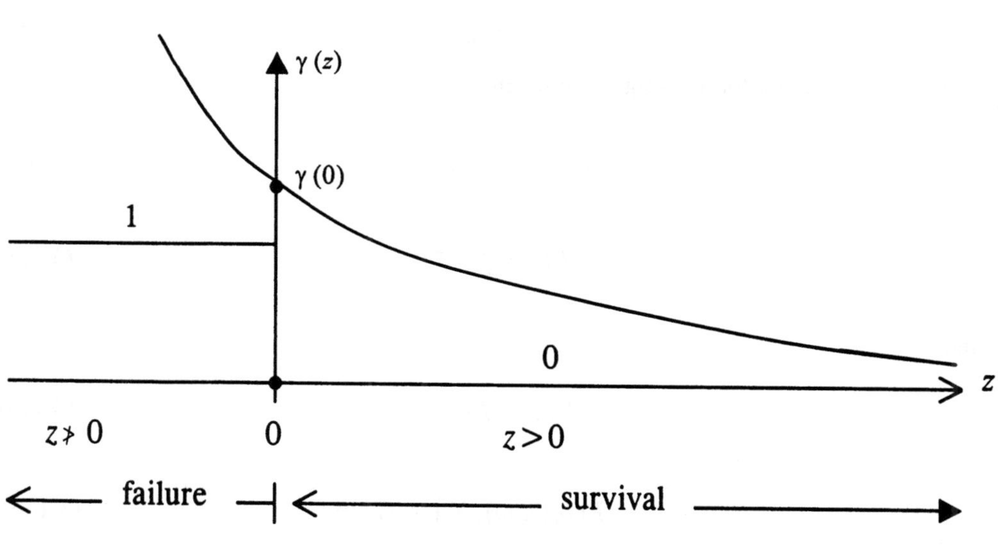


**Figure 3.1** Loss function $\gamma$

*Example 3.1.2.* Considering any nonnegative, nonincreasing function $\gamma = \gamma(z)$ on $\mathbb{R}^{m_M}$ such that

$$\gamma(z) \geq \gamma(0) > 0 \text{ for all } z \in \mathbb{R}^{m_M}, z \not> 0, \tag{3.12a}$$

for the probability of failure we find the following upper bound

$$p_f(X) = P\left(M\Big(a(\omega), X\Big) \not> 0\right) \leq \frac{1}{\gamma(0)} E\gamma\left(M\Big(a(\omega), X\Big)\right), \tag{3.12b}$$

where the right hand side of (3.12b) is obviously an expected cost function of type (3.11a-d).

*Example 3.1.3.* If the loss function $\gamma(z)$ is defined by a vector of individual loss functions for each failure mode

$$\gamma(z) = \Big(\gamma_1(z_1), \ldots, \gamma_{m_M}(z_{m_M})\Big)^T, \tag{3.13a}$$

then

$$\Gamma(X) = (\Gamma_1(X), \ldots, \Gamma_{m_M}(X))^T, \ \Gamma_i(X) := E\gamma_i\left(M_i\Big(a(\omega), X\Big)\right), 1 \leq i \leq m_M, \tag{3.13b}$$

i.e. the $m_M$ failure modes $\left[M_i\Big(a(\omega), X\Big) \leq 0\right], i = 1, \ldots, m_m$, will be treated separately.

Working with the more general expected failure cost function $\Gamma = \Gamma(X)$, instead of (3.8a,b), (3.9a-c) and (3.10a-c) we have the related substitute problems

I)

$$\min c_G EG_0\Big(a(\omega), X\Big) + c_f \Gamma(X), \tag{3.14a}$$

s.t.

$$X \in D \tag{3.14b}$$

where $c_G, c_f$ are (vectorial) weight coefficients.

II)

$$\min EG_0\Big(a(\omega), X\Big) \tag{3.15a}$$

s.t.

$$\Gamma(X) \leq \Gamma^{\max} \tag{3.15b}$$

$$X \in D, \tag{3.15c}$$

and

III)

$$\text{"}\min\text{"}\, \Gamma(X) \tag{3.16a}$$

s.t.

$$EG_0\Big(a(\omega), X\Big) \leq G^{\max} \tag{3.16b}$$

$$X \in D, \tag{3.16c}$$

where "min" indicates again that $\Gamma(X)$ may be a vector valued function.

## 3.2 Quality Engineering

In quality engineering (Park, 1996, Phadke, 1989, and Ross, 1988) the aim is to optimize the performance of a certain product or a certain process, and to make the performance insensitive (robust) to any noise to which the product, the process may be subjected. Moreover, the production costs (development, manufacturing, etc.) should be bounded or minimized.

Hence, the control (input, design) factors $X$ are then to set as to provide

* an optimal mean performance (response)
* a high degree of immunity (robustness) of the performance (response) to noise factors
* low or minimum production costs.

Note. There may be a certain trade-off between the above three criterions.

For this purpose, the quality characteristics or performance (e.g. dimensions, weight, contamination, structural strength, yield, etc.) of the product/process is described (Park, 1996) first by a function

$$y = y\Big(a(\omega), X\Big) \tag{3.17}$$

depending on a $\nu$-vector $a = a(\omega)$ of random varying parameters (noise factors) $a_j = a_j(\omega), j = 1, \ldots, \nu$, causing quality variations, and the $r$-vector $X$ of control/input factors $X_k, k = 1, \ldots, r$.

The **quality loss** of the product/process under consideration is measured then by a certain loss function $\gamma = \gamma(y)$. Standard loss functions $\gamma$ occur in the following decision situations (Ross, 1988):

i) Nominal-is-best
Having a given, finite target or nominal value $y^*$, e.g. a nominal dimension of a product, the quality loss $\gamma(y)$ is defined by

$$\gamma(y) := (y - y^*)^2. \tag{3.18a}$$

ii) Smallest-is-best
If the target or nominal value $y^*$ is zero, i.e., if the absoute value $|y|$ of the product quality function $y = y(a, X)$ should be as small as possible, e.g. a certain product contamination, then the corresponding quality loss is defined by

$$\gamma(y) := y^2. \tag{3.18b}$$

iii) Biggest-is-best
If the absolute value $|y|$ of the product quality function $y = y(a, X)$ should be as large as possible, e.g. the strength of a bar or the yield of a process, then the quality loss is defined by

$$\gamma(y) := \left(\frac{1}{y}\right)^2. \tag{3.18c}$$

In quality engineering, a high mean product/process level or performance at minimum random quality variations and low or bounded production/manufacturing costs is achieved (Phadke, 1989) by minimization of the expected quality loss

$$\Gamma(X) := E\gamma\Big(y\big(a(\omega), X\big)\Big). \tag{3.19a}$$

Obviously, $\Gamma(X)$ can also be minimized by maximizing the so-called "signal-to-noise ratio"

$$SN := -\log\Gamma(X), \tag{3.19b}$$

see Park (1996), Phadke (1989) and Ryan (2000).

If $D \subset \mathbb{R}^r$ denotes the domain of feasible control factors $X$, then one has to solve the following stochastic optimization problem

$$\text{minimize } \Gamma(X) \tag{3.19c}$$
$$\text{s.t. } X \in D. \tag{3.19d}$$

Obviously, (3.18a) and (3.18b) are convex loss functions. Moreover, since the quality or response function $y = y(a, X)$ takes only positive or only negative values $y$ in many practical applications, also (3.18c) is a convex loss function in many cases.

Further decision situations may be modelled by choosing more general convex loss functions $\gamma = \gamma(y)$.

## 3.3 Approximations of Substitute Problems

The main problem in solving the substitute problems defined above is that the probability and expected cost functions $p_f = p_f(X), \Gamma = \Gamma(X), X \in \mathbb{R}^r$, are defined by means of multiple integrals over a $\tilde{\nu}$-dimensional space.

Thus, the substitute problems may be solved in practice only by some approximative numerical methods. In the following we consider approximations for substitute problems based on general expected cost functions $\Gamma = \Gamma(X)$ having a real valued convex loss function $\gamma(z), \gamma(y)$. Note that the probability of failure function $p_f = p_f(X)$ may be approximated from above, see (3.12a,b), by expected cost functions $\Gamma = \Gamma(X)$ having a nonincreasing, continuously differentiable convex loss function $\gamma = \gamma(z)$.

### 3.3.1 Analytical Approximates in Optimal Structural Design

**Continuously differentiable loss function $\gamma$.** Suppose here that $\gamma = \gamma(z)$ is a continuously differentiable, convex loss function on $\mathbb{R}^{m_M}$. Let denote then

$$\overline{M}(X) := EM\big(a(\omega), X\big) = \Big(EM_1\big(a(\omega), X\big), \ldots, EM_{m_M}\big(a(\omega), X\big)\Big)^T \tag{3.20}$$

the expectation of the safety margins or limit state functions $M_i = M_i\big(a(\omega), X\big), i = 1, \ldots, m_M$.

For an arbitrary continuously differentiable, convex loss function $\gamma$ we have that

$$\gamma\Big(M\big(a(\omega),X\big)\Big) \geq \gamma\big(\overline{M}(x)\big)+\nabla\gamma\big(\overline{M}(X)\big)^T\Big(M\big(a(\omega),X\big)-\overline{M}(X)\Big). \tag{3.21a}$$

Thus, taking expectations in (3.21a), we find Jensen's inequality

$$\Gamma(X) = E\gamma\Big(M\big(a(\omega),X\big)\Big) \geq \gamma\big(\overline{M}(X)\big) \tag{3.21b}$$

which holds for any convex function $\gamma$. Using the mean value theorem, we have that

$$\gamma(M) = \gamma(\overline{M})+\nabla\gamma(\widehat{M})^T(M-\overline{M}), \tag{3.21c}$$

where $\widehat{M}$ is a point on the line segment $[\overline{M}, M]$ between $\overline{M}$ and $M$. By means of (3.21b,c) we get

$$0 \leq \Gamma(X)-\gamma\big(\overline{M}(X)\big) \leq E\Big\|\nabla\gamma\Big(\widehat{M}\big(a(\omega),X\big)\Big)\Big\|\cdot\Big\|M\big(a(\omega),X\big)-\overline{M}(X)\Big\|. \tag{3.21d}$$

a) Bounded gradient

If the gradient $\nabla\gamma$ is bounded on the (convex) support of $M = M\big(a(\omega),X\big)$, i.e.

$$\Big\|\nabla\gamma\Big(M\big(a(\omega),X\big)\Big)\Big\| \leq \vartheta^{\max} \text{ a.s. (almost sure) for each } X \in D, \tag{3.22a}$$

then

$$0 \leq \Gamma(X)-\gamma\big(\overline{M}(X)\big) \leq \vartheta^{\max}E\Big\|M\big(a(\omega),X\big)-\overline{M}(X)\Big\|, X \in D. \tag{3.22b}$$

Since $t \to \sqrt{t}, t \geq 0$, is a concave function, we get

$$0 \leq \Gamma(X)-\gamma\big(\overline{M}(X)\big) \leq \vartheta^{\max}\sqrt{E\Big\|M\big(a(\omega),X\big)-\overline{M}(X)\Big\|^2} = \vartheta^{\max}\sqrt{q(X)}, \tag{3.22c}$$

where

$$q(X) := E\Big\|M\big(a(\omega),X\big)-\overline{M}(X)\Big\|^2 = trQ(X) \tag{3.22d}$$

is the generalized variance, and

$$Q(X) := \operatorname{cov}\Big(M\big(a(\cdot),X\big)\Big) \tag{3.22e}$$

denotes the covariance matix of the random vector $M = M\big(a(\omega),X\big)$. Consequently, the expected loss funtion $\Gamma(X)$ can be approximated from above by

$$\Gamma(X) \leq \gamma\big(\overline{M}(X)\big) + \vartheta^{\max}q(X) \text{ for } X \in D. \tag{3.22f}$$

b) Bounded eigenvalues of the Hessian
Considering also second order expansions, we find

$$\gamma(M) - \gamma(\overline{M}) = \nabla\gamma(\overline{M})^T(M - \overline{M}) + \frac{1}{2}(M - \overline{M})^T\nabla^2\gamma(\widetilde{M})(M - \overline{M}). \tag{3.23a}$$

Supposing that the eigenvalues $\lambda$ of $\nabla^2\gamma(M)$ are bounded from below and above for each $M = M\Big(a(\omega), X\Big), X \in D$, a.s., i.e.

$$0 < \lambda^{\min} \leq \lambda\left(\nabla^2\gamma\left(M\Big(a(\omega), X\Big)\right)\right) \leq \lambda^{\max} < +\infty, X \in D,\ a.s., \tag{3.23b}$$

by taking expectations in (3.23a) we get

$$\gamma\Big(\overline{M}(X)\Big) + \lambda^{\min}q(X) \leq \Gamma(X) \leq \gamma\Big(\overline{M}(X)\Big) + \lambda^{\max}q(X), X \in D. \tag{3.23c}$$

Consequently, using (3.22f) or (3.23c), various approximations for (3.14a,b), (3.15a-c), (3.16a-c), resp., may be obtained.

We mention the following two problems which are well known in *robust optimal design:*

i)

$$\min \gamma\Big(\overline{M}(X)\Big) + c_0 q(X) \tag{3.24a}$$

s.t.

$$EG_0\Big(a(\omega), X\Big) \leq G^{\max} \tag{3.24b}$$

$$X \in D, \tag{3.24c}$$

where $c_0$ is a scale factor, cf. (3.22f) and (3.23c);

ii)

$$\min EG_0\Big(a(\omega), X\Big) \tag{3.25a}$$

s.t.

$$\gamma\Big(\overline{M}(X)\Big) + c_0 q(X) \leq \gamma^{\max} \tag{3.25b}$$

$$X \in D \tag{3.25c}$$

where $\gamma^{\max}$ is an upper loss bound.

Obviously, by means of (3.24a-c) or (3.25a-c) feasible designs $X^*$ are achieved which

* yield a high mean performance of the structure/structural system
* are minimally sensitive or have a limited sensitivity with respect to random parameter variations (material, load, manufacturing, process, etc.) and
* cause only limited costs for design, construction, maintenance, etc.

The numerical solution is simplified considerably if one could work with one single limit state function $M = M(a, X)$. Formally, this is possible by defining the function

$$M^{\min}(a, X) := \min_{1 \leq i \leq m_M} M_i(a, X). \tag{3.26a}$$

Indeed, according to (3.2a,b) the failure of the structural system can be represented by the condition

$$M^{\min}(a, X) \leq 0. \tag{3.26b}$$

Thus, the structural failure can be evaluated numerically by the function

$$\Gamma(X) := E\gamma\left(M^{\min}\left(a(\omega), X\right)\right) \tag{3.26c}$$

with a nonincreasing loss function $\gamma : \mathbb{R} \to \mathbb{R}_+$, see Fig. 3.1.

Unfortunately, the "min"-operator in (3.26a) yields a nonsmooth function $M^{\min} = M^{\min}(a, X)$ in general, and the computation of the mean and covariance function

$$\overline{M^{\min}}(X) := EM^{\min}\left(a(\omega), X\right) \tag{3.26d}$$

$$Q_{M^{\min}}(X) := \operatorname{cov}\left(M^{\min}\left(a(\cdot), X\right)\right) \tag{3.26e}$$

by means of Taylor expansion with respect to the model parameter vector $a$ at $\overline{a} = Ea(\omega)$ is not possible in general.

Hence, using e.g. regression techniques, $M^{\min}(a, X)$ may be approximated by functions $\widetilde{M} = \widetilde{M}(a, X)$ being smooth with respect to the parameter vector $a$, see Figure 3.2 and Figure 3.3.

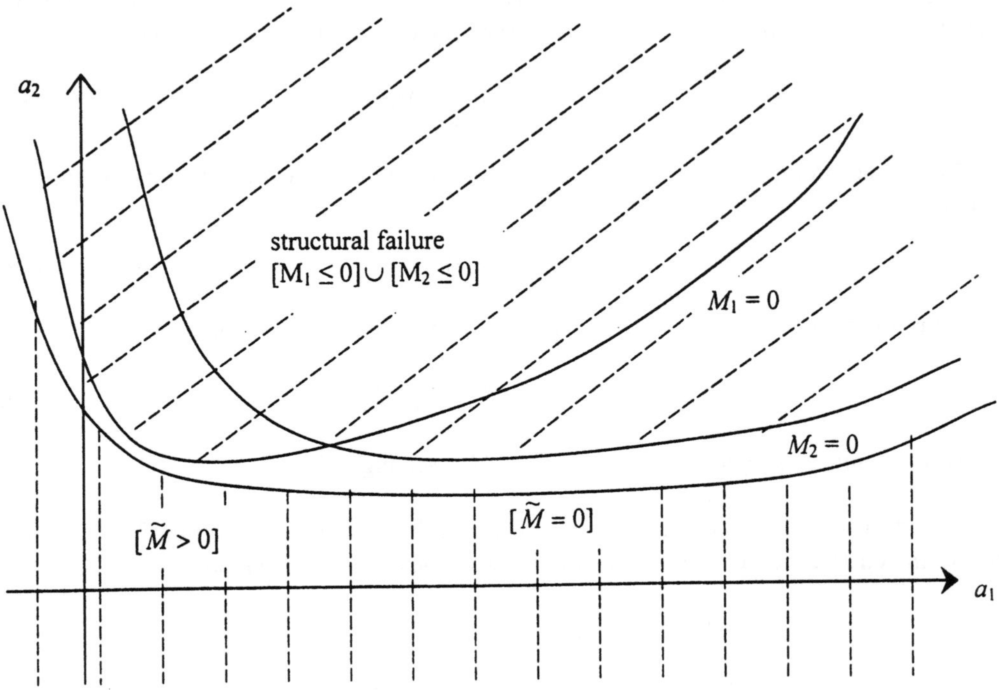


**Figure 3.2** Approximate smooth failure domain $\left[\widetilde{M}(a, X) \leq 0\right]$

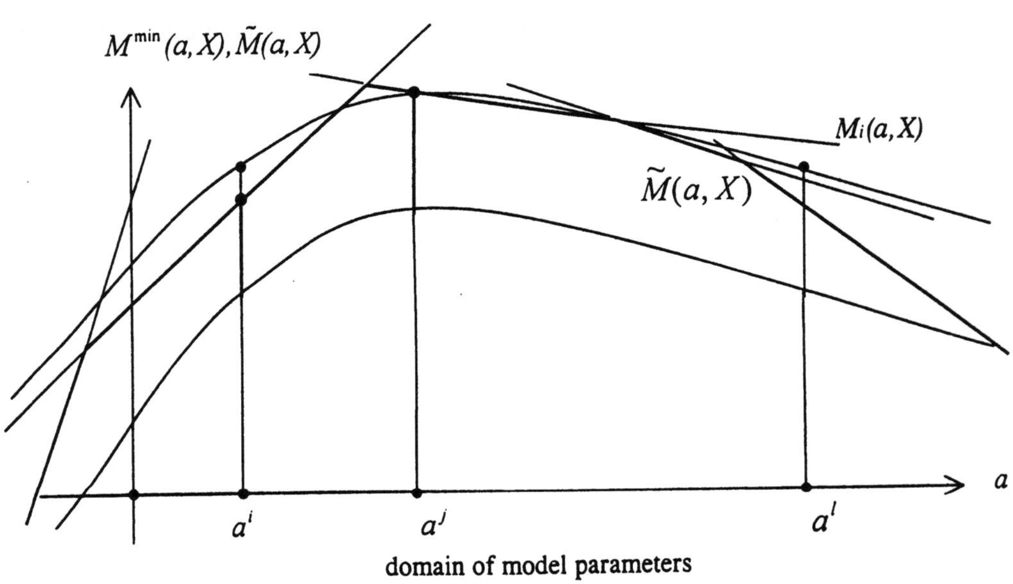


**Figure 3.3** Approximations of $M^{\min}(a, X)$

In many important cases, for each $i = 1, \dots, m_M$, the safety margins

$$(a, X) \longrightarrow M_i(a, X)$$

are bilinear functions. Thus, in this case $M^{\min} = M^{\min}(a, X)$ is a piecewise linear function with respect to $a$. Fitting e.g. a quadratic Response Surface Model

$$\widetilde{M}(a, X) = c(X) + q(X)'a + a'Q(X)a \tag{3.26f}$$

to $a \to M^{\min}(a, X)$, after the selection of appropriate reference points

$$a^{(1)}, a^{(2)}, \dots, a^{(N)} (\in \mathbb{R}^{m+\nu}), \tag{3.26g}$$

the unknown coefficients $c = c(X), q = q(X)$ and $Q = Q(X)$ are obtained by minimizing the mean square error

$$\rho(c, q, Q) := \sum_{j=1}^{N} \left( \widetilde{M}(a^j, X) - M^{\min}(a^j, X) \right)^2 \tag{3.26h}$$

with respect to $(c, q, Q)$, which yields explicit formulas for the optimal coefficients

$$c^* = c^*(X), q^* = q^*(X), Q^* = Q^*(X), \tag{3.26i}$$

since the model (3.26f) depends linearly on $(c, q, Q)$.

**Taylor expansions.** As can be seen before, cf. (3.9a), (3.11a), (3.14a), (3.19a), etc., the functions arising in the objective and/or in the constraints of substitute problems for optimization problems under stochastic uncertainty are mean value functions of the type

$$\Psi(X) := E\psi\left( g\left( (\omega), X \right) \right), \tag{3.27a}$$

where $\psi = \psi(z)$ is a certain real function on $\mathbb{R}^{m_g}$, and $g = g(a, X)$ denotes an $m_g$-vector-valued function on $\mathbb{R}^{\tilde{\nu}} \times \mathbb{R}^r$; moreover, $a = a(\omega)$ is a $\tilde{\nu}$-random vector. If $h = h(a, X)$ is defined by

$$h(a, X) := \psi\left( g(a, X) \right), \tag{3.27b}$$

then, under appropriate differentiability assumptions, by Taylor expansion of $h(a, X)$ with respect to $a$ at $\overline{a} := Ea(\omega)$ we have that

$$\begin{aligned} h(a, X) &:= h(\overline{a}, X) + \nabla_a h(\overline{a}, X)^T (a - \overline{a}) \\ &+ \tfrac{1}{2}(a - \overline{a})^T \nabla_a^2 h(\overline{a}, X)(a - \overline{a}) + \tfrac{1}{3!} \nabla_a^3 h(\tilde{a}, X) \cdot (a - \overline{a})^3, \end{aligned} \tag{3.27c}$$

where $\tilde{a} := \overline{a} + \vartheta(a - \overline{a}), 0 < \vartheta < 1$, and

$$h(\overline{a}, X) = \varphi\left( g(\overline{a}, X) \right) \tag{3.27d}$$

$$\nabla_a h(\overline{a}, X) = \left( \nabla_a g(\overline{a}, X) \right)^T \nabla\psi\left( g(\overline{a}, X) \right) \tag{3.27e}$$

$$\begin{aligned} \nabla_a^2 h(\overline{a}, X) &= \left( \nabla_a g(\overline{a}, X) \right)^T \nabla^2\psi\left( g(\overline{a}, X) \right) \nabla_a g(\overline{a}, X) \\ &+ \sum_{i=1}^{m} \frac{\partial \psi}{\partial g_i}\left( g(\overline{a}, X) \right) \nabla_a^2 g_i(\overline{a}, X), \end{aligned} \tag{3.27f}$$

etc., where $\nabla_a g(a, X)$ denotes the Jacobian of $g = g(a, X)$ with respect to $a$. Thus, taking expectations in (3.27c), $\Psi(X)$ can be approximated by

$$\widetilde{\Psi}(X) := \psi\Big(g(\bar{a}, X)\Big) + \frac{1}{2}E\Big(a(\omega) - \bar{a}\Big)^T \nabla_a^2 h(\bar{a}, X)\Big(a(\omega) - \bar{a}\Big), \tag{3.28a}$$

with an error estimated from above by

$$R(x) := \frac{1}{3!}c(X)E\|a(\omega) - \bar{a}\|^3, \tag{3.28b}$$

provided that

$$\|\nabla_a^3 h\Big(a(\omega), X\Big)\| \leq c(X) < +\infty \text{ a.s.} \tag{3.28c}$$

*Remark 3.3.1.* In the important special case $m_g = 1$ and $\psi(z) = z$ we have that $h(a, X) = g(a, X)$. Of course, also higher order Taylor expansions may be taken into account. Moreover, the above deescribed method can be extended to the case of vector valued functions $\psi(z) = \Big(\psi_1(z), \ldots, \psi_{m_\psi}(s)\Big)^T$.

**FORM/SORM (First/Second Order Reliability Methods).** Special approximation methods were developed (Thoft-Christensen et al., 1986) for the probability of failure $p_f = p_f(X)$ given by (3.8c). Applying certain probability inequalities (Marti, 1997), the case with $m_>1$ limit state functions is reduced first to the case with one limit state function. A first approximation is given below, for more details see, e.g., Marti (1997) and Breitung (1995).

According to (3.8c) we have that

$$\begin{aligned} p_f(x) &= P\left(M_i\Big(a(\omega), X\Big) \leq 0 \text{ for at least one index } i, 1 \leq i \leq m_M\right) \\ &\leq \textstyle\sum_{i=1}^{m_M} P\left(M_i\Big(a(\omega), X\Big) \leq 0\right) = \textstyle\sum_{i=1}^{m_M} p_{f_i}(X), \end{aligned} \tag{3.29a}$$

where

$$p_{f_i}(X) := P\left(M_i\Big(a(\omega), X\Big) \leq 0\right). \tag{3.29b}$$

Assume now that the random $(m + \nu)$-vector $a = a(\omega)$ can be represented by

$$a(\omega) = Z\Big(\tilde{a}(\omega)\Big), \tag{3.30a}$$

where $Z : \mathbb{R}^{m+\nu} \to \mathbb{R}^{m+\nu}$ is a certain transformation in $\mathbb{R}^{m+\nu}$ and $\tilde{a} = \tilde{a}(\omega)$ is an $N(0, I)$-normal distributed random $(m + \nu)$-vector, i.e. with mean vector zero and covariance matrix equal to the identity matrix $I$. In the following, let $X$ be a fixed $r$-vector. Defining

$$\widetilde{M}_i(\tilde{a}, X) := M_i\Big(Z(\tilde{a}), X\Big), \tag{3.30b}$$

we get

$$p_{f_i}(X) = P\left(\widetilde{M}_i\Big(\tilde{a}(\omega), X\Big) \leq 0\right), i = 1, \ldots, m_M. \tag{3.30c}$$

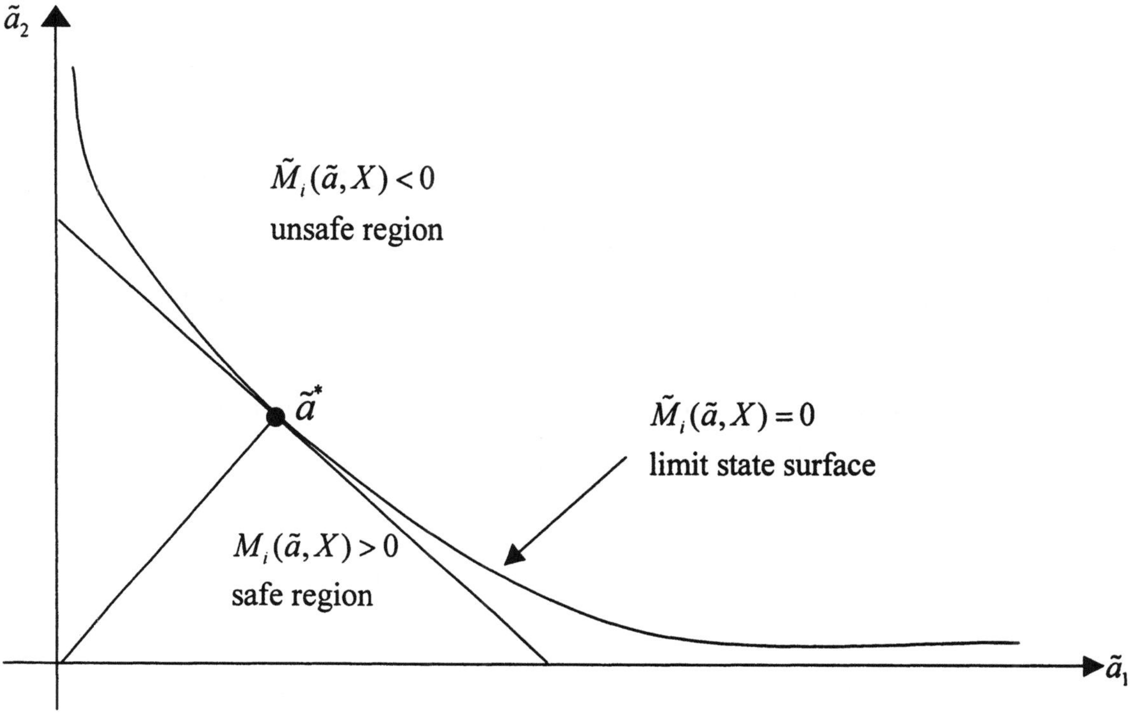


**Figure 3.4** FORM

In the First Order Reliability Method (FORM) the limit state function $\widetilde{M}_i = \widetilde{M}_i(\tilde{a}, X)$ is linearized at a so-called design point $\tilde{a}^* = \tilde{a}^*(X)$ which is obtained by projection of the origin $\tilde{a} = 0$ onto the limit state surface $\widetilde{M}_i(\tilde{a}, X) = 0$, see Fig. 3.4. Since $\widetilde{M}_i(\tilde{a}^*, X) = 0$, we get

$$\begin{aligned}\widetilde{M}_i(a, X) &\approx \widetilde{M}_i(\tilde{a}^*, X) + \left(\nabla_{\tilde{a}}\widetilde{M}_i(\tilde{a}^*, X)\right)^T (a - \tilde{a}^*)\\ &= \left(\nabla_{\tilde{a}}\widetilde{M}_i(\tilde{a}^*, X)\right)^T (a - \tilde{a}^*)\end{aligned} \tag{3.30d}$$

and therefore

$$p_{f_i}(X) \approx P\left(\left(\nabla_{\tilde{a}}\widetilde{M}_i(\tilde{a}^*, X)\right)^T (a(\omega) - \tilde{a}^*) \le 0\right) = \Phi\Big(-\beta(X)\Big), \tag{3.30e}$$

where $\Phi$ denotes the distribution function of the univeriate $N(0,1)$-normal distribution, and the reliability index $\beta = \beta(X)$ is given by

$$\beta(X) := -\frac{\left(\nabla_{\tilde{a}}\widetilde{M}_i(\tilde{a}^*, X)\right)^T \tilde{a}^*}{\|\nabla_{\tilde{a}}\widetilde{M}_i(\tilde{a}^*, X)\|}. \tag{3.30f}$$

Note that $\beta(X)$ is the minimum distance from $\tilde{a} = 0$ to the limit state surface $\widetilde{M}_i(\tilde{a}^*, X) = 0$.

For more details and Second Order Reliability Methods (SORM) based on second order Taylor expansions, see, e.g., Breitung (1995).

### 3.3.2 Analytical Approximates in Quality Engineering

Applying the approximative techniques described in Section 3.3.1 to the objective function

$$\Gamma(X) = E\gamma\left(y\left(a(\omega), X\right)\right)$$

of the basic robust design problem (3.19a-d) with an arbitrary continuously differentiable convex loss function $\gamma = \gamma(y)$, we find the following approximations:

a) Bounded 1st order derivative
In case that $\gamma$ has a bounded derivative on the support of the quality function, i.e., if

$$\left|y'\left(a(\omega), X\right)\right| \leq \vartheta^{\max} < +\infty \text{ a.s. }, X \in D, \tag{3.31a}$$

then

$$\Gamma(X) \leq \gamma\left(\overline{y}(X)\right) + \vartheta^{\max}\sqrt{q(X)}, x \in D, \tag{3.31b}$$

where

$$\overline{y}(X) := Ey\left(a(\omega), X\right), \tag{3.31c}$$
$$q(X) := V\left(y\left(a(\cdot), X\right)\right) = E\left(y\left(a(\omega), X\right) - \overline{y}(X)\right)^2 \tag{3.31d}$$

denote the mean and variance of the quality function $y = y\left(a(\omega), X\right)$.

b) Bounded 2nd order derivative
If the second order derivative $\gamma''$ of $\gamma$ is bounded from above and below, i.e., if

$$0 < \lambda^{\min} \leq \gamma''\left(y\left(a(\omega), X\right)\right) \leq \lambda^{\max} < +\infty \text{ a.s. }, X \in D, \tag{3.32a}$$

then

$$\gamma\left(\overline{y}(X)\right) + \lambda^{\min} q(X) \leq \Gamma(X) \leq \gamma\left(\overline{y}(X)\right) + \lambda^{\max} q(X), X \in D. \tag{3.32b}$$

Corresponding to (3.24a-c), (3.25a-c), here we have then the following approximate problems:

i)

$$\min \gamma\left(\overline{y}(X)\right) + c_0 q(X) \tag{3.33a}$$

s.t.

$$EG_0\left(a(\omega), X\right) \leq G^{\max} \tag{3.33b}$$
$$X \in D, \tag{3.33c}$$

where $c_0$ is again a certain scale factor, see (3.31a,b), (3.32a,b), and $G_0 = G_0\left(a(\omega), X\right)$ represents the production costs having an upper bound $G^{\max}$.

ii)

$$\min EG_0\Big(a(\omega), X\Big) \tag{3.34a}$$

s.t.

$$\gamma\Big(\overline{y}(X)\Big) + c_0 q(X) \le \gamma^{\max} \tag{3.34b}$$

$$X \in D, \tag{3.34c}$$

where $\gamma^{\max}$ denotes an upper loss bound.

### 3.3.3 Numerical Techniques

Solving the substitute problems (3.14a,b), (3.15a-c), (3.16a-c), (3.19a-d), resp., described in Sections 3.1 and 3.2, or their approximates (3.24a-c), (3.25a-c), (3.33a-c), (3.34a-c), resp., given in the present Section 3.3, one has to evaluate numerically always certain mean value, probability and variance functions

$$\overline{G_0}(X) = EG_0\Big(a(\omega), X\Big), p_f = p_f(X), \Gamma = \Gamma(X), \overline{M} = \overline{M}(X),$$
$$\overline{y} = \overline{y}(X), q = q(X).$$

According to the analytical properties of the corresponding integrands, the appearing expectations can be evaluated numerically by

i) Taylor expansions with respect to the random parameter vector $a$ at its mean $\overline{a} = Ea(\omega)$, see Section 3.1.
In this case, the substitute problems are approximated (Haubach, 1998, Reinhart, 1997) by a nonlinear parameter optimization problem depending on certain moments of the random vector $a = a(\omega)$.
ii) Statistical optimization techniques, as e.g. regression and Response Surface Methods (RSM) in combination with Design of Experiment (DOE) methods, stochastic approximation (gradient) methods, simulation techniques (Biles and Swain, 1979, Box and Wilson, 1951, and Jacobsen and Schruben, 1989).
Using regression techniques or Response Surface Methods, some form of experimentation is required for the evaluation of the effects of noise factors or random parameters in the cost or performance characteristics of a design or product specified by a given vector $X$ of values $X_k, k = 1, \dots, n$ for the control variables.
In order to limit the numer of necessary experimentation, in (DOE) alternative levels for the controllable design variables (factors) are considered; generally two or three different levels or settings are selected for each factor. Usually, one of these settings $X = (X_1, \dots, X_r)'$ is taken to be the initial condition. These levels should be taken sufficiently far apart so that the chance is increased for capturing any nonlinearity of the relationship between the control factors or design variables. Of course, all of these settings $X$ should satisfy the remaining deterministic constraints " $X \in D$". In the same way, for the evaluation of the expectations and probabilities, hence, for the evaluation of expected losses, expected quality characteristics, moments of loss and quality functions, etc., also some alternative levels of the noise

factors or random parameters are chosen. By using matrix experiments with so called fractional factorial designs and other orthogonal arrays (all factor levels occur, and factor levels occur in equal number) of settings $X$ of design variables or control factors and random parameters or noise factors $a = a(\omega)$, based on the design of experiment (DOE) theory (Box and Draper, 1987, and Myers, 1971), a maximum of efficiency is obtained at a small number of experiments.

iii) Stochastic linear programming techniques.
Stochastic linear programming (SLP) techniques for solving optimal structural design problems under stochastic uncertainty are developed in the next Section 3.4. However, using subadditive, positively homogeneous loss functions $\gamma = \gamma(y)$ in (3.19a), these SLP methods can be applied also to problems in quality engineering.

## 3.4 Stochastic Linear Programming Techniques in Optimal Structural Design

As can be seen from the above development, the approach based directly on the limit state functions or safety margins $M_i = M_i(a, X), i = 1, \ldots, m_M$, may be very complicated or unprecise because of

* the large number $m_M$ of limit state functions $M_i, i = 1, \ldots, m_M$,
* the large computational expenses in the generation of all or the "most relevant" safety margins, and
* the approximation error which occurs due to the inevitable functional simplifications.

Hence, in the following another method is presented which is based on the very powerful linear programming (LP) techniques.

Returning to the fundamental mechanical conditions for the survival of a mechanical structure, based on the static (lower bound or safe) theorem and the kinematic (upper bound or unsafe) theorem, which are equivalent by means of duality theory (see, e.g., Charnes et al., 1959), we have the following two criterions

$$CF = R(\omega) \tag{3.35a}$$
$$HF \leq F_0\Big(\sigma_y(\omega), X\Big), \tag{3.35b}$$

where (3.35a) is the equilibrium equation involving the $m \times n$ equilibrium matrix $C$ with full rank $C = m$ and the random vector of external loads (forces, moments) $R = R(\omega)$. Moreover, (3.35b) is a system of linear inequalities resulting from an appropriate linearization of the convex yield condition (Cocchetti and Maier, 1988, and Lawo, 1987). Here, $H$ is a given $\mu_H \times n$ matrix, and the right hand side $F_0$ of (3.35b) depends on the random vector of yield stresses $\sigma_y = \sigma_y(\omega)$ and the vector $X$ of design variables, like the dimensions of the elements of the structure. In case of trusses (Kirsch, 1993) we have that

$$H := \begin{pmatrix} I \\ -I \end{pmatrix}, \; F_0\Big(\sigma_y(\omega), X\Big) := \begin{pmatrix} \sigma_d^U(\omega)A(X) \\ -\sigma_d^L(\omega)A(X) \end{pmatrix}, \tag{3.36}$$

where $I$ is the $n \times n$ identity matrix, $\sigma^L = \sigma^L(\omega), \sigma^U = \sigma^U(\omega)$ denote the $n$-vector of yield stresses of the elements in compression (negative values) and tension (positive values), $\sigma_d^L, \sigma_d^U$

are the related diagonal matrices having the components of $\sigma^L, \sigma^U$, resp., on its diagonal, and $A = A(X)$ is the vector of cross-sectional areas of the members of the structure.

Consequently, we have the following optimal design problem under stochastic uncertainty

$$\min G_0\Big(\eta(\omega), X\Big) \tag{3.37a}$$

s.t.

$$CF = R(\omega) \tag{3.37b}$$
$$HF \leq F_0\Big(\sigma_y(\omega), X\Big) \tag{3.37c}$$
$$A_0 X = b_0 \tag{3.37d}$$
$$X \geq 0. \tag{3.37e}$$

Here, $(A_0, b_0)$ is a given matrix such that the convex polyhedron

$$D = \{X \in \mathbb{R}^r; A_0 X = b_0, X \geq 0\} \tag{3.37f}$$

represents the remaining deterministic constraints, e.g. box constraints, of the structural optimization problem. Moreover, $\eta = \eta(\omega)$ is a vector of cost coefficients.

For given parameter vector $(\sigma_y, R)$ and design vector $X$, let then denote $s^* = s^*\Big((\sigma_y, R), X\Big)$ the minimum value of the linear program (with respect to $(s, F)$)

$$\min s \tag{3.38a}$$

s.t.

$$CF = R \tag{3.38b}$$
$$HF - F_0(\sigma_y, X) \leq s1, \tag{3.38c}$$

where $1 := (1, \ldots, 1)'$. Obviously, according to the static/kinematic theorem, the structural systems fails if and only if $s^*\Big((\sigma y, R), X\Big) > 0$ or

$$t^*\Big((\sigma_y, R), X\Big) := -s^*\Big((\sigma_y, R), X\Big) < 0. \tag{3.39a}$$

Moreover, $t^* = t^*\Big((\sigma_y, R), X\Big)$ can be represented, cf. (3.1b), (3.2a,b), (3.26a,b), by

$$t^*\Big((\sigma_y, R), X\Big) = \min_{1 \leq l \leq l_0} M_l\Big((\sigma_y, R), X\Big), \tag{3.39b}$$

where the limit state functions $M_l = M_l\Big((\sigma_y, R), X\Big), l = 1, \ldots, l_0$, are determined (e.g., Charnes et al., 1959) by the extreme points of the admissible domain of the kinematic linear program (LP), i.e. the dual of the LP (3.38a-c), see also (3.57a-c), (3.58a-c).

Since

$s^* = -t^*$ or any other nonincreasing function $\gamma(t^*)$ of $t^*$
indicates the costs of structural failure,

using the reliability based optimal design problems (3.8a,b), (3.9a-c), (3.10a-c) and (3.14a,b), (3.15a-c), (3.16a-c), for the optimal design problem (3.37a-e) we have the following substitute problem

$$\min c_G EG_0\Big(\eta(\omega), X\Big) + c_f \Gamma(X), \tag{3.40a}$$

s.t.

$$X \in D \tag{3.40b}$$

where $EG_0\Big(\eta(\omega), X\Big)$ are the expected costs of design, construction, maintenance, etc., and

$$\begin{aligned} \Gamma(X) &:= E\gamma\left(t^*\Big(\big(\sigma_y(\omega), R(\omega)\big), X\Big)\right) \\ &= E\gamma\left(-s^*\Big(\big(\sigma_y(\omega), R(\omega)\big), X\Big)\right) \end{aligned} \tag{3.40c}$$

denote the expected costs of failure, damage, repair, etc., where $\gamma = \gamma(t)$ is a nonincreasing loss function. Moreover, $c_G, c_f > 0$ are certain weight factors.

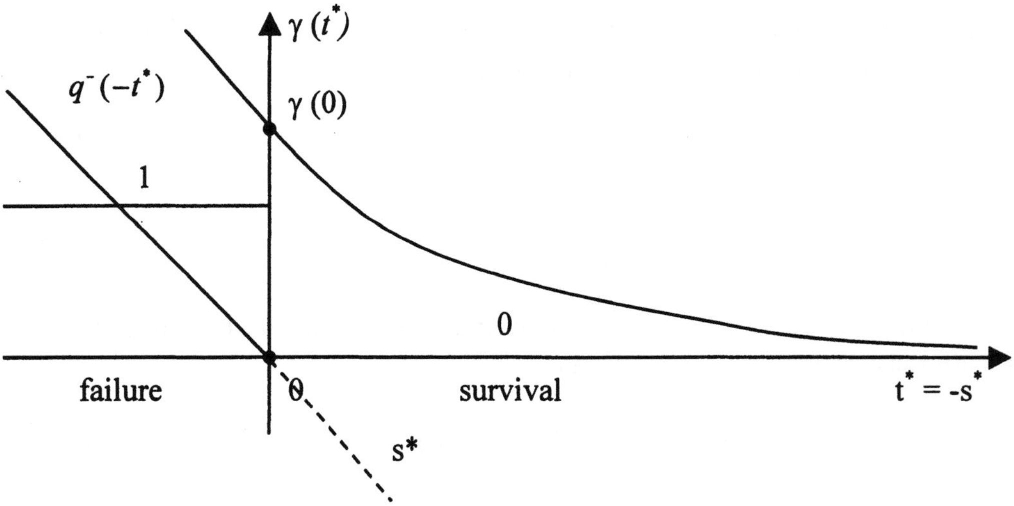


**Figure 3.5** Loss functions

Corresponding to Example 3.1.1 and 3.1.2, here we mention the following examples:

*Example 3.4.1.* If $\gamma(t) = 1$ for $t < 0$ and $\gamma(t) = 0$ for $t \geq 0$, then

$$\Gamma(X) = p_f(X) = P\left(t^*\Big(\big(\sigma_y(\omega), R(\omega)\big), X\Big) < 0\right). \tag{3.41a}$$

*Example 3.4.2.* For any nonnegative function $\gamma = \gamma(t)$ such that $\gamma(t) \geq \gamma(0) > 0$ for $t < 0$ we have that

$$p_f(X) \leq \frac{1}{\gamma(0)} \Gamma(X), \tag{3.41b}$$

see (3.12a,b).

*Example 3.4.3.* If $\gamma(t) = q^-(-t)$ for $t < 0$ and $\gamma(t) = 0$ for $t \geq 0$ with a cost factor $q^- > 0$, then instead of (3.41b) for any negative threshold $t_0 < 0$ we get the inequality

$$p_f^{(t_0)}(X) := P\left(t^*\left(\left(\sigma_y(\omega), R(\omega)\right), X\right) < t_0\right) \leq \frac{1}{q^-|t_0|}\Gamma(X), \tag{3.41c}$$

where $p_f^{(t_0)}(X) \leq p_f(X)$ can be interpreted as the probability of failure with respect to a somewhat weaker failure criterion $t^* < t_0$.

Suppose now that $\gamma = \gamma(t)$ is a nonincreasing loss function on $\mathbb{R}$. Since the minimum $s^*$ is attained in the LP (3.38a-c), we find that that the minimum failure costs

$$\gamma\left(t^*\left((\sigma_y, R), X\right)\right) = \gamma\left(-s^*\left((\sigma_y, R), X\right)\right)$$

can be represented by the minimum value of the program

$$\min \gamma(-s) \tag{3.42a}$$

s.t.

$$CF = R \tag{3.42b}$$

$$HF - s1 \leq F_0(\sigma_y, X). \tag{3.42c}$$

Hence, the substitute problem (3.40a,b) may be represented by the following 2-stage stochastic optimization problem

$$\min E\left(c_G G_0\left(\eta(\omega), X\right) + c_f \gamma(-s)\right) \tag{3.43a}$$

s.t.

$$CF = R(\omega) \text{ a.s.} \tag{3.43b}$$

$$HF - s1 \leq F_0\left(\sigma_y(\omega), X\right) \text{ a.s.} \tag{3.43c}$$

$$X \in D. \tag{3.43d}$$

Consider now the class of sublinear loss functions $\gamma = \gamma(t)$ defined (Marti, 1975, 1979) by

$$\gamma(t) := \begin{cases} q^-(-t), & t \leq 0 \\ q^+t, & t \geq 0, \end{cases} \tag{3.44a}$$

where $q^-, q^+$ are cost factors such that

$$q^- > 0, 0 \geq q^+ \geq -q^-. \tag{3.44b}$$

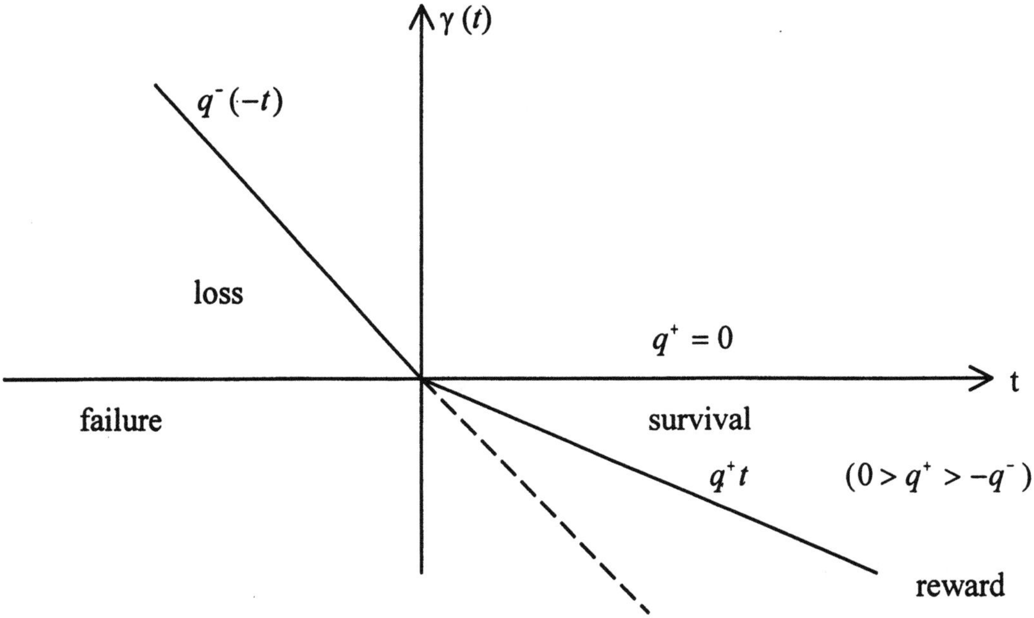


**Figure 3.6** Sublinear loss function

Since a sublinear function $\gamma(t)$, may be represented by the minimum value of the LP

$$\min q^- y^- + q^+ y^+ \tag{3.45a}$$

s.t.

$$y^+ - y^- = t \tag{3.45b}$$
$$y^+, y^- \geq 0, \tag{3.45c}$$

in case (3.44a,b) problem (3.43a-d) can be represented by the so-called **stochastic (linear) program with complete fixed recourse** (Kall, 1976, and Kall and Wallace, 1994)

$$\min E\left(c_G G_0\left(\eta(\omega), X\right) + c_f q^- y^- + c_f q^+ y^+\right) \tag{3.46a}$$

s.t.

$$CF = R(\omega) \quad \text{a.s.} \tag{3.46b}$$
$$HF + (y^+ - y^-)1 \leq F_0\left(\sigma_y(\omega), X\right) \quad \text{a.s.} \tag{3.46c}$$
$$X \in D, y^+, y^- \geq 0 \quad \text{a.s.} \tag{3.46d}$$

Note that the most general substitute problem for (3.37a-e) of this type reads

$$\min E\left(c_G G_0\left(\eta(\omega), X\right) + c_f q_f{}' y_I\right) \tag{3.47a}$$

s.t.

$$CF = R(\omega) \text{ a.s.} \tag{3.47b}$$
$$-F_0\left(\sigma_y(\omega), X\right) + My_I + HF = 0 \quad \text{a.s.} \tag{3.47c}$$
$$X \in D, y_I \geq 0 \text{ a.s.} \tag{3.47d}$$

Here, $q_f$ is a vector of cost coefficients, and $M$ is a matrix such that

$$\{My_I : y_I \geq 0\} = \mathbb{R}^{\mu_H}. \tag{3.48}$$

Obviously, for given design vector $X$, vector of yield stress $\sigma_y$, load vector $R$, and with $F_0 = F_0(\sigma_y, X)$, the costs of a structural failure, damage, for repair, etc., are evaluated by means of the function $Q(z), z = \binom{R}{F_0}$, where

$$Q = Q(z) = \min \{q^T y : Wy = z, y_I \geq 0\} \tag{3.49}$$

is the minimum value of the 2nd-stage LP

$$\min q^T y \tag{3.50a}$$

s.t.

$$Wy = z, z \in \mathbb{R}^{m+\mu_H} \tag{3.50b}$$
$$y_I \geq 0, \tag{3.50c}$$

having variables and data given by

$$y := \begin{pmatrix} y_I \\ F \end{pmatrix}, W := \begin{pmatrix} 0 & C \\ M & H \end{pmatrix}, q := \begin{pmatrix} q_f \\ 0 \end{pmatrix}. \tag{3.50d}$$

In the following we consider now the case of "simple recourse" with respect to $y_I$, given (Kall, 1976, and Kall and Wallace, 1994) by

$$M = (I, -I),\ q_f := \begin{pmatrix} q^+(\omega) \\ q^-(\omega) \end{pmatrix}, y_I = \begin{pmatrix} \Delta^+(\omega) \\ \Delta^-(\omega) \end{pmatrix}, \tag{3.51a}$$

where $I$ denotes the $\mu_H \times \mu_H$ identity matrix, $q^+(\omega), q^-(\omega)$ are vectors of cost factors such that

$$q^-(\omega) > 0, 0 \geq q^+(\omega) \geq -q^-(\omega) \text{ a.s.} \tag{3.51b}$$

cf. (3.44b), and

$$\Delta(\omega) = \Delta^+(\omega) - \Delta^-(\omega) = F_0 - HF \tag{3.51c}$$

is the discrepancy between $F_0$ and $HF$.

Defining now

$$T(\omega, X) := \begin{pmatrix} 0 \\ -F_0\big(\sigma_y(\omega), X\big) \end{pmatrix}, h(\omega) := \begin{pmatrix} R(\omega) \\ 0 \end{pmatrix} \tag{3.52a}$$

$$W := \begin{pmatrix} 0 & 0 & C \\ I & -I & H \end{pmatrix} \quad \text{(recourse matrix)} \tag{3.52b}$$

$$y(\omega) := \begin{pmatrix} \Delta^+(\omega) \\ \Delta^-(\omega) \\ F(\omega) \end{pmatrix},\ y_I(\omega) := \begin{pmatrix} \Delta^+(\omega) \\ \Delta^-(\omega) \end{pmatrix}, q(\omega) := \begin{pmatrix} q^+(\omega) \\ q^-(\omega) \\ 0 \end{pmatrix}, \tag{3.52c}$$

the above problem (3.47a-d) can be represented also by

$$\min E\left(G_0\left(\eta(\omega), X\right) + q(\omega)^T y(\omega)\right) \tag{3.53a}$$

s.t.

$$T(\omega, X) + Wy(\omega) = h(\omega) \text{ a.s.} \tag{3.53b}$$
$$A_0 X = b_0 \tag{3.53c}$$
$$X \geq 0, y_I(\omega) \geq 0 \text{ a.s.} \tag{3.53d}$$

However, this is a so-called "*stochastic program with complete fixed recourse*" (Kall, 1976, Kall and Wallace, 1994, Marti, 1979, and Mayer, 1998). In the case that the function, the mapping, resp.,

$$G_0\left(\eta(\omega), X\right) := c_0(\omega) + c(\omega)^T X \tag{3.54a}$$
$$T(\omega, X) := v_0(\omega) + T(\omega)X \tag{3.54b}$$

is linear with respect to $X$, as e.g. for trusses (Haftka et al., 1985, Hemp, 1973, and Kirsch, 1993) or after a linearization step (Marti, 1979), (3.53a-d) is a "*stochastic linear program with complete fixed recourse*" (Kall, 1976, Kall and Wallace, 1994, and Mayer, 1998).

*Remark 3.4.1.* If $X_i = A_i(X), i = 1, \ldots, B$, denotes the cross-sectional area of the $i$-th rod element of a truss, and the 1st stage cost function $G_0$ is the volume or weight of the truss, then we have that

$$c_0(\omega) := 0, c_i(\omega) := \eta_i(\omega) L_i, i = 1, \ldots, B, \tag{3.55a}$$

where $L_i$ is the length, and $\eta_i = \eta_i(\omega)$ is a weigth factor of the $i$-th element. Furthermore, for $v_0(\omega), T(\omega)$ we find that, cf. (3.36),

$$v_0(\omega) := 0, T(\omega) := \begin{pmatrix} 0 \\ -\sigma_d^U(\omega) \\ \sigma_d^L(\omega) \end{pmatrix}, \tag{3.55b}$$

where $\sigma^L, \sigma^U$ are the vectors of yield stresses in compression and tension, and $\sigma_d^L, \sigma_d^U$ denote the related diagonal matrices.

For stochastic programs with recourse there exists a large amount of theoretical results (Kall, 1976, Kall and Wallace, 1994, and Marti, 1979) and efficient numerical solution procedures (Birge, 1997, Kall and Wallace, 1994, Marti, 1979, and Mayer, 1998) based on

* discretization techniques (for $P_{a(\cdot)}$) and sequential linearization/convex approximation (with respect to $X$)
* special purpose LP-solvers: Special purpose LP-solvers use extreme point methods (special factorization techniques), interior point methods in combination with factorization schemes, column splitting, decomposition approaches (L-shaped methods, decomposition with sampling, stochastic decomposition), Lagrangian-based approaches (Lagrangian dual ascent, augemented Lagrangian method).

We still mention here that for the efficient computation of the equilibrium matrix $C$, recently also a special subroutine to the FEM software package "NASTRAN" is available (MSC, 1998).

In the following, several important properties of the stochastic program (3.53a-d) are presented.

### 3.4.1 The Second-Stage Linear Program

For a given design vector $X$, a so-called 1st stage decision vector, and a given realization $\Big(q(\omega), T(\omega, X), h(\omega)\Big)$ of the random variables $\Big(q(\cdot), T(\cdot, X), h(\cdot)\Big)$, here we consider the *second-stage linear program* (3.50a-d):

$$\min q(\omega)^T y$$

s.t.

$$\begin{aligned} Wy &= z\Big(z = h(\omega) - T(\omega, X)\Big) \\ y_I &\geq 0. \end{aligned}$$

According to (3.49), let then $Q = Q\Big(q(\omega), z\Big), z \in \mathbb{R}^{\overline{m}}, \overline{m} := m + \mu_H$, denote the optimal value of the LP (3.50a-c); this optimal value function is also called "recourse function" (Kall, 1976, and Kall and Wallace, 1994). Since

$$\{Wy : y_I \geq 0\} = \mathbb{R}^{\overline{m}}, \tag{3.56a}$$

and we still suppose

$$\text{dual feasibility of the LP (3.50a-c)}, \tag{3.56b}$$

which holds e.g. in case $q_f \geq 0$, we have the following result:

**Lemma 3.4.1.** *Suppose that $M, q_f, y_I$ are given by (3.51a) and (3.56b) holds. Then, a) The linear program (3.50a-c) has an optimal solution $y^* = (\Delta^{+*}, \Delta^{-*}, F^*)$ for each $z \in \mathbb{R}^{\overline{m}}$; b) The optimal value function $Q = Q\Big(q(\omega), z\Big)$ of (3.50a-c) is a positively homogeneous, subadditive functional with respect to $z$.*

Proof. See Kall and Wallace (1994), and Marti (1975).

Obviously, for $z := h(\omega) - T(\omega, X) = \begin{pmatrix} R(\omega) \\ F_0\Big(\sigma_y(\omega), X\Big) \end{pmatrix}$, the second-stage or static linear program (3.50a-c) reads

$$\min q^+(\omega)^T \Delta^+ + q^-(\omega)^T \Delta^- \tag{3.57a}$$

s.t.

$$CF = R(\omega)$$

$$\Delta^+ - \Delta^- + HF = F_0\Big(\sigma_y(\omega), X\Big) \tag{3.57b}$$

$$\Delta^+ \geq 0, \Delta^- \geq 0. \tag{3.57c}$$

The dual LP of the static linear program (3.57a-c) is the kinematic LP (Augusti et al., 1984, and Marti, 1997)

$$\max R(\omega)^T u - F_0\Big(\sigma_y(\omega), X\Big)^T \tilde{u} \tag{3.58a}$$

s.t.

$$C^T u = H^T \tilde{u} \tag{3.58b}$$
$$-q^+(\omega) \le \tilde{u} \le q^-(\omega). \tag{3.58c}$$

An important consequence of Lemma 3.4.1 and the subsequent considerations is mentioned next:

**Corollary 3.4.1.** *a) The static LP (3.57a-c) yields the most favorable internal loading* $F^* = F^*\big(q(\omega), h(\omega) - T(\omega, X)\big)$ *of the structure for given design vector* $X$ *and data* $\big(q(\omega), R(\omega), \sigma_y(\omega)\big)$*.*
*b) Let* $q^+ = 0$ *and* $q^- > 0$*. Then the loaded structure represented by* $\big(X, \sigma_y(\omega), R(\omega)\big)$ *fails if and only if* $Q > 0$*, i.e. if the minimum value* $Q$ *of the LP (3.57a-c) is positive.*

By means of the optimal value functions $Q(q, z)$ of the second-stage LP (3.50a-c), the stochastic optimization problem (3.53a-d) can be represented also by

$$\min E\left(G_0\Big(\eta(\omega), X\Big) + Q\Big(q(\omega), h(\omega) - T(\omega, X)\Big)\right) \tag{3.59a}$$

s.t.

$$A_0 X = b_0 \tag{3.59b}$$
$$X \ge 0. \tag{3.59c}$$

This yields the following result, cf. (3.54a,b):
**Theorem 3.4.1.** *If* $X \to G_0\Big(\gamma(\omega), X\Big)$ *is convex and* $X \to T(\omega, X)$ *is linear, then (3.59a-c) is a convex program.*
Proof. See Kall (1976), and Kall and Wallace (1994).

**Selection of the cost factors $q^-, q^+$.** According to (3.57a-c), an appropriate, simple choice for $q^+$ is

$$q^+ = 0. \tag{3.60a}$$

Moreover, for trusses, $q^-(\omega) = \Big(q_j^-(\omega)\Big)$ may be defined by

$$q_j^-(\omega) := \eta_j^q(\omega)\frac{L_j}{E_j(\omega)}, j = 1, \ldots, n, \tag{3.60b}$$

where $L_j$ is the length, $E_j = E_j(\omega)$ denotes the modulus of elasticity, and $\eta_j^q = \eta_j^q(\omega)$ is a scale factor of the $j$-th bar of the truss. Since $y_I = \binom{\Delta^+}{\Delta^-}$ contains force components, cf. (3.57b), (3.36), $\Delta V_j = \frac{L_j}{E_j(\omega)}\Delta_j^-$ is the volume-change caused by the force $\Delta_j^-$.

In case of frames, the problem is treated (Lawo, 1987, Smith et al., 1976, Tin–Loi, 1990, 1995, and Zeman et al., 1986) usually iteratively by working first, in an inner optimization loop, with fixed ratios

$$c_{pj} = \left( \frac{T_{pj}}{N_{pj}}, \frac{M^y_{pj}}{N_{pj}}, \frac{M^z_{pj}}{N_{pj}} \right)^T \tag{3.60c}$$

of twisting/bending and principal axial capacities (see Marti, 1997, and Spillers, 1972). For the inner optimization loop the (vectorial) cost factor $q_j^-$ may be defined by

$$q_j^-(\omega) := \eta_j^q(\omega) c_{pj}, \tag{3.60d}$$

where $\eta_j^q(\omega)$ is again a cost factor.

### 3.4.2 Discretely Distributed Problems

Assume here that the total random parameter vector $a(\omega) = \Big(\eta(\omega), \sigma_y(\omega), R(\omega), q(\omega)\Big)$ has a joint *discrete probability distribution*

$$P\Big(a(\omega) = a^j\Big) = \alpha_j, j = 1, \dots, s, \tag{3.61a}$$

having realizations $a^j$ taken with probabilities $\alpha_j > 0, j = 1, \dots, s$. This holds if

- there are only a finite number $s$ of material, cost, loading, ... *scenarios* $a^j, j = 1, \dots, s$, which may occur with certain probabilities $\alpha_1, \dots, \alpha_s$, or
- (3.61a) results from a *discretization* of an original (continuous) distribution $P^{(0)}_{a(\cdot)}$.

Assume now that $G_0\Big(\eta(\omega), X\Big), T(\omega, X)$ have the linear representation (3.54a,b), as for trusses, see (3.36) and Remark 3.4.1, or after linearization with respect to $X$. Define then

$$h_v(\omega) := h(\omega) - v_0(\omega), \tag{3.61b}$$

and let denote

$$T^j, h_v^j := h^j - v_0^j; q^j, j = 1, \dots, s, \tag{3.61c}$$

the realizations of $T(\omega), h_v(\omega), q(\omega)$ related to the realizations (scenarios) $a^j, 1 \le j \le s$, of $a(\omega)$.

A central role in the various numerical solution techniques for the stochastic optimization problem (3.53a-d) is played by the following result.

**Theorem 3.4.2.** *Assume that the linearity condition (3.54a,b) holds and the distribution assumption (3.61a-c) is fulfilled. Then the stochastic optimization problem (3.53a-d) can be represented by a linear program with a "dual decomposition data structure" as shown in Figure 3.7*

Proof. See Kall, 1976, and Kall and Wallace, 1994.

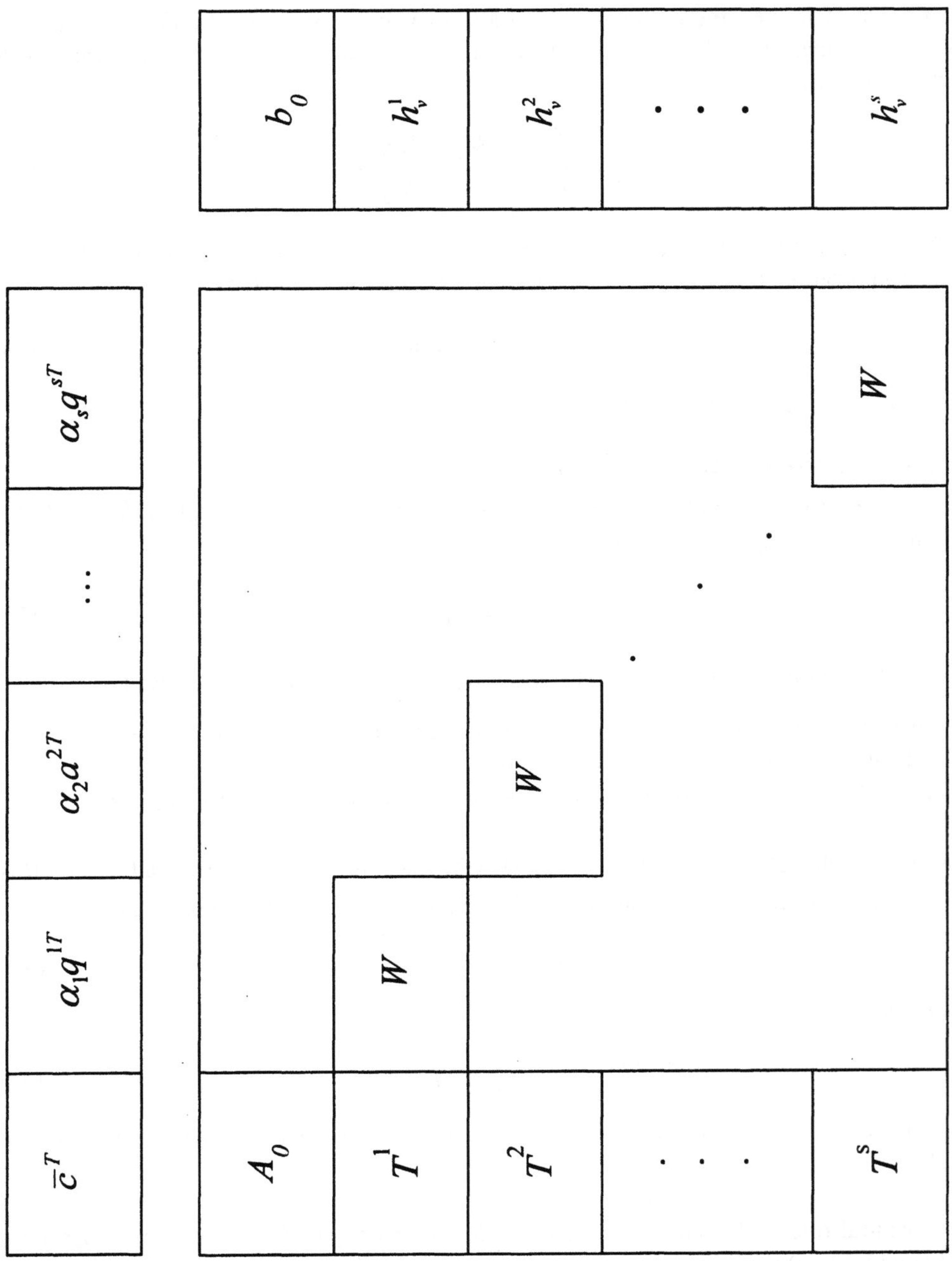


**Figure 3.7** LP with "Dual Decomposition Data Structure"

## 3.5 Basic Solution Techniques for Stochastic Linear Programs

### 3.5.1 Stochastic Linear Programs Having a Discrete Distribution

In this section we assume that the linear representation (3.54a,b) holds and $a(\omega)$ has a discrete distribution as described by (3.61a-c). Furthermore, for simplification suppose a fixed second-stage cost factor

$$q(\omega) = q_0 \text{ a.s.} \tag{3.62}$$

with a given vector $q_0$. According to (3.59a-c) we have then to solve the convex optimization problem

$$\min E\Big(c(\omega)'X + Q\Big(q_0, h_v(\omega) - T(\omega)X\Big)\Big) \tag{3.63a}$$

s.t.

$$A_0 X = b_0 \tag{3.63b}$$
$$X \geq 0, \tag{3.63c}$$

where $c(\omega), T(\omega)$ and $h_v(\omega)$ are defined according to (3.54a,b), (3.61b,c). Based on the above assumptions, the objective function $G = G(X)$ of (3.63a-c) is given by

$$G(X) := \bar{c}^T X + \sum_{j=1}^{s} \alpha_j Q(q_0, h_v^j - T^j X), \tag{3.64a}$$

where

$$\bar{c} := Ec(\omega). \tag{3.64b}$$

Introducing auxiliary variables $\theta = (\theta_1, \ldots, \theta_j, \ldots, \theta_s)'$, program (3.63a-c) is equivalent to

$$\min \bar{c}^T X + \sum_{j=1}^{s} \alpha_j \theta_j \tag{3.65a}$$

s.t.

$$A_0 X = b_0 \tag{3.65b}$$
$$\theta_j \geq Q(q_0, h_v^j - T^j X), j = 1, \ldots, s \tag{3.65c}$$
$$X \geq 0, \theta_j \geq 0, j = 1, \ldots, s. \tag{3.65d}$$

Note that the nonnegativity constraint $\theta \geq 0$ is due to the fact that we may assume that $Q \geq 0$, see (3.49), (3.50a-d), (3.51a,b).

The key to the *dual decomposition method* for solving (3.63a-c) lies in the fact that the values $Q(q_0, h_v^j - T^j X), 1 \leq j \leq s$, of the recourse function $Q$ can be represented, using LP-duality theory, by

$$Q\Big(q_0, h_v^j - T^j X\Big) = \max_{w \in U} w^T (h_v^j - T^j X), \tag{3.66a}$$

cf. (3.52a-c), (3.50a-c), (3.57a-c), (3.58a-c), where the feasible domain $U$ of the dual LP reads

$$U := \left\{ w = \begin{pmatrix} u \\ \tilde{u} \end{pmatrix} : C^T u = H^T \tilde{u}, -q_0^+ \leq \tilde{u} \leq q_0^- \right\}. \tag{3.66b}$$

Thus, representing (3.65c) by

$$\theta_j \geq w^T (h_v^j - T^j X), w \in U, j = 1, \ldots, s, \tag{3.65c'}$$

(3.65a-d) can be written also as a semi-infinite LP. Approximations of the recourse function $Q$ can be obtained then by

$$Q(q_0, h_v^j - T^j X) \approx \tilde{Q}(q_0, h_v^j - T^j X) := \max_{w \in U_{j\nu}} w^T (h_v^j - T^j X), \tag{3.67}$$

where $U_{j\nu} \subset U, \nu = 0, 1, \ldots,$ are finite sets which are defined recursively.

In detail, the dual decomposition method proceeds as follows:

*STEP 1: Initialization*

Since $Q \geq 0$, solve the approximative LP, cf. (3.65a-d),

$$\min \bar{c}^T X + \sum_{j=1}^{s} \alpha_j \theta_j \tag{3.68a}$$

s.t.

$$A_0 X = b_0 \tag{3.68b}$$
$$X \geq 0, \theta \geq 0. \tag{3.68c}$$

Let denote $\left(\hat{X}, \hat{\theta} = (\hat{\theta}_j)\right)$ an optimal solution of this LP.

*STEP 2: Optimality cuts*

For $z^j := h_v^j - T^j \hat{X}, j = 1, \ldots, s$, solve the second-stage program (3.50a-c) producing a primal, dual optimal solution $\hat{y}^j, \hat{w}^j$, resp., $j = 1, \ldots, s$. Hence,

$$Q(q_0, h_v^j - T^j \hat{X}) = \hat{w}^{jT} (h_v^j - T^j \hat{X}), j = 1, \ldots, s, \tag{3.69}$$

cf. (3.66a,b). Because of (3.65c) we consider now the constraints

$$\theta_j \geq \hat{w}^{jT} (h_v^j - T^j X), j = 1, \ldots, s. \tag{3.70}$$

*Case 2A:* $(\hat{X}, \hat{\theta})$ fulfills (3.70)

Because of (3.69), (3.66a), $\hat{X}$ is optimal in (3.63a-c), and the algorithms stops. Note that this argument remains true also in later steps, since only constraints of the type (3.70) are added to the preceding approximate LP, see (3.65c').

*Case 2B:* $(\hat{X}, \hat{\theta})$ violates (3.70)

In this case there is at least one index $j_0, 1 \le j_0 \le s$, such that

$$Q(q_0, h_v^{j_0} - T^{j_0}\hat{X}) = q_0^T \hat{y}^{j_0} = \hat{w}^{j_0 T}(h_v^{j_0} - T^{j_0}\hat{X}) \tag{3.71}$$

$$= \hat{u}^{j_0 T} R^{j_0} - \hat{\bar{u}}^{j_0 T} F_0(\sigma_y^{j_0}, \hat{X}) := -M_{\lambda_0}(a^{j_0}, \hat{X}) > \hat{\theta}_{j_0} \ge 0,$$

where $M_{\lambda_0} = M_{\lambda_0}(a^{j_0}, X), 1 \le \lambda_0 \le l_0$, denotes the corresponding limit state function, see (3.39b), (3.58a-c). This means that $\hat{X}$ lies in a certain failure domain with positive probability, and the related minimum failure costs exceed the actual optimal cost level $\hat{\theta}_{j_0}$.

Hence, add to the preceding LP the so-called *optimality cuts* (3.70), cutting off the nonoptimal solution $(\hat{X}, \hat{\theta})$.

*STEP 3:* Update of the approximate LP
If $\hat{D}$ denotes the feasible domain of the current approximate LP, redefine

$$\tilde{D} := \left\{ (X, \theta) \in \hat{D} : \quad (X, \theta) \text{ fulfills } (3.70) \right\}, \tag{3.72}$$

and solve the updated LP

$$\min \bar{c}^T X + \sum_{j=1}^{s} \alpha_j \theta_j \tag{3.73a}$$

s.t.

$$(X, \theta) \in \tilde{D}, \tag{3.73b}$$

yielding an optimal solution $(\tilde{X}, \tilde{\theta})$. With $(\hat{X}, \hat{\theta}) := (\tilde{X}, \tilde{\theta})$, return to *STEP 2*.

Under weak assumptions, this algorithms finds an optimal solution $X^*$ of (3.63a-c) after finitely many steps:

**Theorem 3.5.1.** *If* $\{X : A_0 X = b_0, X \ge 0\}$ *is bounded, then the dual decomposition method yields an optimal design point* $X^*$ *after finitely many steps.*

Proof. See Kall and Wallace (1994).

### 3.5.2 Sequential Discretization

For simplification, suppose again that (3.62) holds, i.e. the second-stage cost coefficients are fixed. For an arbitrary random matrix $\big(T(\omega), h_v(\omega)\big)$ in (3.60a-c), by discretization we can find a sequence

$$\Big(T^1(\omega), h_v^1(\omega)\Big), \Big(T^2(\omega), h_v^2(\omega)\Big), \ldots, \Big(T^N(\omega), h_v^N(\omega)\Big), \ldots \tag{3.74a}$$

of piecewise constant random matrices $\big(T^N(\cdot), h_v^N(\cdot)\big), N = 1, 2, \ldots$, converging to $\big(T(\cdot), h_v(\cdot)\big)$ in some probabilistic sense. Consequently, we obtain then the approximative objective function

$$G^N(X) := \bar{c}^T X + EQ\Big(q_0, h_v^N(\omega) - T^n(\omega)N\Big), \tag{3.74b}$$

cf (3.64a). Under weak assumptions (Marti, 1975) we have that

$$G_N(x) \to G(X), N \to \infty, \tag{3.74c}$$

uniformly on compact subsets of $\mathbb{R}^r$. Moreover, the approximative program

$$\min G^N(X) \tag{3.75a}$$

s.t.

$$A_0 X = b_0 \tag{3.75b}$$
$$X \geq 0 \tag{3.75c}$$

has the following properties.

**Theorem 3.5.2.** *a) If $\left(T^N(\cdot), h_v^N(\cdot)\right)$ is a piecewise constant random matrix having finitely many realizations then (3.75a-c) can be represented by a linear program having a dual decomposition data structure as described by Figure 3.7.*
*b) Suppose that $D_0 = \{X \in \mathbb{R}^r : A_0 X = b_0, X \geq 0\}$ is a compact set, and let denote $X^{N*}$ an optimal solution of the approximate problem (3.75a-c). Then the optimal value $G^{N*}$ of (3.75a-c) converges to the optimal value $G^{N*}$, as $N \to \infty$, and every accumulation point $X^*$ of $(X^{N*})$ is an optimal solution of the original problem (3.63a-c).*
Proof. See Kall (1976), and Marti (1975).

*Remark 3.5.1.* By special iterative refining techniques applied in the sequential discretization

$$\left(T^N(\cdot), h_v^N(\cdot)\right) \to \left(T^{N+1}(\cdot), h_v^{N+1}(\cdot)\right), N = 1, 2, \ldots,$$

of $\left(T(\cdot), h_v(\cdot)\right)$, the increase of the size of the resulting (large scale) LP's (3.75a-c) can be kept moderately.

More details about the properties of the sequential discretization procedure are given in Section 3.7.

### 3.5.3 Convex Approximation by Partial Linearization

If $X \to G_0\left(\eta(\omega), X\right)$ and/or $X \to T(\omega, X)$ are nonlinear mappings, then by linearization of the function(s)

$$G_0\left(\eta(\omega), X\right) \approx G_0\left(\gamma(\omega), X_0\right) + \nabla_x G_0\left(\gamma(\omega), X_0\right)^T (X - X_0), \tag{3.76a}$$
$$T(\omega, X) \approx T(\omega, X_0) + \frac{\partial T}{\partial X}(\omega, X_0)(X - X_0) \tag{3.76b}$$

with respect to $X$ at a certain reference point $X_0$, the original problem (3.63a-c) can be reduced to a convex problem fulfilling (3.54a,b); let $X_1$ denote its optimal solution. Proceeding iteratively, a minimizing sequence $X_0, X_1, \ldots$ for the original problem (3.63a-c) is obtained. More details can be found in Marti (1979).

## 3.6 Adaptive Structural Designs

In this section we assume that we have the following *adaptive decision situation*, (see, e.g., Hartmann, 2000, and Schüler, 1997).

*STAGE I at time $t_0$:* An *initial or pre-design $X$* must be selected at a time $t_0$ based on limited information on the model parameters and structural loading represented by the probability distribution $P_{a(\cdot)}$ of the total parameter vector $a(\cdot)$.

*STAGE II at a later time $t_1 > t_0$:* The random vector $a(\cdot)$ is realized $a = a(\omega)$ at a later time instant $t_1 > t_0$, which enables then a correction, completion $\xi = \xi(a)$ of the initial design, or a strengthening, retrofitting/repair of the present structure, etc., by a further design action $\xi$. Besides the costs $G_0 = G_0\Big(\eta(\omega), X\Big)$ caused by the inital or pre-design $X$, we have now additional costs

$$G_1 = G_1\Big(\zeta(\omega), \xi(a)\Big), a := \Big(\sigma_y, \sigma^L, \sigma^U, E, \eta, \zeta, R, \ldots\Big), \tag{3.77}$$

caused by the second-stage design action $\xi$. Here, $\zeta(\omega)$ is again a vector of cost coefficients.

Working again with the yield condition (3.35b), the vector

$$F_0 = F_0(\sigma_y, X, \xi)$$

depends now on the initial design vector $X = X(P_{a(\cdot)})$ *and* on the correction/completion or strengthening design vector $\xi = \xi(a)$ which can be taken at time $t_1 > t_0$ under full information about the realization $a = a(\omega)$ of $a(\cdot)$. For the definition of $F_0$ in this two-stage decision situation, we mention the following models:

a) If $\xi = \xi(a)$ is a vector of design corrections for the initial design $X$, then

$$F_0(\sigma_y, X, \xi) = F_0(\sigma_y, X + \xi), \tag{3.78a}$$

where the right hand side of (3.78a) is defined as in (3.35b).

b) If $\xi = \xi(a)$ represents certain pre-planned completions at time $t_1$ of the structure built using the initial design $X$ at time $t_0$, then $F_0$ may be given by the additive model

$$F_0(\sigma_y, X, \xi) := F_0^I(\sigma_y^I, X) + F_0^{II}(\sigma_y^{II}, \xi), \tag{3.78b}$$

where the vectors $F_0^I, F_0^{II}$ are defined again as in (3.35b).

Corresponding to program (3.37a-e) in Section 3.4, for the optimal selection of the initial design $X = X(P_{a(\cdot)})$ and the second-stage design vector $\xi = \xi(a)$ we have then this random program

$$\min G_0\Big(\eta(\omega), X\Big) + G_1\Big(\zeta(\omega), \xi\big(a(\omega)\big)\Big) \tag{3.79a}$$

s.t.

$$CF = R(\omega) \tag{3.79b}$$

$$HF \le F_0\Big(\sigma(\omega), X, \xi\big(a(\omega)\big)\Big) \tag{3.79c}$$

$$A_0 X = b_0 \tag{3.79d}$$

$$A_1 \xi\Big(a(\omega)\Big) = b_1 \tag{3.79e}$$

$$X \ge 0, \xi^L \le \xi \le \xi^U, \tag{3.79f}$$

where $\xi^L, \xi^U$, are given bounds for $\xi$ which may depend on $X$, and $(A_1, b_1)$ is a given matrix.

The transformation of (3.79a-f) into a *two-stage* stochastic (linear) program can be done now by the stochastic optimization tools described in Section 3.4.

In case (3.78a), and assuming then that

$$G_0\Big(\eta(\omega), X\Big) := c(\omega)^T X,\ G_1\Big(\zeta(\omega), \xi\Big) := d(\omega)^T \xi, \tag{3.80a}$$

$$F_0\Big(\sigma_y(\omega), X + \xi\Big) := S(\omega)(X + \xi) = S(\omega)X + S(\omega)\xi \tag{3.80b}$$

with random vectors $c(\omega), d(\omega)$ and a random matrix $S(\omega)$, cf. (3.54a,b), for the minimization of the expected total costs we get a stochastic linear program with recourse of the type (3.53a-d) having the following main components, cf. (3.52a-c):

$$T(\omega, X) = T(\omega)X := \begin{pmatrix} 0 \\ -S(\omega)X \end{pmatrix},\ h(\omega) := \begin{pmatrix} R(\omega) \\ 0, \end{pmatrix} \tag{3.81a}$$

$$W := \begin{pmatrix} 0 & 0 & 0 & C \\ I & -I & -S(\omega) & H \end{pmatrix} \text{ (recourse matrix)} \tag{3.81b}$$

$$y(\omega) := \begin{pmatrix} \Delta^+(\omega) \\ \Delta^-(\omega) \\ \xi(\omega) \\ F(\omega) \end{pmatrix}, y_I(\omega) := \begin{pmatrix} \Delta^+(\omega) \\ \Delta^-(\omega) \\ \xi(\omega) \end{pmatrix}, q(\omega) := \begin{pmatrix} q^+(\omega) \\ q^-(\omega) \\ d(\omega) \\ 0 \end{pmatrix}. \tag{3.81c}$$

## 3.7 Solutions of Stochastic Linear Programs by Discretization Methods

### 3.7.1 Stochastic Linear Programs with Recourse

A well known method to handle linear programs

$$\text{minimize } c(\omega)^T x \text{ s.t. } T(\omega)x = h(\omega), x \in D \tag{3.82}$$

with random data $\Big(c(\omega), T(\omega), h(\omega)\Big)$ on a probability space $(\Omega, \mathcal{A}, P)$ is to replace (3.82) by the stochastic optimization problem, cf. (3.53a-d), (3.59a-c), (3.63a-c),

$$\text{minimize } F(x) \text{ s.t. } x \in D, \tag{3.83a}$$

where

$$F(x) = E\Big(c(\omega)^T x + p\Big(h(\omega) - T(\omega)x\Big)\Big) \tag{3.83b}$$

and $p = p(z)$ denote the so-called second stage or recourse costs defined by

$$p(z) = \inf\{q^T y : Wy = z, y \geq 0\}, z \in \mathbb{R}^m. \tag{3.83c}$$

Here, $D$ is a fixed convex polyhedron in $\mathbb{R}^n$, hence, "$x \in D$" represents the deterministic constraints in (3.82), and "E" denotes the expectation operator. In the following we suppose that the $m \times n_1$ matrix $W$ and the $m$ vector $q$ are related such that

$$\{Wy : x \geq 0\} = \mathbb{R}^m. \tag{3.84a}$$
$$\{v : W^T v \leq q\} \neq \emptyset. \tag{3.84b}$$

Thus, the loss function $p$ is defined on the whole $\mathbb{R}^m$. If $q \geq 0$, then (3.84a) holds, and $p$ is nonnegative. According to Marti (1975) we know that $p$ is a sublinear function on $\mathbb{R}^m$, thus,

$$p(z+w) \leq p(z) + p(w) \text{for all } z, w \in \mathbb{R}^m \tag{3.85a}$$
$$p(\lambda z) = \lambda p(z) \text{for all } z \in \mathbb{R}^m, \lambda \geq 0. \tag{3.85a}$$

Consequently, $p$ is a convex function on $\mathbb{R}^m$, and we have that

$$p(0) = 0 \tag{3.85c}$$
$$-p(-z) \leq p(z) \text{ for all } z \in \mathbb{R}^m \tag{3.85d}$$
$$-p(z-w) \leq p(w) - p(z) \leq p(w-z) \text{ for all } z, w \in \mathbb{R}^m. \tag{3.85e}$$

Moreover, if

$$\|p\| = \sup_{\|z\|_E \leq 1} \left| p(z) \right| = \sup_{\|z\|_E = 1} \left| p(z) \right| \; (< +\infty) \tag{3.85f}$$

denotes the norm of the sublinear function $p$, then

$$\left| p(z) \right| \leq \|p\| \, \|z\| \text{ for all } z \in \mathbb{R}^m, \tag{3.85g}$$
$$\left| p(w) - p(z) \right| \leq \|p\| \, \|w - z\| \text{ for all } z, w \in \mathbb{R}^m. \tag{3.85h}$$

Denoting $Ec(\omega)$ by $\bar{c}$, (3.83b) reads

$$F(x) = \bar{c}^T x + Ep\Big(h(\omega) - T(\omega)x\Big). \tag{3.86}$$

### 3.7.2 Discretization and Error Bounds

Approximating now the $m \times (n+1)$ random matrix $\Big(T(\omega), h(\omega)\Big)$ by a certain sequence of random matrices

$$\Big(T^1(\omega), h^1(\omega)\Big), \Big(T^2(\omega), h^2)(\omega)\Big), \ldots, \Big(T^N(\omega), h^N(\omega)\Big), \ldots,$$

converging in some probabilistic sense to $\Big(T(\omega), h(\omega)\Big)$, we obtain the approximative objective functions

$$F^N(x) = \bar{c}^T x + Ep\Big(h^N(\omega) - T^N(\omega)x\Big), \quad N = 1, 2, \ldots. \tag{3.87}$$

Using (3.85d), for $F(x) - F^N(x)$ we obtain, (see Marti, 1975 and 1979).

$$-\vartheta^N(x) \leq F(x) - F^N(x) \leq \eta^N(x), \tag{3.88a}$$

where the lower, upper error term $\vartheta^N(x), \eta^N(x)$, resp., is defined by

$$\vartheta^N(x) := Ep\left(\left(T(\omega) - T^N(\omega)\right)x + \left(h^N(\omega) - h(\omega)\right)\right) \tag{3.88b}$$

$$\eta^N(x) := Ep\left(\left(T^N(\omega) - T(\omega)\right)x + \left(h(\omega) - h^N(\omega)\right)\right). \tag{3.88c}$$

Using discretization methods, the approximations $\left(T^N(\omega), h^N(\omega)\right)$ are piecewise constant random variables, hence

$$\left(T^N(\omega), h^N(\omega)\right) = \left(T^{N,j}, h^{N,j}\right) \text{ for all } \omega \in \Omega^{N,j}, j = 1, 2, \ldots, r_N, \tag{3.89a}$$

where

$$\Omega^{N,1}, \Omega^{N,2}, \ldots, \Omega^{N,j}, \ldots, \Omega^{N,r_N} \text{ is a partition of } \Omega. \tag{3.89b}$$

Consequently, $F^N(x)$, given by (3.87), reads

$$F^N(x) = \bar{c}^T x + \sum_{j=1}^{r_N} P(\Omega^{N,j}) p(h^{N,j} - T^{N,j} x). \tag{3.89c}$$

Using Jensen's inequality, for the error estimate $\vartheta^N(x)$ we get

$$\begin{aligned} \vartheta^N(x) &= \int p\left(\left(T(\omega) - T^N(\omega)\right)x + \left(h^N(\omega) - h(\omega)\right)\right) P(d\omega) \\ &= \sum_{j=1}^{r_N} \int_{\Omega^{N,j}} p\left(\left(T(\omega) - T^{N,j}\right)x + \left(h^{N,j} - h(\omega)\right)\right) P(d\omega) \\ &\geq \sum_{j=1}^{r_N} P(\Omega^{N,j}) p\left(\left(\overline{T}^{\Omega^{N,j}} - T^{N,j}\right)x + \left(n^{N,j} - \overline{h}^{\Omega^{N,j}}\right)\right), \end{aligned} \tag{3.90a}$$

where

$$\overline{T}^{\Omega^{N,j}} = \frac{1}{P(\Omega^{N,j})} \int_{\Omega^{N,j}} T(\omega) P(d\omega), \tag{3.90b}$$

$$\overline{h}^{\Omega^{N,j}} = \frac{1}{P(\Omega^{N,j})} \int_{\Omega^{N,j}} h(\omega) P(d\omega) \tag{3.90c}$$

are the conditional expectation of $T(\omega), h(\omega)$, resp., with respect to $\Omega^{N,j}$. Obviously, for $\eta^N(x)$ we obtain

$$\eta^N(x) \geq \sum_{j=1}^{r_N} P(\Omega^{N,j}) p\left(\left(T^{N,j} - \overline{T}^{\Omega^{N,j}}\right)x + \left(\overline{h}^{\Omega^{N,j}} - h^{N,j}\right)\right). \tag{3.90d}$$

In several publications (Kall, 1976, Kall and Wallace, 1994, and Mayer, 1998) there is suggested to select the values $T^{N,j}, h^{N,j}$, resp., according to

$$T^{N,j} = \overline{T}^{\Omega^{N,j}} = \frac{1}{P(\Omega^{N,j})} \int_{\Omega^{N,j}} T(\omega) P(d\omega), \tag{3.91a}$$

$$h^{N,j} = \overline{h}^{\Omega^{N,j}} = \frac{1}{P(\Omega^{N,j})} \int_{\Omega^{N,j}} h(\omega) P(d\omega). \tag{3.91b}$$

In this case, (3.90a,d) and (3.85c) immediately yield this result:

**Lemma 3.7.1.** *If the approximation* $\left(T^N(\omega), h^N(\omega)\right)$ *of* $\left(T(\omega), h(\omega)\right)$ *is defined by (3.89a,b) and (3.91a,b), then* $\vartheta^N(x) \geq 0, \eta^N(x) \geq 0$ *for all* $x \in \mathbb{R}^n$.

*If the norm* $\|p\|$ *of the sublinear loss function is known, then the error terms* $\vartheta^N(x), \eta^N(x)$ *can be further estimated from above. Indeed, (3.88b,c) and (3.85f) yield*

$$|\vartheta^N(x)|, |\eta^N(x)| \leq \|p\| \left(E\left\|\left(T(\omega) - T^N(\omega)\right)x\right\| + E\left\|h^N(\omega) - h(\omega)\right\|\right). \tag{3.92a}$$

*Denoting by*

$$P^{N,j}(d\omega) = \frac{1}{P(\Omega^{N,j})} 1_{\Omega^{N,j}} P(d\omega) \tag{3.92b}$$

*the restriction of P to the subdomains* $\Omega^{N,j}$ *of* $\Omega$, *according to (3.89a-c), from (3.92a,b) we obtain*

$$|\vartheta^N(x)|, |\eta^N(x)| \leq \|p\| \sum_{j=1}^{r_N} P(\Omega^{N,j}) \left( \int_{\Omega^{N,j}} \left\|\left(T(\omega) - T^{N,j}\right)x\right\| P^{N,j}(d\omega) \right. \tag{3.92c}$$

$$\left. + \int_{\Omega^{N,j}} \|h^{N,j} - h(\omega)\| P^{N,j}(d\omega) \right).$$

*Remark 3.7.1.* **More general loss functions**. If the sublinear loss function $p$ is replaced by a more general convex loss function $u$, then the inequalities (3.92a,c) remain true if the left hand side in (3.92a,c) is simply replaced by $|F(x) - F^N(x)|$, and $\|p\|$ is replaced by a Lipschitz constant $L > 0$ of $u$, provided that $u$ is Lipschitzian with constand $L$ on the union of the supports of the random $m$-vectors $T(\omega)x - h(\omega)$ and $T^N(\omega)x - h^N(\omega), N = 1, 2, \ldots, x \in D$, cf. Marti (1975).

Because of

$$\left\|\left(T(\omega) - T^{N,j}\right)x\right\| = \left(\|\left(T(\omega) - T^{N,j}\right)x\|^2\right)^{1/2} =$$

$$= \left(x^T\left(T(\omega) - T^{N,j}\right)^T\left(T(\omega) - T^{N,j}\right)x\right)^{1/2}$$

and the concavity of $t \to \sqrt{t}$, we have that

$$\int_{\Omega^{N,j}} \left\|\left(T(\omega) - T^{N,j}\right)x\right\| P^{N,j}(d\omega) \leq \left(\int_{\Omega^{N,j}} \left\|\left(T(\omega) - T^{N,j}\right)x\right\|^2 P^{N,j}(d\omega)\right)^{1/2}$$

$$= \left(x^T\left(\sum_{i=1}^{m} \int_{\Omega^{N,j}} \left(T_i(\omega) - T_i^{N,j}\right)^T\left(T_i(\omega) - T_i^{N,j}\right) P^{N,j}(d\omega)\right)x\right)^{1/2} \tag{3.93a}$$

$$\leq \|x\| \left(\sum_{i=1}^{m}\sum_{k=1}^{n} \int_{\Omega^{N,j}} \left(t_{ik}(\omega) - t_{ik}^{N,j}\right)^2 P^{N,j}(d\omega)\right)^{1/2}$$

where $T_i, T_i^{N,j}$ is the $i$-th row of the $m \times n$ matrices $T(\omega) = \left(t_{ik}(\omega)\right), T^{N,j} = \left(t_{ik}^{N,j}\right)$, respectively. Moreover,

$$\int\limits_{\Omega^{N,j}} \|h^{N,j} - h(\omega)\| P^{N,j}(d\omega) \leq \left( \sum_{i=1}^{m} \int\limits_{\Omega^{N,j}} \left(h_i^{N,j} - h_i(\omega)\right)^2 P^{N,j}(d\omega) \right). \tag{3.93b}$$

Clearly, if (3.10a,b) holds, then

$$\int\limits_{\Omega^{N,j}} \left\| \left(T(\omega) - T^{N,j}\right) x \right\| P^{N,j}(d\omega) \leq \left( x^T \left( \sum_{i=1}^m \operatorname{var}^{N,j} \left(T_i(\cdot)\right)\right) x \right)^{1/2} \tag{3.93c}$$

$$\int\limits_{\Omega^{N,j}} \|h^{N,j} - h(\omega)\| P^{N,j}(d\omega) \leq \left( \sum_{i=1}^m \operatorname{var}^{N,j} \left(h_i(\cdot)\right)\right)^{1/2}, \tag{3.93d}$$

where $\operatorname{var}^{N,j}\left(T_i(\cdot)\right)$ is the covariance matrix of the $i$-th row $T_i(\omega)$ of $T(\omega)$, and $\operatorname{var}^{N,j}\left(h_i(\cdot)\right)$ is the variance of the $i$-th component $h_i(\omega)$ of $h(\omega)$ with respect to the conditional distribution $P^{N,j} = P|_{\Omega^{N,j}}$.

*Remark 3.7.2.* For the general case given by (3.93a,b) we have that

$$\begin{aligned} \int\limits_{\Omega^{N,j}} \left(T_i(\omega) - T_i^{N,j}\right)^T \left(T_i(\omega) - T_i^{N,j}\right) P^{N,j}(d\omega) &= \operatorname{var}^{N,j}\left(T_i(\cdot)\right) \\ &+ \left(\overline{T}_i^{\Omega^{N,j}} - T_i^{N,j}\right)^T \left(\overline{T}^{\Omega^{N,j}} - T_i^{N,j}\right), \end{aligned} \tag{3.93e}$$

$$\int\limits_{\Omega^{N,j}} \left(h_i(\omega) - h_i^{N,j}\right)^2 P^{N,j}(d\omega) = \operatorname{var}^{N,j}\left(h_i(\cdot)\right) + \left(\overline{h}_i^{\Omega^{N,j}} - h_i^{N,j}\right)^2, \tag{3.93f}$$

respectively.

Because of (3.92c) and (3.93a,b) we still have to compute an upper estimate of $\|p\|$. Representing any vector $z \in \mathbb{R}^m$ by $z = \sum_{i=1}^m z_i e_i$, where $z_1, z_2, \ldots, z_m$ are the components of $z$ and $e_1, e_2, \ldots, e_n$ are the unit vectors of the $m$ coordinate directions, because of the sublinearity of $p$ we find

$$\begin{aligned} p(z) &= p\left(\sum_{i=1}^m z_i e_i\right) \leq \sum_{i=1}^m p(z_i e_i) = \sum_{i=1}^m p\left((z_i^+ - z_i^-)e_i\right) \\ &\leq \sum_{i=1}^m \left(p(z_i^+ e_i)\right) + p\left(z_i^-(-e_i)\right) = \sum_{i=1}^m \left(z_i^+ p(e_i) + z_i^- p(-e_i)\right) \\ &\leq \sum_{i=1}^m (z_i^+ + z_i^-) \max\left\{p(e_i), p(-e_i)\right\} = \sum_{i=1}^m \pi_i |z_i|, \end{aligned} \tag{3.94a}$$

where

$$\pi_i = \max\left\{p(e_i), p(-e_i)\right\}, i = 1, 2, \ldots, m. \tag{3.94b}$$

Note that for the computation of the $m$ coefficients $\pi_1, \pi_2, \ldots, \pi_m$ we have to solve the following $2m$ linear programs

$$\min q^T y \text{ s.t. } Wy = \pm e_i, y \geq 0 \text{ for } i = 1, 2, \ldots, m. \tag{3.94c}$$

From (3.94a) we obtain

$$p(z) \le \|\pi\| \cdot \|z\|, \tag{3.94d}$$

where $\|\pi\|, \|z\|$ denote the Euclidean norm of $\pi = (\pi_1, \pi_2, \ldots, \pi_m)^T$ and $z$, respectively. Thus, according to (3.85f) we have the upper norm bound

$$\|p\| \le \|\pi\| = \left(\sum_{i=1}^{m} \pi_i^2\right)^{1/2} = \left(\sum_{i=1}^{m} \left(\max\left\{p(e_i), p(-e_i)\right\}\right)^2\right)^{1/2}, \tag{3.95}$$

where $\|\pi\|$ **can be calculated from the given data**, cf. (3.83c), (3.94c).

Summarizing the above considerations, from (3.92c), (3.93a-f), (3.95) we get:

**Theorem 3.7.1.** *If $\left(T^N(\omega), h^N(\omega)\right)$ is given by (3.89a,b) and (3.84a,b) holds, then for each $x \in \mathbb{R}^n$ we have that*

$$0 \le \vartheta^N(x), \eta^N(x) \le \|\pi\| \sum_{j=1}^{r_N} P(\Omega^{N,j}) \left( \|x\| \left( \sum_{i,k=1}^{m} \int_{\Omega^{N,j}} \left(t_{ik}(\omega) - t_{ik}^{N,j}\right)^2 P^{N,j}(d\omega)\right)\right)^{1/2}$$
$$+ \left( \sum_{i=1}^{m} \int_{\Omega_{N,j}} \left(h_i(\omega) - h_i^{N,j}\right)^2 P^{N,j}(d\omega)\right) \tag{3.96a}$$

Moreover, if equations (3.91a,b) hold, then

$$0 \le \vartheta^N(x), \eta^N(x) \le \|\pi\| \textstyle\sum_{j=1}^{r_N} P(\Omega^{N,j}) \left( \|x\| \left(\sum_{i,k=1}^{m} \text{var}^{N,j}\left(t_{ik}(\cdot)\right)\right)^{1/2} \right. \tag{3.96b}$$
$$\left. + \left(\textstyle\sum_{i=1}^{m} \text{var}^{N,j}\left(h_i(\cdot)\right)\right)^{1/2}\right).$$

**Special representations of the random matrix $\left(T(\cdot), h(\cdot)\right)$.** A common, well-known representation of $\left(T(\omega), h(\omega)\right)$ is given by

$$\left(T(\omega), h(\omega)\right) = \left(T\left(\xi(\omega)\right), h\left(\xi(\omega)\right)\right) = (\overline{T}, \overline{h}) + \sum_{s=1}^{L} \xi_s(\omega)\left(T^{(s)}, h^{(s)}\right), \tag{3.97}$$

where $(\overline{T}, \overline{h})$ denotes the mean of $\left(T(\omega), h(\omega)\right)$, $\left(T^{(s)}, h^{(s)}\right)$, $s = 1, 2, \ldots, L$, are given $m \times (n+1)$ matrices, and $\xi_1(\omega), \xi_2(\omega), \ldots, \xi_L$, are zero-mean, stochastically independent random variables.

Based on representation (3.97), the approximation $\left(T^N(\omega), h^N(\omega)\right)$ of $\left(T(\omega), h(\omega)\right)$ according to (3.89a,b) can be described then by the following piecewise constant approximation $\xi^N(\omega)$ of the random $L$-vector $\xi(\omega) = \left(\xi_1(\omega), \xi_2(\omega), \ldots, \xi_L(\omega)\right)$: Let $\Xi$ denote the support of $\xi(\omega)$ or a set containing the support of $\xi(\omega)$. Moreover, let

$$\Xi^{N,1}, \Xi^{N,2}, \ldots, \Xi^{N,j}, \ldots, \Xi^{N,r_N} \tag{3.98a}$$

be the partition of $\Xi$ generated by the partition (3.89b) of $\Omega$, hence

$$\Xi^{N,j} = \left\{\xi(\omega) : \omega \in \Omega^{N,j}\right\}. \tag{3.98b}$$

Thus, with fixed $L$-vectors $\xi^{N,j} = \left(\xi_1^{N,j}, \ldots, \xi_L^{N,j}\right) \in \Xi^{N,j}$ we have that

$$\xi^N(\omega) = \xi^{N,j} \text{ for all } \omega \in \Omega^{N,j} \tag{3.98c}$$

and therefore

$$(T^{N,j}, h^{N,j}) = (\overline{T}, \overline{h}) + \sum_{s=1}^{L} \xi_s^{N,j}\left(T^{(s)}, h^{(s)}\right). \tag{3.98d}$$

Because of the properties of the random variables $\xi_1(\omega), \ldots, \xi_L(\omega)$ in the representation (3.97) of $\left(T(\omega), h(\omega)\right)$, we suppose that $E\xi_s^N(\omega) = 0$ and $E\xi_s^N(\omega)\xi_t^N(\omega) = 0$ for all $s = 1, \ldots, L, t = 1, \ldots, L, t \neq s$, hence

$$\sum_{j=1}^{r_N} \xi_s^{N,j} P_{\xi(\cdot)}(\Xi^{N,j}) = 0, \ \sum_{j=1}^{r_N} \xi_x^{N,j}\xi_t^{N,j} P_{\xi(\cdot)}\left(\Xi^{N,j}\right) = 0, s, t = 1, \ldots, L, t \neq s. \tag{3.98e}$$

In many cases we may suppose that

$$\Xi = \prod_{s=1}^{L} \Xi_s, \Xi_s = [\alpha_s, \beta_s), \tag{3.99a}$$

is a half-open $L$-dimensional interval. In this case $\Xi$ is then partitioned into certain subintervals $\Xi^{N,j}$. Hence, we set

$$j = (j_1, j_2, \ldots, j_L), r_N = (r_{N1}, r_{N2}, \ldots, r_{NL}), \tag{3.99b}$$

$$\xi^{N,j} = \left(\xi_1^{N,j_1}k, \xi_2^{N,j_2}, \ldots, \xi_L^{N,j_L}\right), \tag{3.99c}$$

where $j_s = 1, 2, \ldots, r_{N^s}, s = 1, \ldots, L$, and a cell $\Xi^{N,j}$ is given by

$$\Xi^{N,j} = \Xi^{N,(j_1,\ldots,j_L)} = \prod_{s=1}^{L} \Xi_s^{N,j_s} \tag{3.99d}$$

with certain half-open subintervals

$$\Xi_s^{N,j_s} = \left[\alpha_s^{N,j_s}, \beta_s^{N,j_s}\right), j_s = 1, 2, \ldots, r_{N^s}, s = 1, 2, \ldots, L \tag{3.99e}$$

with

$$\xi_s^{N,j_s} \in \left[\alpha_s^{N,j_s}, \beta_s^{N,j_s}\right), j_s = 1, 2, \ldots, r_{N^s}, s = 1, 2, \ldots, L. \tag{3.99f}$$

Moreover, (3.98d) reads in the present case

$$\begin{aligned}(T^{N,j},h^{N,j}) &= \left(T^{N,(j_1,\dots,j_L)},h^{N,(j_1,\dots,j_L)}\right) = \\ &= (\overline{T},\overline{h}) + \textstyle\sum_{s=1}^{L} \xi_s^{N,j_s}\left(T^{(s)},h^{(s)}\right)\end{aligned} \tag{3.99g}$$

For the integrals in the error estimation (3.96a), by (3.97), (3.98d), (3.99g) and the cell representation (3.99d), we get

$$\int\limits_{\Omega^{N,j}} \left(t_{ik}(\omega) - t_{ik}^{N,j}\right)^2 P^{N,j}(d\omega) = \frac{1}{P_{\xi(\cdot)}(\Xi^{N,j})} \int\limits_{\xi\in\Xi^{N,j}} \left(t_{ik}(\xi) - t_{ik}^{N,j}\right)^2 P_{\xi(\cdot)}(d\xi)$$

and therefore

$$\begin{aligned}\int\limits_{\Omega^{N,j}} \left(t_{ik}(\omega) - t_{ik}^{N,j}\right)^2 P^{N,j}(d\omega) &= \sum_{\substack{s,\sigma=1\\ s\neq\sigma}}^{L} t_{ik}^{(s)}t_{ik}^{(\sigma)}\left(\overline{\xi}_s^{\Xi_s^{N,j_s}} - \xi_s^{N,j_s}\right)\left(\overline{\xi}_\sigma^{\Xi_\sigma^{N,j_\sigma}} - \xi_\sigma^{N,j_\sigma}\right) \\ &+ \sum_{s=1}^{L} t_{ik}^{(s)2}\frac{1}{P_{\xi_s(\cdot)}(\Xi_s^{N,j_s})}\int\limits_{\xi_s\in\Xi_s^{N,j_s}} (\xi_s - \xi_s^{N,j_s})^2 P_{\xi_s(\cdot)}(d\xi_s)\end{aligned} \tag{3.100a}$$

as well as

$$\begin{aligned}\int\limits_{\Omega^{N,j}} \left(h_i(\omega) - h_i^{N,j}\right)^2 P^{N,j}(d\omega) &= \sum_{\substack{s,\sigma=1\\ s\neq\sigma}}^{L} h_i^{(s)}h_i^{(\sigma)}\left(\overline{\xi}_s^{\Xi_s^{N,j_s}} - \xi_s^{N,j_s}\right)\left(\overline{\xi}_\sigma^{\Xi_\sigma^{N,j_\sigma}} - \xi_\sigma^{N,j_\sigma}\right) \\ &+ \sum_{s=1}^{L} h_i^{(s)2}\frac{1}{P_{\xi_s(\cdot)}(\Xi_s^{N,j_s})}\int\limits_{\xi_s\in\Xi_s^{N,j_s}} (\xi_s - \xi_s^{N,j_s})^2 P_{\xi_s(\cdot)}(d\xi_s),\end{aligned} \tag{3.100b}$$

where

$$\overline{\xi}_s^{\Xi_s^{N,j_s}} = E(\xi_s|\Xi_s^{N,j_s}) := \frac{1}{P_{\xi_s(\cdot)}(\Xi_s^{N,j_s})}\int\limits_{\xi_s\in\Xi_s^{N,j_s}} \xi_s P_{\xi_s(\cdot)}(d\xi_s) \tag{3.100c}$$

is the conditional mean of $\xi_s(\omega)$ with respect to $\Xi_s^{N,j_s}$, cf. (3.90b,c).

Since $\xi_s^{N,j_s} \in \left[\alpha_s^{N,j_s}, \beta_s^{N,j_s}\right)$, cf. (3.99f), and $\bar{\xi}_s^{\Xi_s^{N,j_s}} \in \left[\alpha_s^{N,j_s}, \beta_s^{N,j_s}\right)$, the first terms in (3.100a,b) can be estimated from above as follows:

$$\left|\sum_{\substack{s,\sigma=1\\ s\neq\sigma}}^{L} t_{ik}^{(s)} t_{ik}^{(\sigma)} \left(\bar{\xi}_s^{\Xi_s^{N,j_s}} - \xi_s^{N,j_s}\right)\left(\bar{\xi}_\sigma^{\Xi_\sigma^{N,j_\sigma}} - \xi_\sigma^{N,j_\sigma}\right)\right| \leq$$
$$\sum_{\substack{s,\sigma=1\\ s\neq\sigma}}^{L} \left|t_{ik}^{(s)} t_{ik}^{(\sigma)}\right| \left(\beta_s^{N,j_s} - \alpha_s^{N,j_s}\right)\left(\beta_\sigma^{N,j_\sigma} - \alpha_\sigma^{N,j_\sigma}\right), \tag{3.100d}$$
$$\left|\sum_{s,\sigma=1}^{L} h_i^{(s)} h_i^{(\sigma)} \left(\bar{\xi}_s^{\Xi_s^{N,j_s}} - \xi_s^{N,j_s}\right)\left(\bar{\xi}_\sigma^{\Xi_\sigma^{N,j_\sigma}} - \xi_\sigma^{N,j_\sigma}\right)\right| \leq$$
$$\sum_{\substack{s,\sigma=1\\ s\neq\sigma}}^{L} \left|h_i^{(s)} h_i^{(\sigma)}\right| \left(\beta_s^{N,j_s} - \alpha_s^{N,j_s}\right)\left(\beta_\sigma^{N,j_\sigma} - \alpha_\sigma^{N,j_\sigma}\right). \tag{3.100e}$$

If the interval $\Xi_s = [\alpha_s, \beta_s)$ is partitioned into $r_{Ns}$ equidistant subintervals $[\alpha_s^{N,j_s}, \beta_s^{N,j_s}), j_s = 1, 2, \ldots, r_{Ns}$, then

$$\sum_{\substack{s,\sigma=1\\ s\neq\sigma}}^{L} \left|t_{ik}^{(s)} t_{ik}^{(\sigma)}\right| (\beta_s^{N,j_s} - \alpha_s^{N,j_s})\,(\beta_\sigma^{N,j_\sigma} - \alpha_\sigma^{N,j_\sigma}) = \frac{(\beta_s - \alpha_s)}{r_{Ns}}\frac{(\beta_\sigma - \alpha_\sigma)}{r_{N\sigma}} \sum_{\substack{s,\sigma=1\\ s\neq\sigma}}^{L} \left|t_{ik}^{(s)} t_{ik}^{(\sigma)}\right|, \tag{3.100f}$$
$$\sum_{\substack{s,\sigma=1\\ s\neq\sigma}}^{L} \left|h_i^{(s)} h_i^{(\sigma)}\right| \left(\beta_s^{N,j_s} - \alpha_s^{N,j_s}\right)\left(\beta_\sigma^{N,j_\sigma} - \alpha_\sigma^{N,j_\sigma}\right) = \frac{(\beta_s - \alpha_s)^2}{r_{Ns} r_{N\sigma}} \sum_{\substack{s,\sigma=1\\ s\neq\sigma}}^{L} \left|h_i^{(s)} h_i^{(\sigma)}\right|. \tag{3.100g}$$

If the values $\xi_s^{N,j_s}, j_s = 1, 2, \ldots, r_{Ns}, s = 1, 2, \ldots, L$, are selected such that

$$\xi_s^{N,j_s} = \bar{\xi}_s^{\Xi_s^{N,j_s}}, j_s = 1, 2, \ldots, r_{N}s, s = 1, 2, \ldots, L, \tag{3.101a}$$

see (3.91a,b), then (3.100a,b) is reduced to

$$\int_{\Omega^{N,j}} \left(t_{ik}(\omega) - t_{ik}^{N,j}\right)^2 P^{N,j}(d\omega) = \textstyle\sum_{s=1}^{L} t_{ik}^{(s)2} \operatorname{var}^{N,j_s}\left(\xi_s(\cdot)\right), \tag{3.101b}$$
$$\int_{\Omega^{N,j}} \left(h_i(\omega) - h_i^{N,j}\right)^2 P^{N,j}(d\omega) = \textstyle\sum_{s=1}^{L} h_i^{(s)2} \operatorname{var}^{N,j_s}\left(\xi_s(\cdot)\right), \tag{3.101c}$$

where

$$\operatorname{var}^{N,j_s}\left(\xi_s(\cdot)\right) = \frac{1}{P_{\xi_s(\cdot)}(\Xi_s^{N,j_s})} \int_{\xi_s \in \Xi_s^{N,j_s}} \left(\xi_s - \bar{\xi}_s^{\Xi_s^{N,j_s}}\right)^2 P_{\xi_s(\cdot)}(d\xi_s) \tag{3.101d}$$

is the conditional variance of $\xi_s(\omega)$ with respect to $\Xi_s^{N,j_s}$.

*Example 3.7.1.* Suppose that $\xi_s(\omega)$ has a density $f_s(z)$ such that for all $s = 1, 2, \ldots, L$ and $j_s = 1, 2, \ldots, r_{N,s}$ we have that

$$0 < f_{s,m}^{N,j_s} := \inf\Big\{f_s(z) : z \in \Xi_s^{N,j_s}\Big\} \le f_{s,M}^{N,j_s} := \sup\left\{f_s(z) : z \in \Xi_s^{N,j_s}\right\} < +\infty. \tag{3.102a}$$

This yields

$$P_{\xi_s(\cdot)}(\Xi_s^{N,j_s}) \ge f_{s,m}^{N,j_s}\Big(\beta_s^{N,j_s} - \alpha_s^{N,j_s}\Big), \tag{3.102b}$$

$$\int\limits_{\xi_s \in \Xi_s^{N,j_s}} \Big(\xi_s - \overline{\xi}_s^{\Xi_s^{N,j_s}}\Big)^2 P_{\xi_s(\cdot)}(d\xi_s) \le f_{s,M}^{N,j_s} \int\limits_{\alpha_s^{N,j_s}}^{\beta_s^{N,j_s}} \Big(\xi_s - \overline{\xi}_s^{\Xi_s^{N,j_s}}\Big)^2 d\xi_s \tag{3.102c}$$

and therefore

$$\begin{aligned} \mathrm{var}^{N,j_s}\Big(\xi_s(\cdot)\Big) &\le \frac{f_{s,M}^{N,j_s}}{f_{s,m}^{N,j_s}} \left(\frac{1}{\beta_s^{N,j_s} - \alpha_s^{N,j_s}} \int\limits_{\alpha_s^{N,j_s}}^{\beta_s^{N,j_s}} \Big(\xi_s - \frac{\alpha_s^{N,j_s} + \beta_s^{N,j_s}}{2}\Big)^2 d\xi_s \right. \\ &\qquad \left. + \Big(\frac{\alpha_s^{N,j_s} + \beta_s^{N,j_s}}{2} - \overline{\xi}_s^{\Xi_s^{N,j_s}}\Big)^2\right) \\ &= \frac{f_{s,M}^{N,j_s}}{f_{s,m}^{N,j_s}} \left(\frac{(\beta_s^{N,j_s} - \alpha_s^{N,j_s})^2}{12} + \Big(\frac{\alpha_s^{N,j_s} + \beta_s^{N,j_s}}{2} - \overline{\xi}_s^{\Xi_s^{N,j_s}}\Big)^2\right). \end{aligned} \tag{3.102d}$$

Since $\dfrac{\alpha_s^{N,j_s} + \beta_s^{N,j_s}}{2}$ and $\overline{\xi}_s^{\Xi_s^{N,j_s}}$ are elements of $\Xi_s^{N,j_s} = \Big[\alpha_s^{N,j_s}, \beta_s^{N,j_s}\Big)$, from (3.101d) and (3.102a-d) we obtain

$$\mathrm{var}^{N,j_s}\Big(\xi_s(\cdot)\Big) \le \frac{f_{s,M}^{N,j_s}}{f_{s,m}^{N,j_s}} \frac{13}{12} \Big(\beta_s^{N,j_s} - \alpha_s^{N,j_s}\Big)^2 \tag{3.103a}$$

If each $\Xi_s$ is partitioned into $r_{Ns}$ equidistant subintervals $\Xi_s^{N,j_s}, j_s = 1, 2, \ldots, r_{Ns}$, then $\beta_s^{N,j_s} - \alpha_s^{N,j_s} = \dfrac{1}{r_{Ns}}$ for all $j = 1, 2, \ldots, r_{Ns}$ and therefore

$$\mathrm{var}^{N,j_s}\Big(\xi_s(\cdot)\Big) \le \frac{13}{12} \frac{f_{s,M}^{N,j_s}}{f_{s,m}^{N,j_s}} \cdot \frac{1}{r_{Ns}^2}. \tag{3.103b}$$

### 3.7.3 Approximations of $F$ with a given Error Level $\varepsilon$

According to (3.88a-c) and (3.96a,b) we have that

$$\Big|F(x) - F^N(x)\Big| \le \|\pi\| \sum_{j=1}^{r_N} P(\Omega^{N,j}) \Big(\|x\| V^{N,j}\Big(T(\cdot)\Big) + V^{N,j}\Big(h(\cdot)\Big)\Big), \tag{3.104a}$$

with the estimation errors

$$V^{N,j}\Big(T(\cdot)\Big) = \left(\sum_{i,k=1}^{N} \int\limits_{\Omega^{N,j}} \Big(t_{ik}(\omega) - t_{ik}^{N,j}\Big)^2 P^{N,j}(d\omega)\right)^{1/2}, \tag{3.104b}$$

$$V^{N,j}\Big(h(\cdot)\Big) = \left(\sum_{i,k=1}^{N} \int\limits_{\Omega^{N,j}} \Big(h_i(\omega) - h_i^{N,j}\Big)^2 P^{N,j}(d\omega)\right)^{1/2}. \tag{3.104c}$$

Knowing that $D$ is bounded, hence

$$D \subset \Big\{x \in \mathbb{R}^n : \|x\| \le \rho_0\Big\} \tag{3.105a}$$

for some $\rho_0 > 0$, from (3.104a) we obviously get

$$|F^* - F^{N*}| \le \|\pi\| \sum_{j=1}^{r_N} P(\Omega^{N,j}) \left(\rho_0 V^{N,j}\Big(T(\cdot)\Big) + V^{N,j}\Big(h(\cdot)\Big)\right), \tag{3.105b}$$

where $F^*$ is the optimal value of (3.83a-c) and $F^{N*}$ the optimal value of the approximating problem

$$\min F^N(x) \text{ s.t. } x \in D, \tag{3.105c}$$

which can be represented by a **linear program (LP)**.

Furthermore, if it is known that there is an optimal solution $x^*$ of (3.83a-c) such that with some $p \ge 1$ we have that

$$\|x^*\|_p \le \rho_{0p} \text{ for some given } \rho_{0p} > 0, \tag{3.106a}$$

where $\|x\|_p$ is the $p$-norm of $x$, then again (3.88a-c) and (3.96a,b) yield

$$\Big|F^* - F_{\rho_0}^{N*}| \le \|\pi\| \sum_{j=1}^{r_N} P(\Omega^{N,j}) \left(\rho_0 V^{N,j}\Big(T(\cdot)\Big) + V^{N,j}\Big(h(\cdot)\Big)\right), \tag{3.106b}$$

where $\rho_0 := \rho_{0p} \max\{\|u\| : \|u\|_p \le 1\}$, and $F_{\rho_0}^{N*}$ is the optimal value of the approximation

$$\min F^N(x) \text{ s.t. } x \in D, \|x\|_p \le \rho_{0p}. \tag{3.106c}$$

Note that if $F^N(x)$ is generated by a discretization process of $P_{\left(T(\cdot),h(\cdot)\right)}$, and $p = 1$ or $p = 1$ or $p = +\infty$, then (3.106c) can again be represented by a **linear program**.

The above considerations yield now the following result:

**Theorem 3.7.2.** *Selecting the approximation* $\Big(T^N(\omega), h^N(\omega)\Big)$ *of* $\Big(T(\omega), h(\omega)\Big)$ *such that*

$$\|\pi\| \sum_{j=1}^{r_N} P(\Omega^{N,j}) \left(\rho_0 V^{N,j}\Big(T(\cdot)\Big) + V^{N,j}\Big(h(\cdot)\Big)\right) \le \varepsilon \tag{3.107a}$$

*where $\varepsilon > 0$ is an a priori given error bound, then in cases (3.105a) and (3.106a) we have the* **a priori error bound**

$$|F^* - F^{N*}| < \varepsilon,\ |F^* - F^{N*}_{\rho_0}| < \varepsilon \tag{3.107b}$$

*respectively.*

While (3.105a) is a simple property of the convex polyhedron $D$ which may hold or not, the relation (3.106a) is more involved.

### 3.7.4 Norm Bounds for Optimal Solutions of (3.83a-c)

For finding upper norm bounds $\rho_0$ for an optimal solution $x^*$ of (3.83a-c) we first have to study the growth properties of $F$. These can be obtained if for the loss function $p$, see (3.83c), appropriate lower bounds can be derived.

Using condition (3.84a,b), where we assume that $q$ has components

$$q_1 > 0, q_2 > 0, \ldots, q_\mu > 0, \tag{3.108a}$$

we define now the closed, convex polyhedron $K$ by

$$K = \text{conv}\left\{\frac{1}{q_1}w_1, \frac{1}{q_2}w_2, \ldots, \frac{1}{q_\mu}w_\mu\right\}, \tag{3.108b}$$

where $q_k > 0, k = 1, \ldots, \mu$, are the components of $q$ and $w_1, w_2, \ldots, w_\mu$ are the columns of $W$, where we may assume - without any restrictions - that $w_k \neq 0$ for all $k = 1, \ldots, \mu$. According to Marti (1972, 1975) we know that in this situation the loss function $p$ has the representation

$$p(z) = \inf\left\{\lambda > 0 : \frac{z}{\lambda} \in K\right\}, z \in \mathbb{R}^m. \tag{3.109}$$

Having (3.108a,b) and defining

$$q_0 = \max_{1\leq k\leq\mu} \frac{1}{q_k}\|w_k\|, \tag{3.110a}$$

where $q_0 > 0$, we find

$$\|z\| \leq q_0 \text{ for each } z \in K.$$

Since the relation $\frac{1}{\lambda}z \in K$ obviously implies that $\left\|\frac{1}{\lambda}z\right\| \leq q_0$, for the loss function $p$, having representation (3.109), for each $z \in \mathbb{R}^m$ we have that

$$p(z) \geq \inf\left\{\lambda > 0 : \left\|\frac{z}{\lambda}\right\| \leq q_0\right\} = \frac{1}{q_0}\|z\| = \left(\min_{1\leq k\leq\mu} \frac{q_k}{\|w_k\|}\right)\|z\|. \tag{3.110b}$$

If (3.110b) holds, then (3.86) yields

$$F(x) \geq \bar{c}^T x + \underline{p}E\|T(\omega)x - h(\omega)\|, \tag{3.111a}$$

where

$$\underline{p} = \min_{1 \le k \le \mu} \frac{q_k}{\|w_k\|}. \tag{3.111b}$$

In the following we assume that $\big(T(\omega), h(\omega)\big)$ is bounded w.p.1, hence, there is a constant $\Gamma > 0$ such that

$$\Big\|\big(T(\omega), h(\omega)\big)\Big\| \le \Gamma \text{ w.p. 1.} \tag{3.112}$$

Defining for any $x \in \mathbb{R}^n$

$$\hat{x} = (x^T, 1)^T, e_{\hat{x}} = \frac{\hat{x}}{\|\hat{x}\|},$$

we find

$$\|T(\omega)x - h(\omega)\| = \Big\|\big(T(\omega), h(\omega)\big)\hat{x}\Big\| = \|\hat{x}\| \cdot \Big\|\big(T(\omega), h(\omega\big)e_{\hat{x}}\Big\|;$$

furthermore, we have that

$$\begin{aligned}\Big\|\big(T(\omega), h(\omega)\big)e_{\hat{x}}\Big\|^2 &= \Big\|\big(T(\omega), h(\omega)\big)e_{\hat{x}}\Big\| \cdot \Big\|\big(T(\omega), h(\omega)\big)e_{\hat{x}}\Big\| \\ &\le \Big\|\big(T(\omega), h(\omega)\big)e_{\hat{x}}\Big\| \cdot \Gamma\end{aligned}$$

and therefore

$$\Big\|T(\omega)x - h(\omega)\Big\| \le \frac{1}{\Gamma}\|\hat{x}\| \cdot \Big\|\big(T(\omega), h(\omega)\big)e_{\hat{x}}\Big\|^2, \tag{3.113a}$$

see (3.111e). Taking expectations on both sides of (3.113a), we get

$$E\|T(\omega)x - h(\omega)\| \ge \frac{1}{\Gamma}\|\hat{x}\|e_{\hat{x}}^T E\big(T(\omega), h(\omega)\big)^T\big(T(\omega), h(\omega)\big)e_{\hat{x}}. \tag{3.113b}$$

Denoting by $\lambda_{\min}(Q)$ the minimal eigenvalue of any symmetric matrix $Q$, from (3.113b) we obtain

$$\begin{aligned}E\|T(\omega)x - h(\omega)\| &\ge \tfrac{1}{\Gamma}\|\hat{x}\|\lambda_{\min}\left(E\big(T(\omega), h(\omega)\big)^T\big(T(\omega), h(\omega)\big)\right) \\ &\ge \tfrac{1}{\Gamma}\|\hat{x}\|E\lambda_{\min}\left(\big(T(\omega), h(\omega)\big)^T\big(T(\omega), h(\omega)\big)\right).\end{aligned} \tag{3.113c}$$

Note that

$$E\big(T(\omega), h(\omega)\big)^T\big(T(\omega), h(\omega)\big) = \sum_{i=1}^m \Big(\text{var}\big(T_i(\cdot), h_i(\cdot)\big) + (\overline{T}_i, \overline{h}_i)^T(\overline{T}_i, \overline{h}_i)\Big), \tag{3.113d}$$

where $\text{var}\big(T_i(\cdot), h_i(\cdot)\big)$ designates the covariance matrix of the $i$-th row $(T_i, h_i)$ of $\big(T(\omega), h(\omega)\big)$. If the random matrix $\big(T(\omega), h(\omega)\big)$ is represented by (3.97), then

$$E\big(T(\omega), h(\omega)\big)^T\big(T(\omega), h(\omega)\big) = \overline{T}^T\overline{T} + \sum_{s=1}^L \text{var}\,(\xi_s)T^{(s)^T}T^{(s)}, \tag{3.113e}$$

hence, this matrix can be computed easily. Let then $\lambda_0$ be defined by

$$\lambda_0 := \lambda_{\min}\left(E\Big(T(\omega),h(\omega)\Big)^T\Big(T(\omega),h(\omega)\Big)\right) \text{ or}$$
$$\lambda_0 = E\lambda_{\min}\Big(T(\omega),h(\omega)\Big)^T\Big(T(\omega),h(\omega)\Big). \tag{3.114}$$

Summarizing the above considerations, according to (111a) and (113b) we find the following result:

**Lemma 3.7.2.** *Suppose that conditions (3.84a,b) and (3.112) hold true. If $\underline{p}, \Gamma, \lambda_0$ are defined by (3.111b), (3.112), (3.114), resp., then*

$$F(x) \geq \bar{c}^T x + \frac{\underline{p}\lambda_0}{\Gamma}\|\hat{x}\| \text{ for all } x \in \mathbb{R}^n. \tag{3.115}$$

Consider now any element $x^0$ of the feasible domain $D$ of (3.83a). If $x^*$ is an optimal solution of (3.83a-c), then (3.115) and (3.94d) yield the following inequalities

$$\begin{aligned}\bar{c}^T x^* + \tfrac{\underline{p}\lambda_0}{\Gamma}\|\widehat{x^*}\| &\leq F(x^*) \leq F(x^0) = \bar{c}^T x^0 + \|\pi\| E\|T(\omega)x^0 - h(\omega)\| \\ &\leq F^0 := \bar{c}^T x^0 + \|\pi\|\left(\widehat{x^0} E\Big(T(\omega),h(\omega)\Big)^T\Big(T(\omega),h(\omega)\Big)\widehat{x^0}\right)^{1/2}\end{aligned}, \tag{3.116}$$

where the last inequality is guaranteed by the concavity of the function $z \to \sqrt{z}$. Note that the upper bound $F^0$ can be computed easily, see (3.95), (3.113e). Since $\|\widehat{x^*}\| = \left(1 + \|x^*\|^2\right)^{1/2}$, we have now the following norm bounds for optimal solutions $x^*$ of (3.83a-c).

**Theorem 3.7.3.** *Suppose that the assumptions of Lemma 3.7.2. hold, and let $x^0$ be any feasible solution of (3.83a-c). Moreover, let $F^0$ be defined as in (3.116).*

a) If $\underline{c}^T x \geq 0$ for all $x \in D$, then for any optimal solution $x^*$ of (3.83a-c) we have that

$$\|x^*\| \leq \left(\left(\frac{\Gamma}{\underline{p}\lambda_0 - \Gamma\|\underline{c}\|}F(x^0)\right)^2 - 1\right)^{1/2} \leq \left(\left(\frac{\Gamma F^0}{\underline{p}\lambda_0}\right)^2 - 1\right)^{1/2}. \tag{3.117a}$$

b) If $\|\bar{c}\| < \frac{\underline{p}\lambda_0}{\Gamma}$, then for any optimal solution $x^*$ of (3.83) it holds

$$\|x^*\| \leq \left(\left(\frac{F(x^0)\Gamma}{\underline{p}\lambda_0 - \Gamma\|\bar{c}\|}\right)^2 - 1\right)^{1/2} \leq \left(\left(\frac{F^0\Gamma}{\underline{p}\lambda_0 - \Gamma\|\bar{c}\|}\right)^2 - 1\right)^{1/2}. \tag{3.117b}$$

Proof. a) Here, from (3.115) and (3.116) we get

$$F^0 \geq F(x^0) \geq \bar{c}^T x^* + \frac{\underline{p}\lambda_0}{\Gamma}\|\widehat{x^*}\| \geq \frac{\underline{p}\lambda_0}{\Gamma}\left(1 + \|x^*\|^2\right)^{1/2}$$

which yields the first assertion (3.117a). b) The next assertion (3.117b) follows from

$$F^0 \geq F(x^0) \geq \bar{c}^T x^* + \frac{p\lambda_0}{\Gamma}\|\widehat{x^*}\| \geq -\|\bar{c}\| \cdot \|x^*\| + \frac{p\lambda_0}{\Gamma}\|\widehat{x^*}\|$$
$$\geq \left(-\|\bar{c}\| + \frac{p\lambda_0}{\Gamma}\right)(1 + \|x^*\|^2)^{1/2}.$$

### 3.7.5 Invariant Discretizations

According to Theorem 3.7.1., (3.104a-c), there is a large variety of possible discretizations $(T^N, h^N)$ of $\left(T(\omega), h(\omega)\right)$ guaranteeing a certain given a priori error bound, see (3.107b). Hence, the problem is to find discretizations taking into consideration the special structure of the underlying problem (Marti, 1979). A main idea in stochastic linear programming with recourse is the use of special refining strategies for refining the partitions $\Xi^{N,1}, \dots, \Xi^{N,r_N}$ of $\Xi$, see (3.98a-e), such that only cells $\Xi^{N,j}$ are further partitioned which contribute most to the increase of the accuracy of approximation (see Kall, 1976 and Kall and Wallace, 1994).

Very often the probability distribution $P_{\left(T(\cdot),h(\cdot)\right)}$ has certain symmetry or invariance properties, cf. Marti (1985,1988). Not destroying these invariance properties during the discretization process, in several cases descent discretions can be constructed very easily.

Considering the approximation $\left(T^N(\omega), h^N(\omega)\right)$, given by (3.89a,b) or (3.98a-e), we define $\left(T_0^N(\omega), h_0^N(\omega)\right)$ by

$$\left(T_0^N(\omega), h_0^N(\omega)\right) := \left(T^N(\omega) - \overline{T}^N, h^N(\omega) - \overline{h}^N\right), \tag{3.118a}$$

where $\left(\overline{T}^N, \overline{h}^N\right)$ is the mean of $\left(T^N(\omega), h^N(\omega)\right)$. Using the results of (Marti, 1992), we define the distribution invariance as follows, where the set $\mathcal{B}_\alpha$ of $r_N \times r_N$ matrices $B = (b_{ij})$ is given by

$$\mathcal{B}_\alpha = \{B : 1^T B = 1^T, B\alpha = \alpha, B \geq 0\}. \tag{3.118b}$$

Here, 1 denotes the $r_N$-vector $1 = (1, 1, \dots, 1)$, $\alpha$ is the $r_N$-vector

$$\alpha = \left(P(\Omega^{N,1}), P(\Omega^{N,2}), \dots, P(\Omega^{N,r_N})\right)^T \tag{3.118c}$$

or

$$\alpha = \left(P_{\xi(\cdot)}(\Xi^{N,1}), P_{\xi(\cdot)}(\Xi^{N,2}), \dots, P_{\xi(\cdot)}(\Xi^{N,r_N})\right)^T \tag{3.118d}$$

and $B \geq 0$ means that $b_{ij} \geq 0$ for all elements $b_{ij}$ of $B$.

The distribution $P_{\left(T_0^N(\cdot),h_0^N(\cdot)\right)}$ of $\left(T_0^N(\omega),h_0^N(\omega)\right)$ is called **invariant** if there is a matrix $B \in \mathcal{B}_\alpha$ and an $n \times n$ matrix $C$ such that for each row $i = 1,2,\ldots,m$ we have that

$$B^T \begin{pmatrix} T_{0,i}^{N,1} \\ T_{0,i}^{N,2} \\ \vdots \\ T_{0,i}^{N,r_N} \end{pmatrix} = \begin{pmatrix} T_{0,i}^{N,1} \\ T_{0,i}^{N,2} \\ \vdots \\ T_{0,i}^{N,r_N} \end{pmatrix} C \tag{3.119a}$$

$$B^T \begin{pmatrix} h_{0,i}^{N,1} \\ h_{0,i}^{N,2} \\ \vdots \\ T_{0,i}^{N,r_N} \end{pmatrix} = \begin{pmatrix} h_{0,i}^{N,1} \\ h_{0,i}^{N,2} \\ \vdots \\ h_{0,i}^{N,r_N} \end{pmatrix}. \tag{3.119b}$$

For the general case, we have to introduce some more notations:
Let denote $\tilde{z}$ the $(1+m)$-vector

$$\tilde{z} = \begin{pmatrix} t \\ z \end{pmatrix} \text{ with } t \in \mathbb{R}, z \in \mathbb{R}^m, \tag{3.120a}$$

where we set $\tilde{z} = (\tilde{z}_0, \tilde{z}_1, \ldots, \tilde{z}_m)$ with $\tilde{z}_0 = t, \tilde{z}_i = z_i, i = 1,\ldots,m_i$ furthermore, let $u(\tilde{z})$ denote the total loss function

$$u(\tilde{z}) = t + p(z) \tag{3.120b}$$

of (3.83a-c).

Obviously, the total loss function $u = u(\tilde{z})$ is monotoneous nondecreasing with respect to the component $z_0 = t$. In many cases the loss function $p$ itself has some (partial) monotonicity properties, see, e.g., Marti (1988). Hence, supposing in the following - for example - that $p$ is also **partically nondecreasing**, we have a subset $J \subset \{0,1,\ldots,m\}$ with $0 \in J$ and a corresponding partition

$$\tilde{z} = \begin{pmatrix} \tilde{z}_I \\ \tilde{z}_{II} \end{pmatrix}, \tilde{z}_I = (\tilde{z}_i)_{i\in J}, \tilde{z}_{II} = (z_i)_{i\notin J} \tag{3.121a}$$

of $\tilde{z}$ into two subvectors $\tilde{z}_I, \tilde{z}_{II}$ such that for any vectors $\tilde{z}, \tilde{w} \in \mathbb{R}^{1+m}$ the following relation hold:

$$\tilde{z}_I \le \tilde{w}_I, \tilde{z}_{II} = \tilde{w}_{II} \Rightarrow u(\tilde{z}) \le u(\tilde{w}), \tag{3.121b}$$

where $\tilde{z}_I \le \tilde{w}_I$ means that $z_i \le w_i$ for all $i \in J$. Of course, in many cases we have also this sharper condition

$$\tilde{z}_I \le \tilde{w}_I, \tilde{z}_{II} = \tilde{w}_{II}, \tilde{z}_i < \tilde{w}_i \text{ for at least one } i \in J \Rightarrow u(\tilde{z}) < u(\tilde{w}). \tag{3.121c}$$

Based on the above definitions, the invariance of an arbitrary distribution $P_{\left(A^N(\cdot),b^N(\cdot)\right)}$ with

$$\left(A^N(\omega), b^N(\omega)\right) = \begin{pmatrix} \bar{c}^T & 0 \\ T^N(\omega) & h^N(\omega) \end{pmatrix} \tag{3.122}$$

is stated as follows, where the following inclusion is still assumed:

$$D \subset \mathbb{R}^n_+. \tag{3.123}$$

**Definition 3.7.1.** The probability distribution $P_{\left(A^N(\cdot),b^N(\cdot)\right)}$ of $\left(A^N(\omega), b^N(\omega)\right)$ is called **invariant** it there is a matrix $B \in \mathcal{B}_\alpha$ and an $n \times n$ matrix $C$ such that the following relations hold:

i)

$$\bar{c} \geq \bar{c}^T C \tag{3.124a}$$

ii)

$$\overline{T}_I \leq \overline{T}_I C \tag{3.124b}$$

$$\overline{T}_{II} = \overline{T}_{II} C \tag{3.124c}$$

iii)

(3.119a) and (3.119b) are fulfilled, (3.124d)

where $\overline{T}_I, \overline{T}_{II}$, respectively, is the matrix containing the rows $\overline{T}_i$ with $i \in J, i \notin J$, respectively.

*Remark 3.7.2.* Note that condition (3.124d), hence, relations (3.119a) and (3.119b) can be interpreted as **conditions for the discretization** of the distribution of the centralized random matrix $\left(T_0(\omega), h_0(\omega)\right) = \left(T(\omega) - \overline{T}, h(\omega) - \overline{h}\right)$, where $(\overline{T}, \overline{h})$ is the mean of $\left(T(\omega), h(\omega)\right)$.

The significance of the above invariance concept follows from the following result, cf. Marti (1992).

**Theorem 3.7.4.** *Suppose that $D \subset \mathbb{R}^n_+$. If $\left(A^N(\cdot), b^N(\cdot)\right)$ has an invariant distribution with matrices $B \in \mathcal{B}_\alpha, C$ according to Definition 3.7.1., then*

*a) $F^N(y) \leq F^N(x)$ with $y := Cx$ for every $x \in \mathbb{R}^n$*
*b) $h = y - x$ is a descent direction for $F^N$ at $x$, provided that only $F^N$ is not constant on the line segment $xy$ joining $x$ and $y \neq x$.*

As an important consequence of Theorem 3.7.4. we find the following result:

**Corollary 3.7.1.** *Assume that $P_{\left(A^N(\cdot),b^N(\cdot)\right)}$ is invariant with matrices $B \in \mathcal{B}_\alpha, C$ according to Defintion 3.7.1 . Furthermore, suppose that $F^N(x)$ is not constant on each line segment $xy$ in D. If $x^*$ is an optimal solution of the approximating problem (3.105c), then*

$$Cx^* = x^* \text{ or}$$
$$h = Cx^* - x^* \text{ is not a feasible direction for } D \text{ at } x^*.$$

**Note.** Corollary 3.7.1. holds also under weaker conditions concerning $F^N$.

**Acknowledgement**
The author thanks Mrs. Elisabeth Lößl, UniBw München, for providing the electronic version of the manuscript.

## References


Augusti G., Baratta A. and Casciati F. (1984). Probabilistic Methods in Structural Engineering. London/New York: Chapmann and Hall.

Biles W.E. and Swain J.J. (1979). Mathematical Programming and the Optimization of Computer Simulations. *Math. Programming Studies* 11: 189–207.

Birge J.R. (1997). Stochastic Programming computation and Applications. *INFORMS Journal of Computing* 9 (2): 111-133.

Box G.E.P. and Draper N.R. (1987). Empirical Model-Building and Response Surfaces. New York: Wiley.

Box G.E.P. and Wilson K.G. (1951). On the experimental attainment of optimum conditions. *Journal of the Royal Statistical Soc.* 13: 1-45.

Breitung K. (1995). Asymptotic Approximations for Normal Probability Integrals. In Marti K. and Kall P., eds., *Stochastic Programming: Numerical and Engineering Applications*. Lecture Notes in Economics and Mathematical Systems Vol. 423, Berlin/Heidelberg: Springer-Verlag. 1–7.

Charnes A., Lemke C.E. and Zienkiewicz O.C. (1959). Virtual work, linear programming and plastic limit analysis. *Proceedings of the Royal Soc. A* 25: 110-116.

Cocchetti G. and Maier G. (1988). Static shakedown theorems in piecewise linearized poroplasticity. *Archive of Applied Mechanics* 68: 651-661.

Haftka R.T. and Kamat M.P. (1985). Elements of structural optimization. The Hague/Boston/Lancaster: Martinus Nijhoff Publishers/Kluwer.

Hartmann D. (2000). Modellierung von Informationssystemen. Homepage: http://www.inf.bi.ruhr–uni–bochum.de. Bochum: Ruhr–Universität.

Haubach-Lippman C. (1998). Stochastische Strukturoptimierung flexibler Roboter. *Fortschrittberichte VDI, Reihe 8: Mess-, Steuerungs- und Regelungstechnik* Nr. 706. Düsseldorf: VDI-Verlag.

Hemp W.S. (1973). Optimum Strucutres. Oxford: Oxford University Press.

Heyman J. (1950/51). Plastic Design of Beams and Plane Frames for Minimum Material Consumption. *Quarterly Applied Mathematics* 8: 373-381.

Jacobsen Sh.H. and Schruben L.W. (1989). Techniques for Simulation Response Optimization. *Operations Research Letters* 9: 1-9.

Kall P. (1976). Stochastic Linear Programming. Berlin/Heidelberg/New York: Springer-Verlag.

Kall P. and Wallace S.W. (1994). Stochastic Programming. Chichester: J. Wiley.

Kirsch U. (1993). Structural Optimization. Berlin/Heidelberg/New York: Springer-Verlag.

König J.A. (1987). Shakedown of Elastic-Plastic Structures. Amsterdam, etc.: Elsevier.

Lawo M. (1987). Optimierung im kunstruktiven Ingenieurbau. Braunschweig/Wiesbaden: Vieweg.

Marti K. (1972). Entscheidungsprobleme mit linearem Aktionen- und Ergebnisraum. *Zeitschrift für Wahrscheinlichkeitstheorie verw. Geb.* 23: 133-147.

Marti K. (1975). Approximationen der Entscheidungsprobleme mit linearer Ergebnisfunktion und positiv homogener, subadditiver Verlustfunktion. *Zeitschrift für Wahrscheinlichkeitstheorie verw. Geb.* 31: 203-233.

Marti K. (1979). Diskretisierung stochastischer Programme unter Berücksichtigung der Problemstruktur. *Zeitung für Angewandte Mathematik und Mechanik* 59: T105-T108.

Marti K. (1979). Approximationen stochastischer Optimierungsprobleme. Königstein/Ts.: Verlag Anton Hain Meisenheim GmbH.

Marti K. (1985). Computation of Descent Directions in Stochastic Optimization Problems with Invariant Distributions. *Zeitung für Angewandte Mathematik und Mechanik* 65: 355-378.

Marti K. (1988). Descent Directions and Efficient Solutions in Disretely Distributed Stochastic Programs. *Lecture Notes in Economics and Mathematical Systems* Vol. 299.Berlin/Heidelberg/New York: Springer-Verlag.

Marti K. (1992). Computation of Efficient Solutions of Discretely Distributed Stochastic Optimization Problems. *Mathematical Methods of Operations Research* 36: 259-294.

Marti K. (1997). Approximation and derivatives of probabilities of survival in structural analysis and design. *Structural Optimization* 13: 230-243.

Marti K. (1997). Solving Stochastic Structural Optimization Problems by RSM-Based Stochastic Approximation Methods - Gradient estimation in case of intermediate variables. *Mathematical Methods of Operations Research* 46 (3): 409-434.

Marti K. (1998). Optimal Design of Trusses as a Stochastic Linear Programming Problem. In Nowak, A.S., ed., *Reliability and Optimization of Structural Systems.* Ann Arbor, Michigan: University of Michigan Press. 231–239.

Mayer J. (1998). Stochastic Linear Programming Algorithms. Gordon and Breach.

MSC/NASTRAN (1998). MSC/NASTRAN Module CMG. Munich: MacNeal-Schwendler Corp. Munich.

Myers R.H. (1971). Response Surface Methodology. Boston: Allyn-Bacon.

Nielsen M.P. (1984). Limit Anlaysis and Concrete Plasticity. New Jersey: Prentice Hall, Englewood Cliffs.

Park S.H. (1996). Robust design and analysis for quality engineering. London: Chapman and Hall.

Phadke M.S. (1989). Quality Engineering Using Robust Design – Robuste Prozesse durch Quality Engineering. München: Prentice Hall/gfmt (Gesellschaft für Mangagementtechniken und Technologietraining).

Prager W. and Shield R.T. (1967). A General Theory of Optimal Plastic Design. *Journal Applied Mechanics* 34 (1): 184-186.

Reinhart J. (1997). *Stochastische Optimierung von Faserkunststoffverbundplatten. Adaptive stochastische Approximationsverfahren auf der Basis der Response-Surface-Methode.* Fortschrittberichte VDI, Reihe 5: Grund- und Werkstoffe, Nr. 463. Düsseldorf: VDI-Verlag.

Ross Ph.J. (1988). Taguchi Techniques for Quality Engineering. New York, etc.: McGraw Hill Book Comp.

Rozvany G.I.N. (1989). Structural Design via Optimality Criteria. Dordrecht/Boston/London: Kluwer Academic Publisher.

Rozvany G.I.N. (1997). Topology Optimization in Strucutral Mechanics. Wien/New York: Springer-Verlag.

Ryan Th.P. (2000). Statistical Methods for Quality Improvement. New York, etc.: Wiley.

Schüler H. (1997). *Zur Analyse und Bemessung adaptiver Tragwerke aus Stahlbeton unter dynamischen Einwirkungen.* Dissertation Bauhaus-Universität Wien.

Schuëller G.I. and Gasser M. (1998). Some Basic Principles of Reliability-Based Optimization (RBO) of Structures and Mechanical Components. In Marti K. and Kall P., eds., *Stochastic Programming - Numerical Techniques and Engineering Applications.* Lecture Notes in Economics and Mathematical Systems Vol. 458, Berlin/Heidelberg: Springer-Verlag. 80–103.

Smith D.L. and Munro J. (1976). Plastic Analysis and Synthesis of Frames subject to Multiple Loadings. *Engineering Optimization* 2: 145-157.

Spillers W.R. (1972). Automated structural analysis: An introduction. New York/Toronto/Oxford: Pergamon.

Steffen P.N. (1996). *Elastoplastische Dimensionierung von Stahlbetonplatten mittels Finiter Bemessungselemente und Linearer Programmierung.* Dissertation ETH Zürich, Nr. 11611.

Thoft–Christensen P. and Murotsu Y. (1986). Application of structural systems reliability theory. Berlin a.O.: Springer-Verlag.

Tin–Loi F. (1990). On the optimal plastic synthesis of frames. *Engineering Optimization* 16: 91–108.

Tin–Loi F. (1995). Plastic limit analysis of plane frames and grids using GAMS. *Computers & Structures*54: 15-25.

Zeman P. and Irvine H.M. (1986). Plastic Design. An imposed hinge-rotation approach. London/Boston/Sydney: Allen & Unwin.

# CHAPTER 4

# Response Surface Approximations for Engineering Optimization

A.J.G. Schoofs and J.J.M. Rijpkema

Department of Mechanical Engineering, Eindhoven University of Technology, The Netherlands

**Abstract.** In these lecture notes attention is focused on the use of Response Surface Models as approximation models in engineering optimization. Common strategies are discussed for efficient model construction, based on principles from statistical experimental design theory. Furthermore, modifications of the experimental design theory will be treated, which are necessary and useful on behalf of numerical experimental designs. Finally, guidelines are presented for building and application of response surface models, based on numerical computations.

## 4.1 Introduction

### 4.1.1 Structural Optimization and Experimental Design

The research field of structural optimization is concerned with the design of structures which can meet requirements better, e.g. with respect to the resistance of loads applied to them, or with respect to their performance. Designing consists of an iterative procedure of analysis and synthesis. By means of analysis the behavior of the structure is evaluated, while synthesis is used to modify the structure in a way that it probably meets stated demands better. It is assumed that during the optimization process the behavior and the performance can be described uniquely by a set of $n$ design variables $x_1$, $x_2$, ..., $x_n$. The criterion by which the structure is judged to be better or worse than another one, is described by the so-called objective function and of course also depends on the design variables.

Modern research on numerical structural optimization was started during and after World War II in the aerospace industry due to the need for lightweight aircraft. During the last few decades optimization research has become important in almost any field of structural engineering. As reasons for this development one can see the diminishing energy and material resources, environmental problems concerning air pollution and noise, and last but not least the rapid development of computing facilities which have become a common tool of almost any engineer.

For the analytical task in a structural optimization problem mostly the finite element method (FEM) is used for two good reasons. The first reason is its nice and flexible modeling facilities, by which very different structures can be modeled in essentially the same way. Secondly there is the accuracy of a FEM analysis which can easily be controlled by means of the used element grid. However, FEM analyses of actual engineering structures are often very time-consuming, yielding a serious drawback for application of the method during structural optimization. Therefore, a direct coupling between analysis models and optimization routines may be very inefficient, as during optimization a large number of iterative calls to possibly time-consuming analysis models may be necessary. In those situations it is preferred to uncouple analysis and optimization through the use of so called metamodels or surrogate models: fast to evaluate approximations for objective and constraint functions. These metamodels may also be helpful in large-scale multidisciplinary optimization (MDO), to exchange relevant information between the disciplines involved and to study the effect of design changes on coupled subproblems.

Additionally to Response Surface Approximations (Myers et al., 1995; Schoofs et al., 1992) there exist several other strategies to construct metamodels. Examples are kriging (Sacks et al., 1989; Koehler et al., 1996), Radial Basis Functions (Powell, 1987), Multivariate Adaptive Regression Splines (Friedman, 1991), Smoothing Splines (Lancaster, 1986) and Neural Networks (Bishop, 1995). They all estimate the response for a specific design on the basis of information from the full analysis in a limited number of so called training designs. However, they differ with respect to their underlying conceptual ideas, the calculation effort needed for training and the applicability to specific situations, such as large-scale optimization problems or analysis models based on numerical simulations. Through numerical model studies one tries to find guidelines for their proper use and selection (e.g.: Giunta et al., 1998; Jin et al., 2000; Schoofs et al., 1997).

In structural optimization problems the structure itself is subject of modifications, which is the reason why these problems are often essentially non-linear, and a closed-form solution for the optimum design seldom can be found. As already stated, optimization problems are mostly solved iteratively. In Subsection 4.1.3 we will briefly describe the process.

Experimental design, in the literature also referred to as "Design and Analysis of Experiments", or "Planning of Experiments", provides methods for both the formulation of measurement programs and the analysis of the measurement results. We emphasize that by "design" here is meant the design of a measurement program.

In the scientific research cycle the following stages appear:

- Theory and model building,
- Prediction,
- Experimentation,
- Confrontation of prediction and experiment.

Model modification or refinement may follow the last stage and the cycle is repeated until a satisfactory model is obtained. In many cases experimentation is time and cost consuming and the need for well-organized experiments is clear. The experimental design theory provides tools for this organization.

In conjunction with optimization we are interested in factorial experiments which are designed to study some relationship, for instance:

$$y = f(x_1, x_2, \ldots, x_n) \tag{4.1}$$

showing a response variable $y$ as a function of $n$ independent variables $x_1, x_2, ..., x_n$. In the experimental design theory the variables $x_i$ are called "factors", hence the name "factorial experiments".

In the design and analysis of experiments the following questions have to be resolved:

- Which factors play a role?
- Should the factors be used in their original form, or should they be transformed or coded first?
- Which is the range of interest for these factors?
- Which form of the function $f$ may be suitable to describe the searched relationship?
- In order to run the experiments, discrete values have to be chosen for the factors; how many discrete values, so-called levels, for each factor are needed?
- Making a choice for a certain level of each factor represents a discrete design point in the space spanned by the factors; how many and which design points, in other words which measurement program, should be chosen?
- How can the relationship be estimated, and how can it be tested?

A very valuable aspect of the experimental design theory is that it guides experimentation in a structured way. When the founder of the experimental design theory, R.A. Fisher, in 1919 was appointed as a statistician at the Rothamsted Agriculture Research Station, he found experimental data over a period of 70 years. It proved that no confidential conclusions could be drawn from all these data. For Fisher it was the motive to develop the experimental design theory, the basis for modern experimentation. Answering the above questions is not possible all at once and it is certainly unwise trying to do so. Those questions constitute a circular problem: for instance to define the measurement program, the type of relationship has to be known, but the relationship can only be estimated after a measurement program has been defined and the experimentation done. The solution to this problem is to start with a preliminary measurement program of moderate extent. Analysis of the data emerged from this program will provide better answers to the mentioned questions and a more adequate measurement program can be defined. Box et al. (1978) give a 25% rule, which states that at the outset of an experimental investigation, not more than 25% of the experimental effort (budget) should be invested in a first measurement program. The whole procedure is repeated until finally all questions have been resolved in a satisfactory way.

In Section 4.2 of these lecture notes we will discuss parts of the experimental design theory in more detail.

### 4.1.2 Objectives of this Chapter

In the present chapter we will consider experimental designs in which the experiments are of numerical nature, by applying finite element analyses. We will investigate the derivation of regression models and discuss the combination of experimental design and structural optimization, because we find these methods have much in common:

- The set of design variables in structural optimization plays the same role as the factors in experimental designs.

- In both methods the elementary process steps, that is in experimental design the measurements and in optimization the finite element analyses, are cost and time consuming and therefore their number has to be minimized.
- A finite element analyses of a structure can be considered as a numerical experiment.
- Computer programs for structural optimization using the design variable concept are readily suited to do the "experiments" (finite element analyses) as indicated by the experimental design.
- A regression model can serve as a fast analysis module in a computer program for optimization.
- Such a fast analysis module allows us to perform a large number of runs of the optimization program, every time emerging from a different starting point in the design space in order to find the global optimum for the structure. It is also possible to investigate outcomes of the optimization procedure using a number of different objective functions without being obliged to repeat a lot of finite element calculations.
- By applying a regression model an approximating optimum design can be found; this design can be used as initial design in an optimization based on direct FEM-analyses, thus resulting in a more accurate global optimum design.

In these lecture notes we will give procedural guidelines for the use of experimental designs in model building and structural optimization problems.

### 4.1.3 Problem Formulation

Numerical optimization of a mechanical structure requires the availability of a mathematical model of the structure. Such a model is characterized by a finite number of parameters. These model parameters may be of widely divergent types, such as parameters to specify physical properties, geometrical parameters and topological parameters to specify connections between structural components. The describing equations of the mathematical model can be derived in different ways, for instance, using a finite element method. For a linear, elastic, dynamically loaded structure this results in:

$$\boldsymbol{M}\ddot{\boldsymbol{u}} + \boldsymbol{B}\dot{\boldsymbol{u}} + \boldsymbol{K}\boldsymbol{u} = \boldsymbol{p} \tag{4.2}$$

and an appropriate set of initial conditions. Here $\boldsymbol{M}$ is the mass matrix, $\boldsymbol{B}$ the damping matrix, $\boldsymbol{K}$ the stiffness matrix, $\boldsymbol{u}$ the column of unknown nodal displacements and $\boldsymbol{p}$ the column of external, prescribed nodal loads.

In a straightforward design all model parameters are chosen a priori, for instance based on prior investigations or on experience. In structural optimization, however, all or part of these parameters may depend on a finite number of a priori unknown design variables, which have to be determined during the optimization process. Examples of design variables are the radius of a fillet, the cross-sectional area of a beam, the number of stiffeners on a panel and the thickness of a plate. In these lecture notes only scalar design variables are taken into account. They are denoted by $x_1, x_2, \ldots, x_n$ ($n \geq 1$) and are considered the components of a column $\boldsymbol{x}$:

$$\boldsymbol{x}^T = [x_1 \ x_2 \ \ldots \ x_n] \tag{4.3}$$

The design variables span an $n$-dimensional space, called the "design space" $\boldsymbol{D}$. It should be emphasized that a design is represented by a point $\boldsymbol{x}$ in $\boldsymbol{D}$, even if the design is patently absurd (e.g. negative areas) or inadequate.

A design variable $x_i$ may be discrete or continuous. In the second case, which occurs for instance if $x_i$ is the radius of a fillet, $x_i$ may be any real number in a given interval. The first case occurs, for instance, if $x_i$ is the distance between equally spaced stiffeners on a given panel. A discrete design variable may be of type integer as will be the case if $x_i$ is the number of stiffeners on a panel. Often discrete variables are treated as continuous ones during the optimization process. Then the final calculated value is rounded off to the nearest discrete value a posteriori. In these lecture notes only continuous design variables are taken into account.

In general, design variables are subject to constraints. Here two types of constraints are distinguished:

**Design-constraints** are imposed on design variables for reasons of functionality, production, aesthetics etc. Examples are maximum track of a motorcar, minimum diameter of a hole, maximum and minimum slopes of a roof, etc. A very important characteristic of these constraints is that they can readily be evaluated without the need to solve the mathematical model of the structure.

Design constraints are explicit and/or simple, usually linear implicit relations in x. They can be represented by a set of $md$ ($md \geq 0$) inequalities

$$dci_j(\boldsymbol{x}) \leq 0, \qquad j = 1, 2, \ \ldots, \ md \tag{4.4}$$

and a set of $ld$ ($ld \geq 0$) equalities

$$dce_j(\boldsymbol{x}) = 0, \qquad j = 1, 2, \ \ldots, \ ld \tag{4.5}$$

The equality constraints (4.5) can be used to reduce the number of design variables, such that (4.5) is satisfied identically for the reduced set of variables. However, this is not always recommendable from the viewpoint of a general problem definition,

The constraints (4.4) and (4.5) define a subspace of the design space $\boldsymbol{D}$, which will be called the "$d$-constrained design space" $\boldsymbol{D_d}$. A design $\boldsymbol{x} \in \boldsymbol{D_d}$ is called a "$d$-constrained design" and it is assumed that such a design can be realized physically. The space $\boldsymbol{D_d}$ is of great importance in optimization algorithms because it is a proper subset of $\boldsymbol{D}$: the design constraints reduce the set of points $\boldsymbol{x}$ that might be of interest.

**Behavior constraints** are derived from requirements on the performance or the behavior of the structure. Typical examples are limitations on maximum stresses and displacements, on eigenfrequencies and vibration modes etc. In general these constraints are nonlinear even for linear structures. Elaboration of the behavior constraints requires the solution of the mathematical model of the structure. This is a characteristic and very important difference between behavior and design constraints. Mathematically behavior constraints are represented by a set of $mb$ ($mb \geq 0$) inequalities

$$bci_j(\boldsymbol{x}) \leq 0, \qquad j = 1, 2, \ldots, mb \tag{4.6}$$

and a set of $lb$ ($lb \geq 0$) equalities

$$bce_j(\boldsymbol{x}) = 0, \qquad j = 1, 2, \ldots, lb \tag{4.7}$$

Analogous to (4.5) the equalities (4.7) could be used to reduce the number of design variables. However, often the functions in (4.7) are complicated and implicit, and elimination of design variables may be practically impossible. Sometimes the model equations (4.2) are also incorporated in (4.7). Here (4.2) and (4.7) will be considered as separate sets of equations.

The constraints (4.6) and (4.7) define a subspace $\boldsymbol{D_b}$ of the design space $\boldsymbol{D}$. This subspace will be called the "$b$-constrained design space". The set of all designs $\boldsymbol{x}$ that satisfy both the design constraints and the behavior constraints represents the "feasible design space" $\boldsymbol{D_f}$. A design $\boldsymbol{x} \in \boldsymbol{D_f}$ is called a "feasible design". Mathematically $\boldsymbol{D_f}$ is defined as the intersection of the $d$- and the $b$-constrained design spaces

$$\boldsymbol{D_f} = \boldsymbol{D_d} \cap \boldsymbol{D_b} \tag{4.8}$$

The statement "some designs are better than others" implies the existence of a measure to compare designs. In mathematical optimization this measure is the objective function $F$: $\boldsymbol{D_d} \to \Re$, which is defined on the $d$-constrained design space $\boldsymbol{D_d}$. In many optimization formulations the objective function $F$ is required to be semi-positive definite, i.e.

$$F(\boldsymbol{x}) \geq 0, \ \ \forall\, \boldsymbol{x} \in \boldsymbol{D_d} \tag{4.9}$$

A design $\boldsymbol{x^*} \in \boldsymbol{D_f}$ is considered to be a better design than $\boldsymbol{x} \in \boldsymbol{D_f}$ if and only if $F(\boldsymbol{x^*}) < F(\boldsymbol{x})$. It will be clear that the outcome of the optimization process heavily depends on the choice of the objective function. Therefore this choice should be made with the utmost care. In some cases an obvious objective function exists. For example, in shape optimization of a fillet to reduce the stress concentration factor, it is trivial to choose $F(\boldsymbol{x})$ equal to the maximum Von Mises stress $\sigma_{max}$ in the fillet. For transport systems often $F(\boldsymbol{x})$ is chosen as the weight of the system since many design aspects of those systems are closely related to the weight. In other cases the objective function will be chosen as a weighted sum of functions which can represent totally different aspects, like costs of production and of exploitation, safety and environmental aspects. In such situations choosing an appropriate objective function is not trivial. A related problem occurs in situations where one tries to bring everything into account by means of terms in the objective function. If there is but little consensus about the importance of a certain aspect, that aspect should not be incorporated. The only result would be an increased "noise level" of the objective function.

The type of objective function may influence the flow diagram of the optimization process. It is assumed that $F(\boldsymbol{x})$ is a computable function for each $\boldsymbol{x} \in \boldsymbol{D_d}$. In some problems evaluation of $F(\boldsymbol{x})$ may require a complete elaboration of the mathematical model, for instance in the earlier mentioned case of stress optimization of a fillet. If, on the other hand, the structural weight is chosen as an objective function $F(\boldsymbol{x})$, evaluation of $F(\boldsymbol{x})$ can be done using just the design variables $\boldsymbol{x}$ describing a proposed new design. If the weight proves to be increased with respect to the current weight, analysis of the design $\boldsymbol{x}$ makes no sense. In such a situation the process should be stopped or another design $\boldsymbol{x}$ should be proposed.

In general the behavior constraints and the objective function are highly non-linear functions of the design variables and the optimization problem must be solved iteratively. Usually, structural optimization problems converge to a local optimal design. The proof that this design also is the global optimum design can seldom be delivered. Hence, it is common practice to restart the optimization process with another initial design, and repeat this process until confidence is gained that the global optimum solution has been approached closely enough.

Mathematically the optimization process can now be stated as follows:

$$\text{Find } \boldsymbol{x}^* \in \boldsymbol{D}_f \quad \text{such that } F(\boldsymbol{x}^*) \leq F(\boldsymbol{x}) \quad \text{for all } \boldsymbol{x} \in \boldsymbol{D}_f \tag{4.10}$$

After the previous discussions it will be clear that every phrase of this statement is essential in the optimization process. In other words, each of the steps

- Find a feasible design $\boldsymbol{x} \in \boldsymbol{D}_f$
- Find a (local) minimum for $F(\boldsymbol{x})$ and
- Get confidence that the solution is the global optimum,

generally are sub-problems of the same level of complexity.

### 4.1.4 Finite Element Method and Structural Optimization

In computer programs for structural optimization often the finite element method is used for modeling and analysis of the structure. This offers the possibility to model a great variety of structures in a general, accurate and flexible way. The accuracy can be controlled easily by an appropriate modification of the finite element model. In optimization problems often slight improvements of a structure are of interest, so a realistic model and accurate analysis results are of great importance. However, these advantages must be paid for with some serious drawbacks. First, the computing time is large. Second, the implementation of the design variable concept in finite element packages is non-standard and difficult. Last but not least, the computation of gradients of the objective function and of the behavior constraints with respect to design variables is not trivial and difficult to implement.

For complicated linear structures finite element analyses are straightforward but can be very time consuming. If the structural problem is nonlinear, the finite element analysis has an iterative character and computing times may become tremendous. Furthermore, due to the iterative character of the optimization process several finite element analyses are required to obtain an optimal solution. In general this will be a local optimum and it is therefore recommended to carry out several optimization runs. Hence, in problems where the finite element method is used, the number of iteration steps in the optimization process should be kept as small as possible.

A major problem in structural optimization is the link between design variables and the finite element model. In several more or less dedicated structural optimization programs this link has been realized. In some general-purpose finite element packages the design variable concept has been adopted only with the simplest types of variables, such as cross-sectional properties. The rapid development of computer graphics and solid modeling programs can become of great importance to overcome this shortcoming, because the input needed for an advanced solid modeling program is much more user oriented than a finite element model. Furthermore, the link between solid models and finite element models has already been established.

Implementation of the design variable concept in finite element packages will make computation of gradients much easier. If such an implementation is not realized, gradients can only be computed using numerical differentiation, at the cost of an increased number of finite element analyses. However, if the design variables can be used within the finite element program, then computation of gradients can be carried out in a much more efficient way. This can be illustrated by some examples.

First, consider a linear, elastic, static problem described by

$$\boldsymbol{K}\boldsymbol{u} = \boldsymbol{p} \tag{4.11}$$

Differentiation of (4.11) with respect to design variable $x_i$ yields

$$\boldsymbol{K}\frac{\partial \boldsymbol{u}}{\partial x_i} = \frac{\partial \boldsymbol{p}}{\partial x_i} - \frac{\partial \boldsymbol{K}}{\partial x_i}\boldsymbol{u} \tag{4.12}$$

The gradients on the right hand side can easily be calculated by a perturbation method. Assuming that the matrix $\boldsymbol{K}$ was decomposed during a previous analysis and that $\boldsymbol{u}$ is known from that analysis, the calculation of $\partial \boldsymbol{u}/\partial x_i$ is, in fact, straightforward.

Next a linear, elastic, undamped, dynamically loaded structure is considered. Then it can be shown that the gradients of eigenfrequencies are given by

$$\frac{\partial \omega_j}{\partial x_i} = \frac{\boldsymbol{v}_j^T[\frac{\partial \boldsymbol{K}}{\partial x_i} - \omega_j^2 \frac{\partial \boldsymbol{M}}{\partial x_i}]\boldsymbol{v}_j}{2\omega_j \boldsymbol{v}_j^T \boldsymbol{M} \boldsymbol{v}_j} \tag{4.13}$$

where $\boldsymbol{v}_j$ is the eigenvector coupled with the eigenfrequency $\omega_j$. Again the gradients of $\boldsymbol{K}$ and $\boldsymbol{M}$ can be calculated easily by finite differences. If eigenfrequencies and eigenvectors are available from a previous analysis, then evaluation of (4.13) is straightforward too.

### 4.1.5 The Iterative Optimization Procedure

The solution methods commonly used to obtain the optimum design may be divided into several categories. An important classification is the partition in specific and general methods.

Specific methods are used exclusively in structural optimization. They usually are categorized as optimality criterion methods (Haftka et al., 1992). In these methods one tries to satisfy a well-defined criterion and it is expected that by doing so some measure of the structure will become optimal. However, this measure is not used explicitly to control the optimization process. A typical example of an optimality criterion method is the well-known fully stressed design method. In this method one tries to match the stresses in the structural members to their allowable stresses (the criterion), with the implicit objective to minimize the weight of the structure. In the early stages of development of structural optimization specific methods enjoyed great popularity because they could solve special problems more efficiently than any general method. But the popularity of specific methods is decreasing because of their limitations.

General methods are those which are applicable and commonly are applied to optimization problems in several fields. They are based on linear and nonlinear mathematical programming methods, and efficient computer implementations have been developed. However, despite all powerful software it should be emphasized that good engineering intuition will remain of great importance in solving structural optimization problems.

### 4.1.6 General Optimization Algorithm

Let $\boldsymbol{x}_0 \in \boldsymbol{D}_d$ be a given initial set of design variables. This will not always be a feasible design. Some algorithms exist of two stages: in the first stage a feasible design $\boldsymbol{x} \in \boldsymbol{D}_f$ is determined from $\boldsymbol{x}_0 \in \boldsymbol{D}_d$ and in the second the design is optimized. The algorithm is given by:

begin Boolean CONV;
  {given: initial design $\boldsymbol{x}_0 \in \boldsymbol{D}_d$ , convergence criteria, and max. number of iteration steps $q_{max}$}
  $q:=0$ ;
  $\boldsymbol{x}_q:=\boldsymbol{x}_0$ ;
  CONV: = false
  while not CONV and $q < q_{max}$ do
  begin {analyze design $\boldsymbol{x}_q$ ;
    evaluate behaviour constraints;
    evaluate objective function}
    {evaluate convergence criteria}
    if converged then CONV: = true
    {compute $\alpha_q$, $s_q$}
    {compute $\boldsymbol{x}_{q+1}$}
    $q:=q+1$
    end
  end;

From the various steps in this algorithm only the optimization step will be considered now. In general the updating step can be formulated as:

$$x_{q+1} = x_q + \alpha_q \underline{s}_q \tag{4.14}$$

where $q$ is the iteration number and $s_q$ a search direction column while $\alpha_q$ is a scalar step size factor. So the optimization step (4.14) consists of two parts:

- Finding an appropriate search direction $s_q$. For this purpose the most effective optimization algorithms require partial derivatives of the objective function and of the constraints with respect to the design variables to be calculated.
- Computing the scalar $\alpha_q$ such, that, moving in the direction $s_q$, the objective function is minimized.

In mathematical programming packages a great number of techniques, both linear and non-linear, are available to solve optimization problems as described above. Treatment of such techniques is beyond the scope of this lecture.

## 4.2 Experimental Design Theory

### 4.2.1 Introduction: Survey of the Theory

Advanced scientific and technological research requires comprehensive and expensive experiments and the need for careful planning of the experiments is quite clear. On the one hand it is required to minimize the number of experiments, but on the other it is desired to gather as much information as possible about the relevant aspects of the system under consideration.

The experimental design theory (EDT) consists of two main parts. The first part, discussed in the Subsections 4.2.2, 4.2.3 and 4.2.4, concerns the planning of experiments and ends up with a list of experiments to be carried out. This list is called the experimental design, abbreviated to ED. In the second part the experimental results are analyzed and fitted to some mathematical relationship. This is discussed in Subsection 4.2.5.

In these lecture notes we are primarily interested in the use of (a special version of) EDT for the planning of numerical, FEM analyses. Such analyses can be regarded as numerical experiments: the investigator specifies a number of input parameters and as a result a number of response quantities emerge from the analysis program. The computations can be described as deterministic processes. Repeated computations using the same set of input parameters invariably result in exactly the same response quantity values. In general physical experiments show a stochastic character. Even if it were possible in physical experiments to adjust the input parameters to exactly the same values, repeated experiments always will show random variations of the response quantities. EDT has been developed for such stochastic processes. We conform as much as possible to the common theory for stochastic processes, which is discussed first in the next four sections. A separate section, Subsection 4.2.6, is devoted to the consequences of the use of EDT in deterministic processes. It turns out that much of the common theory may still be used.

In Subsection 4.2.4 special attention is given to the use of gradients in EDT. The use of gradients proves to be very advantageous in numerical experiments. In the remainder of this subsection some definitions are given and the problem of experimental design is formulated.

We presume that there exists a true physical relationship, $\boldsymbol{\kappa}$, between a response quantity $u$ and input quantities $\boldsymbol{\xi}$ via physical constants $\boldsymbol{\theta}$:

$$u = \boldsymbol{\kappa}(\boldsymbol{\xi}, \boldsymbol{\theta}) \tag{4.15}$$

If the true physical relationship is known, then we only need to estimate the physical constants $\boldsymbol{\theta}$. Often relation (4.15) is unknown or too complex and (4.15) is approximated locally by some interpolating function $g(\boldsymbol{x}, \boldsymbol{\beta})$:

$$u = g(\boldsymbol{x}, \boldsymbol{\beta}) \tag{4.16}$$

where $\boldsymbol{x}$ is a set of control variables obtained by coding or standardizing the input quantities $\boldsymbol{\xi}$, and $\boldsymbol{\beta}$ are the unknown coefficients in the interpolating function. The name "control variable" will be used throughout this chapter because it expresses very well our objectives: we want to control the behavior of a system by means of manipulating the control variables. The concept of control variables is closely related to the concept of design variables discussed in Section 4.1.

The control variables are assumed to be mutually independent. They are denoted by $x_1, x_2, \ldots, x_n$ $(n \geq 1)$ and are considered the components of a column $\boldsymbol{x}$

$$\boldsymbol{x} = [x_1 \; x_2 \; \ldots \; x_n]^T \tag{4.17}$$

In these lecture notes only control variables will be considered which are continuous quantities. In EDT, however, it is common practice that each control variable is adjusted to a finite number of discrete values, called levels. A column $\boldsymbol{x}$ containing one level for each of the control variables is called a treatment or experimental run. The set of all possible treatments for which observations can be made (later on also called 'set of candidate points'), is denoted by $\Gamma$. The number of elements of $\Gamma$, i.e. the number of possible treatments, is denoted by $r$. In a so-called complete factorial experiment all treatments $\boldsymbol{x} \in \Gamma$ are considered, whereas in a fractional factorial experiment only treatments in a subset of $\Gamma$ are used.

If observations are repeated for a treatment $\boldsymbol{x}$ then the observed results will vary about $\mu$ due to the stochastic nature of the experiments. We shall assume a normal distribution for the observed result, which we will call $\underline{y}$. The underscore denotes that $y$ is a stochastic variable. The expected value of $\underline{y}$ equals $\mu$, that is, $E(\underline{y})=\mu$ .

Because of the stochastic nature of the experiments and because of the approximation of the true relationship by an interpolating function, two types of errors will occur:

1. Random error: $\underline{e} = \underline{y} - E(\underline{y})$ ,
2. Systematic error or bias: $E(\underline{y}) - g(\boldsymbol{x}, \boldsymbol{\beta})$ .

Thus, the following relation exists:

$$\underline{y} = g(\boldsymbol{x},\boldsymbol{\beta}) + \underline{e} + \text{bias} \tag{4.18}$$

We shall deal with bias in Subsection 4.2.3.6, where an experimental design technique is presented with which designs can be constructed that account for bias. For the present it is assumed that the interpolating function exactly matches the true physical relationship and only random errors exist:

$$\begin{aligned}\underline{y} &= g(\boldsymbol{x},\boldsymbol{\beta}) + \underline{e} \\ &= \boldsymbol{f}^T(\boldsymbol{x})\boldsymbol{\beta} + \underline{e} \\ &= \beta_1 f_1(\boldsymbol{x}) + \ldots + \beta_k f_k(\boldsymbol{x}) + \underline{e}\end{aligned} \tag{4.19}$$

where the components $f_1, f_2, \ldots, f_k$ of the column $\boldsymbol{f}$ are known, postulated, mutual independent functions of $\boldsymbol{x}$ while the components $\beta_1, \beta_2, \ldots, \beta_k$ of the column $\boldsymbol{\beta}$ are unknown parameters. Often $f_1(\boldsymbol{x})$ is chosen equal to 1 for all $\boldsymbol{x}$. The postulated functions $f_1, f_2, \ldots, f_k$ are called model functions or regressor variables. Relation (4.19) is the so-called regression equation. For the present only linear regression equations (i.e. linear in the components of $\boldsymbol{\beta}$) are considered. In Section 4.3 some comments on nonlinear regression are given.

The unknown parameters $\boldsymbol{\beta}$ have to be determined from the results of a series of $N$ observations characterized by the treatments $\boldsymbol{x}_1, \boldsymbol{x}_2, \ldots, \boldsymbol{x}_N$. All these treatments need not necessarily be distinct; observations may be repeated, i.e. carried out under the same treatment. It is the objective of the design of experiments to select the most suitable set of $N$ treatments from the set $\chi$ of all candidate treatments such that the obtained results for $\boldsymbol{\beta}$ are as accurate and reliable as possible for the given number $N$ ($N \geq 1$) of treatments.

For the treatments $\boldsymbol{x}_1, \boldsymbol{x}_2, \ldots, \boldsymbol{x}_N$ the sample model is:

$$\underline{\boldsymbol{y}} = \boldsymbol{X}\boldsymbol{\beta} + \underline{\boldsymbol{e}} \tag{4.20}$$

where $\underline{\boldsymbol{y}} = [\underline{y}_1 \; \underline{y}_2 \; \cdots \; \underline{y}_N]^T$ is the column of response quantities, $\underline{\boldsymbol{e}}$ is the column of errors and $\boldsymbol{X}$ is an $(N * k)$-matrix, the design matrix, which is given by:

$$\boldsymbol{X}^T = [\boldsymbol{f}(\boldsymbol{x}_1) \; \boldsymbol{f}(\boldsymbol{x}_2) \; \ldots \; \boldsymbol{f}(\boldsymbol{x}_N)] \tag{4.21}$$

Unbiased estimates for $\boldsymbol{\beta}$ can be computed if the following assumptions for $\underline{\boldsymbol{e}}$ are made:

1. The expected value $E(\underline{\boldsymbol{e}})$ is zero:

$$E(\underline{\boldsymbol{e}}) = \boldsymbol{0} \tag{4.22}$$

2. The variance of the responses and hence of the errors are the same and equal to $\sigma^2$ for all treatments:

$$E(\underline{e}\,\underline{e}^T) = V(\underline{e}) = \sigma^2 \boldsymbol{I} \qquad (4.23)$$

In general, the number of treatments in the experiment should exceed the number of the parameters $\boldsymbol{\beta}$. Then an estimate of $\boldsymbol{\beta}$ (denoted by $\hat{\boldsymbol{\beta}}$) can be computed using a least-squares technique. $\hat{\boldsymbol{\beta}}$ is calculated from the requirement that the residual sum of squares; $KS_r$ defined by

$$KS_r = (\boldsymbol{y} - \boldsymbol{X}\hat{\boldsymbol{\beta}})^T (\boldsymbol{y} - \boldsymbol{X}\hat{\boldsymbol{\beta}}) \qquad (4.24)$$

is minimal. Here $\boldsymbol{y}$ represents a column with $N$ measured response quantity values. This results in a set of $k$ so-called normal equations:

$$\boldsymbol{X}^T \boldsymbol{X}\hat{\boldsymbol{\beta}} = \boldsymbol{X}^T \boldsymbol{y} \qquad (4.25)$$

For each appropriate experimental design, i.e. each appropriate set of treatments $\boldsymbol{x}_1$, $\boldsymbol{x}_2$, ..., $\boldsymbol{x}_N$ the matrix $\boldsymbol{X}^T\boldsymbol{X}$ will be regular and hence positive definite. Therefore $\hat{\boldsymbol{\beta}}$ can be solved from (4.25) , yielding

$$\hat{\boldsymbol{\beta}} = (\boldsymbol{X}^T \boldsymbol{X})^{-1} \boldsymbol{X}^T \boldsymbol{y} \qquad (4.26)$$

For each treatment $\boldsymbol{x}$ estimates $\hat{y}$ of the response variable $y$ can be calculated from

$$\hat{y}(\boldsymbol{x}) = \boldsymbol{f}^T(\boldsymbol{x})\,\hat{\boldsymbol{\beta}} \qquad (4.27)$$

This relation represents an operational regression model of the mathematical model (4.16). A measure for the accuracy of the estimator $\underline{\hat{\boldsymbol{\beta}}}$ is the variance- covariance matrix $\boldsymbol{V}(\underline{\hat{\boldsymbol{\beta}}})$:

$$\boldsymbol{V}(\hat{\boldsymbol{\beta}}) = E((\hat{\boldsymbol{\beta}} - \boldsymbol{\beta})(\hat{\boldsymbol{\beta}} - \boldsymbol{\beta})^T) = \sigma^2 (\boldsymbol{X}^T \boldsymbol{X})^{-1} \qquad (4.28)$$

Furthermore, correlations between the elements of $\underline{\hat{\boldsymbol{\beta}}}$ are determined by the correlation coefficient matrix $\boldsymbol{\rho}$, the elements of which are given by

$$\rho_{ij} = \frac{V(\underline{\hat{\boldsymbol{\beta}}})_{ij}}{(V(\underline{\hat{\boldsymbol{\beta}}})_{ii} V(\underline{\hat{\boldsymbol{\beta}}})_{jj})^{1/2}} \qquad \text{for } i, j = 1, 2, ..., k \qquad (4.29)$$

For the response estimator $\hat{\underline{y}}(\boldsymbol{x})$ the variance $V(\hat{\underline{y}}(\boldsymbol{x}))$ is used as a measure for its accuracy. From (4.27) and (4.28) it follows:

$$V(\hat{\underline{y}}(\boldsymbol{x})) = \boldsymbol{f}^T(\boldsymbol{x})(\boldsymbol{X}^T\boldsymbol{X})^{-1}\boldsymbol{f}(\boldsymbol{x})\sigma^2 \tag{4.30}$$

Often the variance $\sigma^2$ of the measured responses is unknown. An estimate for $\sigma^2$ can be calculated from

$$\sigma^2 = \frac{1}{n-k}KS_r \tag{4.31}$$

Using this result estimates $\hat{V}(\hat{\underline{\beta}})$, $\rho$ and $\hat{V}(\hat{\underline{y}}(\boldsymbol{x}))$ for $V(\hat{\underline{\beta}})$, $\rho$ and $V(\hat{\underline{y}}(\boldsymbol{x}))$ can be calculated.

As stated before it is the objective of the design of experiments to select an optimal set of $N$ treatments from the set of candidate points, $\Gamma$, such that the obtained estimation $\hat{\boldsymbol{\beta}}$ for $\boldsymbol{\beta}$ is as accurate and reliable as possible. In order to distinguish between candidate points and points actually present in the experimental design, candidate points are denoted by $\chi_j$, $j = 1, 2, \ldots, r$, whereas design points are denoted by $\boldsymbol{x}_j$, $j = 1, 2, \ldots, N$. In a numerical experiment the selected treatments $\boldsymbol{x}_1, \boldsymbol{x}_2, \ldots, \boldsymbol{x}_N$ will all be different. In physical experiments however replicating observations may result in a more accurate estimation of $\boldsymbol{\beta}$. Given the number $N$ of treatments and given the set $\Gamma$ of candidate treatments the objective of the design of experiments is to determine for each candidate treatment $\chi_j \in \Gamma$, $j = 1, 2, \ldots, r$, the number $N_j$ $(N_j \geq 0)$ of observations such that $\hat{\boldsymbol{\beta}}$ is as accurate and reliable as possible and such that the total number of observations is equal to $N$, i.e.

$$N = \sum_{j=1}^{r} N_j \tag{4.32}$$

For this purpose several methods are available. We will treat two of these methods. In the next section a more or less classical method, resulting in so-called $2^n$-designs, is discussed. The optimal design theory is the subject of the Subsections 4.2.3 and 4.2.4.

### 4.2.2 The $2^n$ Factorial Design

#### 4.2.2.1 The Concept of $2^n$ Designs

The design of a factorial experiment, where each of the n control variables is varied on two levels, is called a $2^n$ factorial design. Such designs are very popular and are quite suitable to develop regression models. Furthermore, they are useful in optimal design methods since they can constitute a set of candidate treatments. There exists a wealth of literature on $2^n$ designs (e.g. Box et al., 1978; Montgomery, 1996) and only the main topics of the method are discussed here.

In $2^n$ factorial designs a special notation is adopted. The control variables, i.e. the factors, are indicated by capitals: *A, B, C* ...etc., and can be adjusted to two levels: "high" and "low". "High" and "low" are just names; their specific meaning must be properly defined. A treatment is indicated by a string of lowercase letters. This string only contains those characters that correspond to the factors at the "high" level. The observation of the response quantity for a given treatment is denoted by a string of the same lowercase letters as the string denoting that treatment. For example, observation *ab* is the result of the treatment *ab*, where the factors *A* and *B* are both at the "high" level. Fig. 4.1 illustrates the possible treatments in a complete $2^3$ experiment with the factors *A, B* and *C*. Treatment (1) in Fig. 4.1, represents the treatment with all factors at the "low" level.

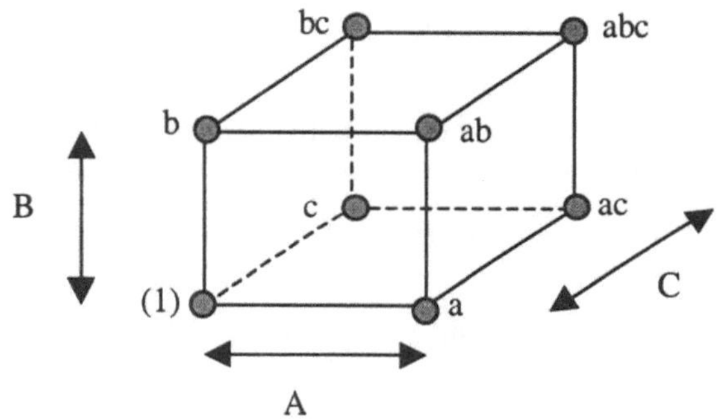


**Figure 4.1** Eight possible treatments in a $2^3$ experiment.

A complete $2^3$ experiment contains eight different treatments, which enables us to estimate up to a maximum of eight parameters in a regression equation, for instance:

$$y = \beta_1 + \beta_A x_A + \beta_B x_B + \beta_C x_C + \beta_{AB} x_A x_B + \beta_{AC} x_A x_C + \beta_{BC} x_B x_C + \beta_{ABC} x_A x_B x_C \tag{4.33}$$

The parameters are often called effects and are usually indicated by the same upper case characters as the factors: the effect $\beta_1$ is indicated by *I*, $\beta_A$ by *A* etc. Main effects are those effects which are indicated by just one letter, i.e. the effects *I, A, B* and *C* in the considered example. Effects indicated by more than one letter, for example *AB, AC*, etc., are called interactions or interactive effects.

In most cases it is possible to code the levels of the factors to +1 and -1. If for the $2^3$ experiment at every treatment one observation is made, the design matrix $\boldsymbol{X}$ is given by:

$$\begin{array}{c} \text{Effects} \qquad\qquad\qquad \text{Observations } \mathbf{y} \\ \begin{array}{cccccccc} I & A & B & C & AB & AC & BC & ABC \end{array} \end{array}$$

$$\mathbf{X} = \begin{bmatrix} +1 & -1 & -1 & -1 & +1 & +1 & +1 & -1 \\ +1 & +1 & -1 & -1 & -1 & -1 & +1 & +1 \\ +1 & -1 & +1 & -1 & -1 & +1 & -1 & +1 \\ +1 & -1 & -1 & +1 & +1 & -1 & -1 & +1 \\ +1 & +1 & +1 & -1 & +1 & -1 & -1 & -1 \\ +1 & +1 & -1 & +1 & -1 & +1 & -1 & -1 \\ +1 & -1 & +1 & +1 & -1 & -1 & +1 & -1 \\ +1 & +1 & +1 & +1 & +1 & +1 & +1 & +1 \end{bmatrix} \begin{bmatrix} (1) \\ a \\ b \\ c \\ ab \\ ac \\ bc \\ abc \end{bmatrix} \tag{4.34}$$

and the parameters$\beta_1$, through $\beta_{ABC}$ can be estimated using (4.26). Another algorithm to estimate a certain effect can be explained with the help of the above-mentioned example. For instance an estimate $\hat{AC}$ (= $\hat{\beta}_{AC}$) of the effect $AC$ can be calculated as follows:

1. Summing all the observations with $x_A * x_C = +1$ ("high"):
   $h = (1) + b + ac + abc$

2. Summing all the observations with $x_A * x_C = -1$ ("low"):
   $l = a + c + ab + bc$

3. Dividing the difference $h$-$l$ by the total number of observations, results in the estimate

$$AC = \frac{1}{8}\{((1) + b + ac + abc) - (a + c + ab + bc)\} \tag{4.35}$$

For the other effects similarly simple relations can be derived. The right hand side of (4.35) is called the contrast associated with the effect $AC$.

#### 4.2.2.2 Blocks and Fractional $2^n$ Experiments

In order to estimate all possible effects from a $2^n$ design a complete set of $2^n$ treatments should be used, and under each treatment at least one observation should be made. For increasing $n$ the measurement program will become impractical large and the number of measurements must be reduced. This is possible since in many problems the great majority of effects are interactions of little importance. To estimate the effects of interest usually much fewer observations are necessary than in a complete $2^n$ factorial design. To account for this a (small) number of chosen treatments are grouped in a so-called block and observations are made only for the treatments in this block. The crucial point in this procedure is the block definition, i.e. the choice of the treatments in the block. Since the number of observations is reduced, it is not possible to estimate all effects separately: some effects will be "confounded". The effects related to the block coincide with other effects and only the "sum" of those effects can be estimated. If observations are made only for one block, then the effects, which are confounded with the block effect, cannot be estimated (e.g. Box et al., 1978).

### 4.2.3 Optimal Design Theory

#### 4.2.3.1 Introduction

The central theme in the design of experiments is the formulation of experimental designs such that suitable regression models can be determined with a minimum of experimental effort. In the optimal design theory direct evaluation criteria are used: the accuracy of the estimated parameters itself and/or the variance in the estimated response and iterative algorithms are used to search for optimal designs (Atkinson, 1992). Optimal design techniques are useful in those situations where classic designs are unsuitable or unavailable, that is when:

- the experimental region is irregularly shaped due to constraints on the control variables,
- it is needed to augment or repair an existing design,
- the number of levels for the control variables varies considerably,
- designs have to be constructed for special models, i.e. other than polynomial models,
- designs have to be constructed for simultaneous observation of several responses.

Smith (1918) initiated the optimal design theory. Significant; progress was stimulated by the development of computers. Kiefer (1959) and Fedorov (1972) made important contributions to the methods. Nagtegaal (1987) gave some generalizations for the construction of experimental designs in case several regression models are being used simultaneously.

#### 4.2.3.2 Optimality Criteria

In the optimal design of experiments a number of $N$, not necessarily distinct treatments $\boldsymbol{x}_1, \boldsymbol{x}_2$ ..., $\boldsymbol{x}_N$ are determined from an a priori chosen and fixed set $\Gamma$ of $r$ discrete candidate treatments $\chi_1, \chi_2, \quad, \chi_r$. The control variables $x_1, x_2, ..., x_n$ are not allowed to vary continuously. It is assumed that somehow an appropriate choice for $\Gamma$ can be made. The main reason for this restriction is to simplify optimization algorithms, thus resulting in less computing time for the determination of optimal EDs. It should be emphasized that in the sequel optimality of designs is always subject to these restrictions. The outcomes of optimization algorithms depend on the choice of the set of candidate points. For each of the treatments $\boldsymbol{x}_1, \boldsymbol{x}_2$ ..., $\boldsymbol{x}_N$, often called "design points" or shortly "points", an observation will be made so these points define an ED and the objective is to optimize that ED. The problem is now to choose $N$ treatments from the $r$ candidate treatments, resulting in the best $N$-point design.

Experimental designs can be evaluated, using the variances of the parameter estimator $V(\underline{\hat{\beta}})$ or the variances of the response estimator $V(\underline{\hat{y}}(\boldsymbol{x}))$ as a measure, see (4.28) and (4.30) respectively. In both cases the quality of the ED is a function of the matrix $(\boldsymbol{X}^T\boldsymbol{X})^{-1}$ and the objective is to determine that ED among all possible $N$-point EDs which makes $(\boldsymbol{X}^T\boldsymbol{X})^{-1}$ minimal. However, the minimum of a matrix is not a well-defined concept and a number of operational criteria have been developed. The most important of these criteria are:

1. D-optimality, which is achieved if $\det(\boldsymbol{X}^T\boldsymbol{X})^{-1}$ is minimal. i.e. if the product of the eigenvalues of $(\boldsymbol{X}^T\boldsymbol{X})^{-1}$ is minimal.
2. A-optimality, which is achieved if $\mathrm{tr}(\boldsymbol{X}^T\boldsymbol{X})^{-1}$ is minimal, i.e. if the sum of the eigenvalues of $(\boldsymbol{X}^T\boldsymbol{X})^{-1}$ is minimal.
3. E-optimality, which is achieved if the largest eigenvalue of $(\boldsymbol{X}^T\boldsymbol{X})^{-1}$ is minimal.
4. G-optimality, which is achieved if the maximum over all candidate points of the estimated response variance $V(\underline{\hat{y}}(\boldsymbol{\chi}))$ is minimal.
5. V-optimality, which is achieved if the estimated response variance, averaged over all candidate points, $\frac{1}{r}\sum_{j=1}^{r} V(\underline{\hat{y}}(\boldsymbol{\chi}_j))$ is minimal.

#### 4.2.3.3 Definition of the Experimental Design

An $N$-point ED is completely specified by the choice of the candidate treatments and of the number $N_j$ $(N_j > 0)$ of observations under candidate treatment $\boldsymbol{\chi}_j$ $(j = 1, 2, \ldots, r)$, such that (4.32) holds. The numbers $N_1, N_2, \ldots, N_r$, are the elements of a column $\boldsymbol{N}$:

$$\boldsymbol{N} = [N_1 \ \ldots \ N_r\,]^T \tag{4.36}$$

For each of the numbers $N_1, \ldots, N_r$ upper and lower bounds can be specified, apart from the trivial condition $0 \le N_j \le N$ for $j = 1, 2, \ldots, r$. Some typical conditions are:

$N_j \le 1$ if point $\boldsymbol{\chi}_j$ is allowed only once ,
$1 \le N_j$ if point $\boldsymbol{\chi}_j$ should appear at least once,
$N_j^* \le N_j$ if an existing ED has to be augmented and already $N_j^*$ observations have been made in point $\boldsymbol{\chi}_j$.

In practice the number $r$ of candidate treatments and each of the numbers $N_1, \ldots, N_r$ are finite and integer. For numerical purposes, however, it may be advantageous to allow these numbers to be real. An ED with integer numbers is called an exact ED. Otherwise the ED is called continuous.

#### 4.2.3.4 The Information Matrix and Normalized Optimality Criteria

The matrix $\boldsymbol{X}^T\boldsymbol{X}$ can be normalized, resulting in the information matrix $\boldsymbol{M}$, where:

$$\boldsymbol{M}(\boldsymbol{N}) = \frac{1}{N}\sum_{j=1}^{r} N_j\,\boldsymbol{f}(\boldsymbol{\chi}_j)\boldsymbol{f}^T(\boldsymbol{\chi}_j) \tag{4.37}$$

In order to determine an optimal ED we try to find a column $\boldsymbol{N}^*$ for which a given criterion $C(\boldsymbol{N})$ is minimal and for which the constraints are not violated. This criterion may be any of the earlier given criteria. Written in terms of $M$ instead of $X^TX$ these criteria take the following forms :

$$C(N) = \det(M^{-1}(N)) \quad \text{for D-optimality} \tag{4.38}$$

$$C(N) = \mathrm{tr}(M^{-1}(N)) \quad \text{for A-optimality} \tag{4.39}$$

$$C(N) = \lambda_{max}(N) \quad \text{for E-optimality} \tag{4.40}$$

where $\lambda_{max}(N)$ is the largest eigenvalue of $M^{-1}(N)$

$$C(N) = \max_{j=1} d(\chi_j, N) \quad \text{for G-optimality} \tag{4.41}$$

$$C(N) = \frac{1}{r}\sum_{j=1}^{r} d(\chi_j, N) \quad \text{for V-optimality} \tag{4.42}$$

In the last two criteria $d(\chi, N)$ is the variance of the response estimator in the point $x$, normalized for $\sigma^2$/N:

$$d(\chi, N) = f^T(\chi)\, M^{-1}(N)\, f(\chi) \tag{4.43}$$

**4.2.3.5 Optimization Algorithms**

Four classes of optimization algorithms can be distinguished. The first class contains algorithms, which generate and evaluate all possible $N$-point designs. This is very time consuming and only feasible for small values of $N$. As a variant Welch (1982) developed a "branch and bound"-algorithm, in which a binary tree of minimization problems is generated. Not all designs are generated and evaluated. By exploiting bounds on the minimization only branches which might contain D-optimal designs are created. The algorithm guarantees global D-optimal designs, but computing costs are extremely high.

The second class of algorithms uses mathematical optimization techniques for a direct maximization of $\det(M(N))$, which is equivalent to minimization of $\det(M^{-1}(N))$. Due to the complexity of $\det(M(N))$ these methods are not successful.

The third class of algorithms neglects the integer character of the components of $N$ and, in consequence, results in a discrete ED. Subsequently this ED is rounded off to an exact ED. This may result in a good approximation if the number of points, $N$, is sufficiently higher than the number of parameters $k$. For this reason discrete designs are sometimes called "approximate designs". Such a design can be used as an initial design in exchange algorithms, see next paragraph. Algorithms for the construction of optimal discrete designs depend on the optimality criterion to be used (Fedorov, 1972).

Finally, the fourth class concerns exchange algorithms. Efficient and well-known is the algorithm called DETMAX (Mitchell, 1974). This algorithm starts with an initial $m$-point ED; the final goal is an optimal $N$-point ED. During each iteration step that candidate point, which results in the largest increase of det($\boldsymbol{M}$), is added to the design, and subsequently that point, which results in the smallest decrease of det($\boldsymbol{M}$), is removed from the design. The number $m$ of points in the initial design may be larger or smaller than $N$. If necessary, the algorithm first adds (if $m < N$) or rejects (if $m > N$) points until the number of points in the ED is equal to $N$. In order to avoid local optima the algorithm is able to perform "excursions", in which several points are added at one go and subsequently the number of points is reduced to $N$. If the resulting $N$-point ED has not been improved, another excursion will be made from the same initial design. If the excursion is successful the resulting ED will be used as starting ED in a further attempt to maximize det($\boldsymbol{M}$). The algorithm terminates when, after several excursions, no better ED is found. The algorithm generates high quality EDs against relatively low computing costs.

#### 4.2.3.6 Robust Experimental Designs

An ED is called "robust" if parameters and responses can be adequately estimated, even when the basic assumptions, for the construction of the ED, are violated to some extent. We might, for example assume that the considered response quantity is stochastic with a N(0, $\sigma^2$) distribution. In order to construct EDs we must specify a regression model in advance. However, the goodness of fit of the model can be checked only after the observations have been made. The model may prove to be biased and is said to be not exact.

Optimal EDs based on the criteria discussed earlier, show a dependency on the postulated regression model. They do not result in good estimates of the parameters when the real model differs significantly from the assumed model.

In order to account for bias, the so-called J-criterion can be used (Atkinson, 1992). This criterion is defined as the average mean squared error over all candidate points, normalized for $\sigma^2/N$

$$J = \frac{N}{\sigma^2}\frac{1}{r}\sum_{j=1}^{r} E[(\hat{\underline{y}}(\boldsymbol{\chi}_j) - y(\boldsymbol{\chi}_j))^2] \tag{4.44}$$

$J$ can be expressed as the sum of a variance and a bias term:

$$J = V + B, \tag{4.45}$$

where $V$ is the normalized average variance:

$$V = \frac{1}{r}\sum_{j=1}^{r} d(\boldsymbol{\chi}_j, N) = \frac{1}{r}\sum_{j=1}^{r} \boldsymbol{f}^T(\boldsymbol{\chi}_j)\,\boldsymbol{M}^{-1}(N)\,\boldsymbol{f}(\boldsymbol{\chi}_j) \tag{4.46}$$

and $B$ is the normalized average squared bias given by:

$$B = \frac{N}{\sigma^2}\frac{1}{r}\sum_{j=1}^{r} (E(\hat{\underline{y}}(\boldsymbol{\chi})) - y(\boldsymbol{\chi}))^2 \tag{4.47}$$

A problem is that $B$ depends on $\sigma^2$ and on the unknown real responses. However, Welch (1983) developed a method for the construction of robust designs based on the $J$-criterion. Nagtegaal (1987) generalized the method to the case of simultaneous observations of several response quantities.

Whether or not a postulated model is close to reality is seldom known beforehand and one may be tempted to use robust designs in each case. This is not recommended, because in experiments where the postulated model is close to reality a robust design would increase the variance much more than it would reduce the bias. In this case a robust design will result in worse parameter and response estimates, compared to designs based on one of the earlier mentioned criteria. Robust designs should be used only in those cases where the adequacy of the postulated model is seriously doubted.

A first choice of the regression model can be based on experience, on preliminary measurements or on theoretical considerations about the system under study. In general it is worthwhile to invest relatively much effort in such examinations.

### 4.2.4 Experimental Designs in Case of Simultaneous Observations of Several Quantities. Use of Gradients in Experimental Design Theory.

#### 4.2.4.1 Introduction

Sometimes more than one response quantity is relevant. Each of these quantities will have its own functional relationship with the control variables. Often they will be correlated, because they emerge from the same system.

A possible approach in the case of several responses would be to perform an experimental investigation for each response quantity separately. From economical point of view it is much more favorable to construct one optimal ED, which takes into account all response quantities simultaneously.

In Section 4.1 it was shown how partial derivatives of response quantities such as displacements and eigenfrequencies could be derived in FEM-formulations. Generalization to the case of several responses allows us to make use of these derivatives in experimental design and model building. The models for the partial derivatives of a response quantity with respect to the control variables follow by differentiation of the postulated model of that response quantity. In the following subsections some generalized definitions and optimization criteria are presented. Generalizations of optimization algorithms can be found in Fedorov (1972) (for D-optimality only) and Nagtegaal (1987).

#### 4.2.4.2 Definitions

Let $m$ be the number of response quantities $y_1, \ldots, y_m$ and let $\boldsymbol{u}$ be the column with these variables as elements:

$$\boldsymbol{u} = [y_1 \ y_2 \ \ldots \ y_m]^T \tag{4.48}$$

Since only linear models are used, response quantity $y_j$ ($j$ = 1, 2, ..., $m$) is assumed to depend linearly on $k_j$ model parameters, being the elements of a column $\boldsymbol{\beta}_j$. Then the model equations are given by:

$$y_j(\boldsymbol{x}) = \boldsymbol{f}_j^T(\boldsymbol{x})\,\boldsymbol{\beta}_j \qquad \text{for } j = 1, 2, \dots, m \tag{4.49}$$

where $\boldsymbol{f}_j(\boldsymbol{x})$ is a column whose elements are given functions, the so-called model functions or regressor variables. The postulated models (4.49) can be represented in a compact form as follows. The columns containing the model parameters $\boldsymbol{\beta}_1, \boldsymbol{\beta}_2, \dots, \boldsymbol{\beta}_m$, are filed in a column $\boldsymbol{\beta}$ given by:

$$\boldsymbol{\beta} = [\boldsymbol{\beta}_1^T \boldsymbol{\beta}_2^T \dots \boldsymbol{\beta}_m^T]^T \tag{4.50}$$

Column $\boldsymbol{\beta}$ contains $k$ elements, where $k$ is given by:

$$k = \sum_{j=1}^{m} k_j \tag{4.51}$$

Next, each column $\boldsymbol{f}_j(\boldsymbol{x})$ is extended with $m$-$1$ columns $z_i$ ($i$ = 1, 2, ..., $m$, $i \neq j$), each containing $k_i$ zeros, to a column $\boldsymbol{f}_j^*(\boldsymbol{x})$ given by:

$$\boldsymbol{f}_j^*(\boldsymbol{x}) = [z_1^T\ z_2^T \dots z_{j-1}^T \boldsymbol{f}_j^T(\boldsymbol{x}) z_{j+1}^T \dots z_m^T]^T \tag{4.52}$$

Using these definitions the model equations (4.49) can be written as:

$$\boldsymbol{u}(\boldsymbol{x}) = \boldsymbol{F}^T(\boldsymbol{x})\boldsymbol{\beta} \tag{4.53}$$

where $\boldsymbol{F}(\boldsymbol{x})$ is a ($k$*$m$) matrix, such that column $j$ of $\boldsymbol{F}(\boldsymbol{x})$ is equal to $\boldsymbol{f}_j^*(\boldsymbol{x})$:

$$\boldsymbol{F}(\boldsymbol{x}) = [\boldsymbol{f}_1^*(\boldsymbol{x})\ \boldsymbol{f}_2^*(\boldsymbol{x}) \dots \boldsymbol{f}_m^*(\boldsymbol{x})] \tag{4.54}$$

In this subsection our attention is focused on the special case in which elements of $\boldsymbol{u}(\boldsymbol{x})$, say $y_i(\boldsymbol{x})$, $1 \leq i \leq m$-$n$, are response quantities, while the elements $y_{i+1}, y_{i+2}, \dots y_{i+n}$, are partial derivatives of $y_i(\boldsymbol{x})$ with respect to the elements of $\boldsymbol{x}$. Then only the column $\boldsymbol{f}_i(\boldsymbol{x})$ must be specified, since the columns $\boldsymbol{f}_{i+1}(\boldsymbol{x}) \dots \boldsymbol{f}_{i+n}(\boldsymbol{x})$ follow by differentiation of $\boldsymbol{f}_i(\boldsymbol{x})$ with respect to the elements of $\boldsymbol{x}$. In addition all the columns $\boldsymbol{f}_i(\boldsymbol{x}), \boldsymbol{f}_{i+1}(\boldsymbol{x}) \dots \boldsymbol{f}_{i+n}(\boldsymbol{x})$ are extended in the same way to columns $\boldsymbol{f}_i^*(\boldsymbol{x})$, $\boldsymbol{f}_{i+1}^*(\boldsymbol{x})$, ..., $\boldsymbol{f}_{i+n}^*(\boldsymbol{x})$ given by:

$$\begin{aligned}
f_i^* &= [z_1^T \; \ldots \; z_{i-1}^T f_i^T \; z_{i+1+n}^T \; \ldots \; z_m^T]^T \\
f_{i+1}^* &= [z_1^T \; \ldots \; z_{i-1}^T f_{i+1}^T \; z_{i+1+n}^T \; \ldots \; z_m^T]^T \\
&\;\;\vdots \\
f_{i+n}^* &= [z_1^T \; \ldots \; z_{i-1}^T f_{i+n}^T \; z_{i+1+n}^T \; \ldots \; z_m^T]^T
\end{aligned} \tag{4.55}$$

Each of the columns in (4.55) contains $k$ elements, where $k$ is given by:

$$k = \sum_{j=1}^{i} k_j + \sum_{j=i+n+1}^{m} k_j \tag{4.56}$$

Since the regression equations of response quantity $y_i(\boldsymbol{x})$ and its partial derivatives contain the same unknown parameters $\boldsymbol{\beta}_i$, the parameters in this special case can be filed in a column given by:

$$\boldsymbol{\beta} = [\boldsymbol{\beta}_1^T \ldots \boldsymbol{\beta}_{i-1}^T \; \boldsymbol{\beta}_i^T \; \boldsymbol{\beta}_{i+n+1}^T \ldots \boldsymbol{\beta}_m^T]^T \tag{4.57}$$

Using the definitions (4.54) through (4.57) the regression equations for this case can also be written in the same compact form of (4.53). We illustrate this by means of the following example:

$\boldsymbol{u}(\boldsymbol{x}) = [y_1 \; y_2 \; \ldots \; y_5]$ where, for instance:

$\boldsymbol{x} = [x_1 \;\; x_2]^T$ and

$$y_1 = \beta_1 + \beta_2 x_2$$

$$y_2 = \beta_3 + \beta_4 x_2$$

$$y_3 = \beta_5 + \beta_6 x_1 + \beta_7 x_2 + \beta_8 x_1 x_2$$

$$y_4 = \frac{\partial y_3}{\partial x_1} = \beta_6 + \beta_8 x_2$$

$$y_5 = \frac{\partial y_3}{\partial x_2} = \beta_7 + \beta_8 x_1$$

Then $\boldsymbol{u}(\boldsymbol{x}) = \boldsymbol{F}^T(\boldsymbol{x})\boldsymbol{\beta}$, where $\boldsymbol{F}(\boldsymbol{x})$ and $\boldsymbol{\beta}$ are given by:

$$F^T(x) = \begin{bmatrix} 1 & x_1 & | & 0 & 0 & | & 0 & 0 & 0 & 0 \\ 0 & 0 & | & 1 & x_2 & | & 0 & 0 & 0 & 0 \\ 0 & 0 & | & 0 & 0 & | & 1 & x_1 & x_2 & x_1x_2 \\ 0 & 0 & | & 0 & 0 & | & 0 & 1 & 0 & x_2 \\ 0 & 0 & | & 0 & 0 & | & 0 & 0 & 1 & x_1 \end{bmatrix}$$

$$\boldsymbol{\beta} = [\beta_1 \; \beta_2 \mid \beta_3 \; \beta_4 \mid \beta_5 \; \beta_6 \; \beta_7 \; \beta_8]^T$$

As in the previous section, $N$ not necessarily distinct design points $\boldsymbol{x}_1, \boldsymbol{x}_2, \ldots, \boldsymbol{x}_N$ are to be selected from a set $\Gamma$ of $r$ candidate points. In these design points measurements will be carried out, resulting in values $\boldsymbol{u}(\boldsymbol{x}_1), \ldots, \boldsymbol{u}(\boldsymbol{x}_N)$ for the responses $\underline{\boldsymbol{u}}(\boldsymbol{x}_1), \ldots, \underline{\boldsymbol{u}}(\boldsymbol{x}_N)$. Denoting the difference between $\boldsymbol{u}(\boldsymbol{x}_i)$ and $\underline{\boldsymbol{u}}(\boldsymbol{x}_i)$ by $\underline{\boldsymbol{e}}_i$ it is seen that:

$$\underline{\boldsymbol{u}}(\boldsymbol{x}_i) = \boldsymbol{u}(\boldsymbol{x}_i) + \underline{\boldsymbol{e}}_i = \boldsymbol{F}^T(x)\boldsymbol{\beta} + \underline{\boldsymbol{e}}_i \qquad \text{for } i = 1, 2, \ldots, N \tag{4.58}$$

It is assumed that the elements of $\underline{\boldsymbol{e}}_i$ are uncorrelated random errors with expected value zero and known constant variance. Furthermore it is assumed that $\underline{\boldsymbol{e}}_i$ and $\underline{\boldsymbol{e}}_j$ are uncorrelated if $i \neq j$:

$$E(\underline{\boldsymbol{e}}_i) = \boldsymbol{0}\, ; \quad E(\underline{\boldsymbol{e}}_i\, \underline{\boldsymbol{e}}_j^T) = \delta_{ij}\boldsymbol{S}^2 \quad \text{for } i,\; j = 1, 2, \ldots, N \tag{4.59}$$

Here, $\boldsymbol{S}$ is a known $m*m$ diagonal matrix whose diagonal elements $\sigma_1, \ldots, \sigma_m$ are the standard deviations of the response quantities $y_1, \ldots, y_m$.

Because the error variances are unequal, $\boldsymbol{\beta}$ should be estimated using the method of weighted least squares. By pre-multiplying (4.58) with $\boldsymbol{S}^{-1}$ the new sample model becomes:

$$\underline{\boldsymbol{w}}(\boldsymbol{x}_i) = \boldsymbol{W}^T(\boldsymbol{x}_i)\, \boldsymbol{\beta} + \boldsymbol{\eta}_i \qquad \text{for } i = 1, 2, \ldots, N \tag{4.60}$$

where $\underline{\boldsymbol{w}}(\boldsymbol{x}_i) = \boldsymbol{S}^{-1}\underline{\boldsymbol{u}}(\boldsymbol{x}_i)$, $\boldsymbol{W}^T(\boldsymbol{x}_i) = \boldsymbol{S}^{-1}\boldsymbol{F}^T(\boldsymbol{x}_i)$, and $\boldsymbol{\eta}_i = \boldsymbol{S}^{-1}\underline{\boldsymbol{e}}_i$

The errors $\boldsymbol{\eta}_i$ still have zero expectation:

$$E(\boldsymbol{\eta}_i) = \boldsymbol{S}^{-1}\, E(\underline{\boldsymbol{e}}_i) = \boldsymbol{0}$$

However, the error variance matrix of $\boldsymbol{\eta}_i$, $V(\boldsymbol{\eta})$ now becomes unity:

$$E(\boldsymbol{\eta}_i\, \boldsymbol{\eta}_i^T) = E(\boldsymbol{S}^{-1}\underline{\boldsymbol{e}}_i\, \underline{\boldsymbol{e}}_i^T\boldsymbol{S}^{-1}) = \boldsymbol{S}^{-1}E(\underline{\boldsymbol{e}}_i\, \underline{\boldsymbol{e}}_i^T)\,\boldsymbol{S}^{-1} = \boldsymbol{S}^{-1}\boldsymbol{S}\boldsymbol{S}\,\boldsymbol{S}^{-1} = \boldsymbol{I}$$

Therefore model (4.60) satisfies the standard least squares assumptions and $\boldsymbol{\beta}$ can be estimated by an ordinary least squares fit of the new models

Estimates $\hat{\boldsymbol{\beta}}$ for the model parameters $\boldsymbol{\beta}$ can now be determined from the requirement that the weighted residual sum of squares,

$$KS_r = \sum_{i=1}^{N} [(w(\boldsymbol{x}_i) - \boldsymbol{W}^T(\boldsymbol{x}_i)\boldsymbol{\beta}^T)^T (w(\boldsymbol{x}_i) - \boldsymbol{W}^T(\boldsymbol{x}_i)\boldsymbol{\beta}^T)] \, , \tag{4.61}$$

is minimal. As in Subsection 4.2.1, this results in a set of $k$ equations for $\hat{\boldsymbol{\beta}}$ with the solution

$$\hat{\boldsymbol{\beta}} = [\sum_{i=1}^{N} \boldsymbol{W}(\boldsymbol{x}_i)\, \boldsymbol{W}^T(\boldsymbol{x}_i)]^{-1} \sum_{i=1}^{N} [\boldsymbol{W}(\boldsymbol{x}_i) w(\boldsymbol{x}_i)] \tag{4.62}$$

According to (4.62) a weighted least squares estimator $\hat{\boldsymbol{\beta}}$ can be defined as

$$\underline{\hat{\boldsymbol{\beta}}} = [\sum_{i=1}^{N} \boldsymbol{W}(\boldsymbol{x}_i)\, \boldsymbol{W}^T(\boldsymbol{x}_i)]^{-1} \sum_{i=1}^{N} [\boldsymbol{W}(\boldsymbol{x}_i) \underline{w}(\boldsymbol{x}_i)] \tag{4.63}$$

While

$$\begin{aligned} E(\underline{\hat{\boldsymbol{\beta}}}) &= [\sum_{i=1}^{N} \boldsymbol{W}(\boldsymbol{x}_i)\, \boldsymbol{W}^T(\boldsymbol{x}_i)]^{-1} \sum_{i=1}^{N} \boldsymbol{W}(\boldsymbol{x}_i) E\underline{w}(\boldsymbol{x}_i) \\ &= [\sum_{i=1}^{N} \boldsymbol{W}(\boldsymbol{x}_i)\, \boldsymbol{W}^T(\boldsymbol{x}_i)]^{-1} \sum_{i=1}^{N} \boldsymbol{W}(\boldsymbol{x}_i) \boldsymbol{W}(\boldsymbol{x}_i)\boldsymbol{\beta} \;= \boldsymbol{\beta} \end{aligned} \tag{4.64}$$

Relation (4.63) is an unbiased estimator for the parameters $\boldsymbol{\beta}$. Furthermore it can be shown that the variance-covariance matrix of the parameter estimators, i.e. $E(\underline{\hat{\boldsymbol{\beta}}} - \boldsymbol{\beta})(\underline{\hat{\boldsymbol{\beta}}} - \boldsymbol{\beta})^T$ is given by

$$\begin{aligned} V(\underline{\hat{\boldsymbol{\beta}}}) &= N\boldsymbol{M}^{-1} \sum_{i=1}^{N} [\boldsymbol{W}(\boldsymbol{x}_i) V(\underline{\eta}) \boldsymbol{W}^T(\boldsymbol{x}_i)]\; N\boldsymbol{M}^{-1} \\ &= [\sum_{i=1}^{N} \boldsymbol{W}(\boldsymbol{x}_i)\, \boldsymbol{W}^T(\boldsymbol{x}_i)]^{-1} \;= N\boldsymbol{M}^{-1} \end{aligned} \tag{4.65}$$

where $\boldsymbol{M} = \frac{1}{N} \sum_{i=1}^{N} [\boldsymbol{W}(\boldsymbol{x}_i)\, \boldsymbol{W}^T(\boldsymbol{x}_i)]$ is the information matrix.

If the parameters are estimated by an ordinary least squares fit of the original model (4.58) then

$$\underline{\hat{\boldsymbol{\beta}}}_0 = [\sum_{i=1}^{N} \boldsymbol{F}(\boldsymbol{x}_i)\, \boldsymbol{F}^T(\boldsymbol{x}_i)]^{-1} \sum_{i=1}^{N} [\boldsymbol{F}(\boldsymbol{x}_i) \underline{u}(\boldsymbol{x}_i)] \tag{4.66}$$

is obtained. $\underline{\hat{\boldsymbol{\beta}}}_0$ is unbiased, that is, $E(\underline{\hat{\boldsymbol{\beta}}}_0) = \boldsymbol{\beta}_0$ however,

$$V(\hat{\underline{\beta}}_0) = [\sum_{i=1}^{N} \boldsymbol{F}(\boldsymbol{x}_i)\, \boldsymbol{F}^T(\boldsymbol{x}_i)]^{-1} \sum_{i=1}^{N} [\boldsymbol{F}(\boldsymbol{x}_i)\, \boldsymbol{S}^2 \boldsymbol{F}^T(\boldsymbol{x}_i)][\sum_{i=1}^{N} \boldsymbol{F}(\boldsymbol{x}_i)\, \boldsymbol{F}^T(\boldsymbol{x}_i)]^{-1} \qquad (4.67)$$

If $\boldsymbol{S}^2 = \boldsymbol{I}$ then $V(\hat{\underline{\beta}}_0) = V(\hat{\underline{\beta}})$. If $\boldsymbol{S}^2 \neq \boldsymbol{I}$ then $V(\hat{\underline{\beta}}_0) > V(\hat{\underline{\beta}})$, i.e., $V(\hat{\underline{\beta}}_0)$ is a positive definitive matrix and an ordinary least squares fit of (4.58) results in greater variances than a weighted least squares fit.

In order to estimate $\underline{\beta}$ the error variance matrix $\boldsymbol{S}^2$ should be known beforehand. When $\boldsymbol{S}^2$ is unknown, $\boldsymbol{S}^2$ and $\underline{\beta}$ can be estimated using an iterative reweighted least squares procedure:

Step 1. Compute the ordinary least squares estimations for the original model (4.58)

$$\hat{\underline{\beta}}_0 = [\sum_{i=1}^{N} \boldsymbol{F}(\boldsymbol{x}_i)\, \boldsymbol{F}^T(\boldsymbol{x}_i)]^{-1} \sum_{i=1}^{N} [\boldsymbol{F}(\boldsymbol{x}_i) \underline{\boldsymbol{u}}(\boldsymbol{x}_i)]$$

Step 2. Compute initial estimations of the error variances according to

$$\hat{\sigma}_{j_0}^2 = \frac{1}{N-k} \sum_{i=1}^{N} [y_j(\boldsymbol{x}_i) - \hat{y}_{j_0}(\boldsymbol{x}_i)]^2 \qquad \text{for } j = 1, \dots, m$$

where $\hat{y}_{j_0}(\boldsymbol{x}_i) = \boldsymbol{f}_j^T(\boldsymbol{x}_i) \hat{\underline{\beta}}_0$, $\quad i = 1, \dots, N$

and construct $\hat{S}_0 = \begin{bmatrix} \hat{\sigma}_{1_0} & & 0 \\ & \cdot & \\ 0 & \cdot & \hat{\sigma}_{m_0} \end{bmatrix}$

Step 3. Pre-multiply model (4.58) with $\hat{S}_0^{-1}$. This results in a transformed model:

$$\underline{w}_1(\boldsymbol{x}_i) = \mathbf{W}_1^T(\boldsymbol{x}_i)\underline{\beta} + \underline{\eta}_1(\boldsymbol{x}_i), \quad i = 1, \dots, N \qquad (4.68)$$

Step 4. Applying the ordinary least squares method to the transformed model (4.68) results in new estimations $\hat{\underline{\beta}}_1$, for which holds $V(\hat{\underline{\beta}}_1) \leq V(\hat{\underline{\beta}}_0)$.

Step 5. Compute new estimations of the error variances of the original model (4.58) according to

$$\hat{\sigma}_{j_1}^2 = \frac{1}{N-k} \sum_{i=1}^{N} [y_j(\boldsymbol{x}_i) - \hat{y}_{j1}(\boldsymbol{x}_i)]^2 \qquad \text{for } j = 1, \dots, m$$

where $\hat{y}_{j1}(\boldsymbol{x}_i) = \boldsymbol{f}_j^T(\boldsymbol{x}_i)\underline{\hat{\boldsymbol{\beta}}}_1$ , $i = 1, ..., N$

and construct $$\hat{\boldsymbol{S}}_1 = \begin{bmatrix} \hat{\sigma}_{1_1} & & 0 \\ & \ddots & \\ 0 & & \hat{\sigma}_{m_1} \end{bmatrix}, \quad i = 1, ..., N$$

Step 3 - 5 are repeated with $\hat{\boldsymbol{S}}_1$ instead of $\hat{\boldsymbol{S}}_0$, then with $\hat{\boldsymbol{S}}_2$ and so on, until

$$\left\| \underline{\hat{\boldsymbol{\beta}}}_i - \underline{\hat{\boldsymbol{\beta}}}_{i-1} \right\| \le eps * \left\| \underline{\hat{\boldsymbol{\beta}}}_i \right\| ,$$

where *eps* is the relative machine precision.

Using these results, unbiased estimators $\underline{\hat{\boldsymbol{u}}}(\boldsymbol{x})$ for the response quantities can be derived from:

$$\underline{\hat{\boldsymbol{u}}}(\boldsymbol{x}) = \boldsymbol{F}^T(\boldsymbol{x})\,\underline{\hat{\boldsymbol{\beta}}} \tag{4.69}$$

The variance-covariance matrix $\boldsymbol{V}(\underline{\hat{\boldsymbol{u}}}(\boldsymbol{x}))$ is given by:

$$\boldsymbol{V}(\underline{\hat{\boldsymbol{u}}}(\boldsymbol{x})) = \boldsymbol{F}^T(\boldsymbol{x})\boldsymbol{V}(\underline{\hat{\boldsymbol{\beta}}})\boldsymbol{F}(\boldsymbol{x}) \tag{4.70}$$

From (4.65) it is seen that $\boldsymbol{V}(\underline{\hat{\boldsymbol{\beta}}})$ is determined completely by the statistical properties, $\boldsymbol{S}$, of the measurement errors, the model functions $\boldsymbol{F} = \boldsymbol{F}(\boldsymbol{x})$ and the design points $\boldsymbol{x}_1, \ldots, \boldsymbol{x}_N$. Assuming that the error variance matrix $\boldsymbol{S}$ is known, then, for a given number $N$ of design points the matrix $\boldsymbol{V}(\underline{\hat{\boldsymbol{\beta}}})$ can only be influenced by the choice of the design points $\boldsymbol{x}_1, \ldots, \boldsymbol{x}_N$ from the set $\Gamma$ of candidate points $\chi_1, \ldots, \chi_r$. As in Subsection 4.2.3, the objective of optimal design of experiments for simultaneous responses is to determine the number $N_i$ $(N_i \geq 0)$ of observations at candidate point $\chi_i$ ($i$=1, 2, ..., $r$), such that $N_1 + N_2 + \ldots + N_r$ equals the number $N$ of design points. Using. (4.65) it is readily shown that the variance-covariance matrix $\boldsymbol{V}(\underline{\hat{\boldsymbol{\beta}}})$ can be considered a function of the column $\boldsymbol{N} = [N_1 \;\; N_2 \;\ldots\; N_r]^T$,

$$\boldsymbol{V}(\underline{\hat{\boldsymbol{\beta}}}, \boldsymbol{N}) = N\boldsymbol{M}^{-1}(\boldsymbol{N}) \tag{4.71}$$

where $\boldsymbol{M}(\boldsymbol{N})$ follows from (4.65)

$$M(N) = \frac{1}{N}\sum_{j=1}^{N}[W(x_j)\, W^T(x_j)] = \frac{1}{N}\sum_{i=1}^{r}[N_i W(\chi_i)\, W^T(\chi_i)] \tag{4.72}$$

#### 4.2.4.3 Optimality Criteria

In the previous subsection it has been shown that the accuracies of the parameter estimators $\underline{\hat{\beta}}$ and of the response estimators $\underline{\hat{u}}(x)$ are determined by the variance-covariance matrix $V(\underline{\hat{\beta}}, N)$ and therefore by the elements of $N$. These elements may be subject to constraints. As in subsection 4.2.3.4 this leads to the following definition of an optimal experimental design:

The design of an experiment, characterized by $N^*$, is optimal if the elements of $N^*$ satisfy all specified constraints and a specified criterion $C = C(N)$ is minimal for $N = N^*$

The D-, A- and E- criteria remain the same:

$$C(N) = \det(\hat{V}(\underline{\hat{\beta}}, N)) \quad \text{for D-optimality} \tag{4.73}$$

$$C(N) = tr(\hat{V}(\underline{\hat{\beta}}, N)) \quad \text{for A-optimality} \tag{4.74}$$

$$C(N) = \lambda_{max}(N) \quad \text{for E-optimality} \tag{4.75}$$

where $\lambda_{max}(N)$ is the maximal eigenvalue of $\hat{V}(\underline{\hat{\beta}}, N)$. (4.76)

Nagtegaal (1987) proposes the following G- and V- criteria

$$C(N) = \max_{i=1}^{r}[tr(\hat{V}\,\underline{\hat{u}}(\chi_i))] \quad \text{for G-optimality} \tag{4.77}$$

$$C(N) = \frac{1}{r}\sum_{i=1}^{r}[tr(\hat{V}\,\underline{\hat{u}}(\chi_i))] \quad \text{for V-optimality} \tag{4.78}$$

Note: when the error variance matrix $S^2$ is unknown, successive estimations of $S^2$ can be obtained by applying the iterative model building procedure of Fig. 4.1.

### 4.2.5 Model Fitting and Testing

#### 4.2.5.1 Introduction

The construction of a reliable model for a given system or structure is an iterative process. At the start of each iteration step the number $k$ of model parameters $\beta_1, \ldots, \beta_k$ and the presumed model functions $f_i(\mathbf{x})$, $i = 1, 2, ..., k$ must be available. The iteration step then involves the design of an experiment to collect data, estimation of the parameters from the collected data and evaluation of the model. Evaluation implies answering questions like:

- Is the model valid?
- Are the estimated parameters accurate enough?
- Is the model accurate enough for all relevant values of $\mathbf{x}$?
- Which parameters are relevant and which parameters can be dropped without affecting the predictive ability of the model too much?

In next subsection procedures for selecting regressor variables will be discussed briefly. Testing procedures will be discussed in Subsection 4.2.5.4.

### 4.2.5.2 Selection of Regressor Variables

Experimental design as described in the previous sections, is based on a postulated regression model of the type (4.49). The problem now is to decide which terms or "regressor variables" should be contained in a regression model for a particular response quantity. For the selection of the regressor variables several algorithms have been developed.

The backward elimination algorithm starts with the full model and subsequently removes non-significant variables in successive steps. The forward selection algorithm starts with the most significant regressor variable and adds significant variables successively. A popular and effective algorithm is the stepwise regression algorithm, which can be regarded as a combination of forward selection and backward elimination. A concise description is given below.

In order to decide whether or not a certain parameter $\beta_i$ and its corresponding regressor variable $f_i(\mathbf{x})$ should be preserved in the model or removed from the model a criterion is needed. For this purpose an F-test quantity $\hat{F}_i$ is used, defined by:

$$\hat{F}_i = \frac{\hat{\beta}_i^2}{\hat{V}(\underline{\hat{\beta}})_{ii}} \qquad \text{for i} = 1, 2, ..., k, \tag{4.79}$$

where $\hat{\beta}_i$ is an estimate for the parameter $\beta_i$, and $\hat{V}(\underline{\hat{\beta}})_{ii}$ is the estimated variance of that parameter. The term $\underline{\hat{\beta}}_i f_i(\mathbf{x})$ is said to be more significant than $\underline{\hat{\beta}}_j f_j(\mathbf{x})$ if $\hat{F}_i$ is larger than $\hat{F}_j$. Using these notions the stepwise regression algorithm adds the most significant term $\underline{\hat{\beta}}_j f_j(\mathbf{x})$ to the model for which $\hat{F}_j$ is larger than a predetermined F-value to enter, and subsequently removes that term $\underline{\hat{\beta}}_i f_i(\mathbf{x})$ for which $\hat{F}_i$ is smaller then a predetermined F-value to remove.

### 4.2.5.3 Parameter Estimation

Methods for estimation of the unknown parameters $\boldsymbol{\beta}$ and their variances and for estimation of the response quantities and their variances have already been described in the Subsections

4.2.1 and 4.2.4. Section 4.2.1 treats a single response quantity, whereas Subsection 4.2.4 treats the general case of several response quantities, some of which possibly may be response quantities and their partial derivatives with respect to the control variables $\boldsymbol{x}$.

#### 4.2.5.4 Tests for Model Adequacy

Several tests and criteria for a judgement of the model adequacy are available. It is emphasized that the model should be tested using newly collected data. Tests based on data, which were used to derive the model, can be deceptive and dangerous. From careful inspection of the residuals, i.e. the discrepancies between observed and predicted values, one can learn much about the validity of the model. Assumptions about distributions of errors and the significance of the selected regressor variables can be tested. In e.g. Draper et al. (1998) and Myers et al. (1995) a comprehensive treatment on the subject can be found. Plots of residuals are very illustrative; common plots include normality plots and plots of residuals against predicted values or against control variables.

The variance estimator (4.71) gives a measure of the accuracy of the parameter estimators $\underline{\hat{\boldsymbol{\beta}}}$. If the observations follow a normal distribution, then the parameters $\underline{\hat{\boldsymbol{\beta}}}$ are also normally distributed and confidence intervals for the parameters can be calculated using:

$$\frac{\underline{\hat{\boldsymbol{\beta}}} - \boldsymbol{\beta}_i}{V(\boldsymbol{\beta}_i)} \tag{4.80}$$

which follows an N(0, 1) distribution if $\sigma^2$ and hence $V(\boldsymbol{\beta}_i)$ is known. Based on estimated parameter variances confidence intervals for the parameters can be derived from:

$$\frac{\underline{\hat{\boldsymbol{\beta}}} - \boldsymbol{\beta}_i}{(\underline{\hat{\boldsymbol{V}}}(\underline{\hat{\boldsymbol{\beta}}})_{ii})^{1/2}} = \underline{t_i} \tag{4.81}$$

where $t_i$ is a random variable following a Student's t-distribution with $n$-$k$ degrees of freedom. If the normal equations are ill conditioned, the parameters will become unstable. A slight modification of the observations will result in considerable changes in parameter values. Hence, the condition number of the matrix $\boldsymbol{M}$ provides another criterion for parameter judgement.

For a judgement of the predictive ability of the model the residuals of the responses should be examined. The mean of the squared residuals provides an "overall" measure for the predictive quality. It is an estimate of the response variance. For example, the estimated response variance of $\underline{\hat{y}}$ can be computed from:

$$\underline{\hat{V}}(\hat{y}) = \frac{1}{t}\sum_{i=1}^{t}(\hat{y}_i - y_i)^2 \tag{4.82}$$

As recommended before, here newly collected data should be used in t adequately chosen points. The new set of data points may be generated using a block (or fraction) of a $2^n$-design, which has not yet been used. Experimental design methods render efficient facilities to generate a set of "optimal" testing points, by augmenting the experimental design (Nagtegaal, 1987). When added to the old data, they may result in improved parameter estimates.
If the outcomes of the model testing ask for further model improvement, it is necessary to initiate another model building cycle consisting of design, data collection, model fitting and again testing. The model building process will be considered in Section 4.3.

### 4.2.6 Experimental Design of Deterministic Processes

The experimental design theory has been developed for the planning and evaluation of physical experiments with random errors. The consequences of applying this theory for the planning and evaluation of numerical experiments on very complicated deterministic mathematical models, for example, finite element analyses, are not trivial and will be discussed below.

#### 4.2.6.1 Experimental Design

In general the postulated linear model in an experimental design for some deterministic process is biased and, by definition, does not show a random error. In order to find an optimal design for this situation the J-criterion for robust designs could be used. However, the following approach may yield better results.
In practical situations often a realistic linear model can be postulated using experience and/or preliminary investigations. Then experimental designs can be obtained in the same way as for stochastic processes. Optimization using the common algorithms and criteria may yield designs for which the condition of the set of normal equations will be improved.

#### 4.2.6.2 Parameter Estimation

In deterministic processes estimation of the parameters $\boldsymbol{\beta}$ can be carried out in the same way as in stochastic processes. The measures of accuracy of the estimated parameters and response quantities can be computed as in the stochastic case. However, there is a difference because the variance of a deterministic response quantity is zero. We loosely keep speaking of "variances", although such quantities are not defined in deterministic processes.
The variance $\hat{V}(\hat{\boldsymbol{\beta}})$ of the parameters $\hat{\boldsymbol{\beta}}$ is defined analogously to (4.65) as:

$$\hat{V}(\hat{\boldsymbol{\beta}}) = N\boldsymbol{M}^{-1} \tag{4.83}$$

The variance-covariance matrix of the deterministic response quantities, $\hat{\boldsymbol{V}}(\hat{\boldsymbol{u}}(\boldsymbol{x}))$, can be computed analogously to (4.70) by:

$$\hat{V}(\hat{\boldsymbol{u}}(\boldsymbol{x})) = \boldsymbol{F}^T(\boldsymbol{x})\hat{V}(\hat{\boldsymbol{\beta}})\boldsymbol{F}(\boldsymbol{x}) \tag{4.84}$$

where $\hat{V}(\hat{\boldsymbol{\beta}})$ is computed from (4.83)

#### 4.2.6.3 Selection of Regressor Variables in Deterministic Processes

For the selection of regressor variables F-tests are used which are analogous to (4.79):

$$\hat{F}_i = \frac{\hat{\beta}_i^2}{\hat{V}(\hat{\boldsymbol{\beta}})_{ii}} \quad , \ i = 0, 1, \ \ldots, \ k \tag{4.85}$$

In physical experiments a certain parameter $\beta_i$ is regarded significant ($\beta_i \neq 0$) or not ($\beta_i = 0$), depending on whether or not $\hat{F}_i$ exceeds a critical value taken from an F-table. The test is used to include the corresponding regressor variable in the model if $\beta_i$ is significant, or to remove that variable if $\beta_i$ is not significant.

In deterministic processes we can use this selection procedure as well; however, it cannot be stated that $\beta_i$ is significant or not. In general the correct variables will be selected in the regression model, but this cannot be proved. Tests for model adequacy should be used as a final approval of the regression model.

#### 4.2.6.4 Model testing

The test for model adequacy can be performed using the same procedures as for stochastic processes. However, confidence intervals for estimated parameters cannot be defined.

## 4.3 Building and Application of Regression Models Based on Numerical Computations

### 4.3.1 Building Regression Models

Guidelines for building regression models based on physical experimentation are well known (e.g. Box et al., 1987; Draper et al., 1998; Montgomery et al., 1992, or Myers et al., 1995). In this section we apply these rules to regression models based on numerical computations. First we will make some remarks about linear and nonlinear regression models. Next, aspects are discussed of response quantities, control variables and feasibility, followed by aspects of the preliminary choice of a regression model and the iterative development of that model. Finally, operational aspects of regression models are discussed.

#### 4.3.1.1 Linear and Nonlinear Regression

Regression models can be categorized in so-called functional, control and predictive models. We will give a short explanation of the differences.

Functional models, in literature also indicated as mechanistic models (Box et al., 1978), can be used if there is a fairly detailed knowledge, often a set of differential equations, of the system under consideration. The unknown parameters generally appear in a nonlinear way in the model, and a least squares fitting of the model requires the solution of a nonlinear set of normal equations. Such a solution may be obtained by means of mathematical programming methods. In the present research we will concentrate on the combination of structural optimization and experimental design using less complicated linear regression models.

In structural optimization we are concerned with control of the structural behaviour by means of adjusting the values of the design variables. What we need is a so-called control model, in which the design or "control" variables are under the control of the investigator. With the term "control variables" we mean "design variables" or "factors" as far as structural optimization or experimental design is concerned respectively. For certain classes of structural analysis and optimization problems useful control models can effectively be derived by regression techniques using finite element analyses and experimental designs for linear models.

In many practical situations where experimentation has to be performed on running processes, the factors are not completely under the control of the experimentalist. In such situations experiments cannot be designed, but still one can often obtain a so-called linear predictive model, which at least can reproduce the main effects of the considered response quantity. Regression techniques are much used to construct predictive models. However, since we are concerned with designed (numerical) experiments, we will not discuss this type of model any further.

#### 4.3.1.2 Selection of Response Quantities and Control Variables

In constructing a regression model the first important task is to define the problem as clearly as possible. This definition should include those response quantities and control variables, which play a role in the problem. The pursued regression models should be able to estimate response quantities accurately, with values for the control variables as input data.

With a **response quantity** of a system we mean some behaviour variable of the system which, in our case, can only be evaluated by means of an analysis of the system, given a proper set of design variables. In general structural analyses produce a lot of response quantities. It is important to get clear which of these quantities are needed in the model building process. The relevant response quantities are closely related to the aimed goals of the investigation, so a formulation of these goals has to be available. To obtain afterwards data for an overlooked response quantity usually requires great extra effort or may even be impossible at all. In many cases response quantity values emerge directly from the analysis of the system, but sometimes one or more transformations are necessary to derive such a value from analysis results. In the latter situations two approaches are possible. The first is to perform those transformations first and to use the resulting response value in the model building process. In the second approach regression models are first fitted on the separate analysis results and the transformation is performed on estimates computed using these regression models. The advantage of the first approach is that finally less data (regression parameters) is involved. In the second approach the regression models may be fitted more easily in case the individual regression models are of relatively low order.

The investigator who tries to define the **control variables** often ends up in conflicting situations. On the one hand he wants to take "everything" into account to be sure that nothing important is overlooked. On the other hand the regression model should be as simple as possible, resulting in less model building effort, and in less variance in the model due to not including redundant variables. Nevertheless, the investigator should start with much freedom to consider a certain variable of possible importance. Adequate computer programs can support the selection process, but one should never solely rely on computer outcomes. It will be wise to consult experts both on model building and on the physical system under study.

In many problems the control variables can readily be identified as a set of discrete quantities which define the system and which, of course may be of different individual importance. However, sometimes control variables are defined more complex, for instance as certain classes of functions. The geometry in shape optimization problems may be regarded as such a control variable. It is recommended to transform both the response quantities and the control variables to a set of dimensionless quantities. This can be done by means of a dimensional analysis. Introduction of dimensionless quantities may have the following advantages:

- The number of the elementary dimensions in the problem can reduce the total number of response quantities and control variables.
- Insight into the behaviour of the system may be enlarged.
- Scaling the variables to obtain a well-conditioned regression problem will be more straightforward.

**4.3.1.3 Aspects of Collecting the Data**

Concerning the data collection in numerical experimentation, at an early stage the following aspects have to be considered.

The investigator must build and test an efficient and accurate numerical model, often a finite element model, of the system under study. Generally the control variables for the regression model must be transformed in input data for the finite element model (FEM), and the output of the finite element analysis usually also has to be transformed in order to obtain observations for the required response quantities. Sometimes the link between a certain response quantity and the FEM-output is simple: for instance the lowest eigenfrequency of the structure, or the Von Mises stress in a certain element node. User-friendly FEM-packages have features to gather such information in a manageable file. In more complex situations, one may be obliged to write an interface program between the FEM-output file and the regression analysis program. At this stage of the project the investigator should check whether or not appropriate hardware and software is available to perform the transformations and computations.

**First check on feasibility.** At this stage a rough estimate of the required budget has to be made. Till now, rather little human and computational effort has been invested, but already a considerable knowledge is gathered about the problem. Using this knowledge, a first feasibility study can be carried out. The study should result in one of the following three outcomes:

- The project is infeasible and has to be stopped.
- The project should be revised and checked again.

- The project is allowed to proceed; in this case we proceed according to the next subsection.

**4.3.1.4 A Preliminary Regression Model**

In deriving a preliminary regression model we have the following objectives:

- Gathering more detailed information about the behaviour of the system under study; we use this information to formulate a more established experimental design.
- Making a realistic time planning of the project and producing fairly accurate estimates of the human and computational efforts and of the budget that is required.

The preliminary computations need not necessarily be formulated as an experimental design. If very little is known about the behavior of the system one can start with computations in which the control variables are changed one at a time. A disadvantage of this approach is that interactions between the variables will not be revealed. Even a rather simple regression model, derived from a very first formulation of an experimental design, is a more structured way to achieve the objectives mentioned above. Whatever the approach may be, it is recommended to start tentatively and to spend not more than one quarter (25%-rule of Box et al., 1978) of the budget as estimated at the first feasibility check.

The way in which the preliminary model is derived may be essentially the same as one iteration cycle of the further development of the regression model (Fig. 4.2) and will be discussed in the next subsection.

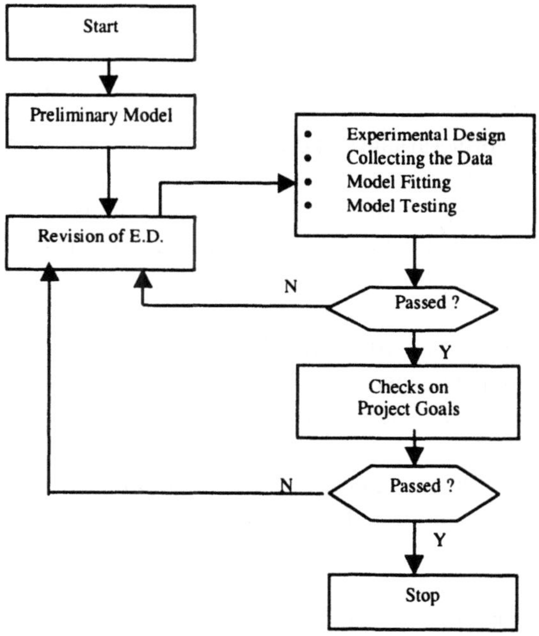


**Figure 4.2** Scheme for model building

**4.3.1.5 Development of the Regression Model**

The development of the preliminary model into the ultimate regression model globally proceeds according to the scheme shown in Fig. 4.2.

The preliminary regression model can be used to perform a second screening of the control variables in two ways.

First, the $X^T X$-matrix of the underlying experimental design can be written in the so-called correlation form (Draper et al., 1998). Marquardt (1970) gives the rule that the diagonal elements of the inverse of this correlation matrix should be larger than 1.0 but certainly smaller than 10.0. Violation of the latter limit indicates that the current data are not appropriate to produce valuable estimates for the regression coefficients. It. may be an indication to modify the underlying linear model.

Next, for every response variable the correlation with the control variables is checked. Every response should show one or two strong correlations. If this is not the case, perhaps an important variable was overlooked or the range of variation of one or more of the variables is too small, resulting in poor predictive properties of the regression model.

Once the experimental design is more established, more reliable estimates can be made for the financial budget, human effort; and needs for hardware and software. The schedule can be updated and the whole project has to be submitted for approval. The possible outcomes of this second checkpoint are similar as before:

- The project is stopped.
- The project is revised and submitted again.
- We proceed the project with the further development of the regression model.

**Revision of the experimental design.**

Using the information obtained from the preliminary or the current regression model, the experimental design can now be revised with respect to the following aspects:

- The set of relevant control variables and their ranges and the regression equations must be adapted.
- Depending on the outcomes of the previous point, for every control variable a suitable number of discrete levels and values for the levels must be chosen. These choices depend on whether or not partial derivatives of the response quantities will be used to estimate the model parameters.
- If more than one response quantity is involved, one may decide to treat these quantities one at a time and formulate experimental designs for every individual response quantity. Generally such an approach is very inefficient and one of the following two approaches is used:
    1. By inspection we formulate a linear model incorporating the expected models of the individual response quantities. This model is used to formulate the experimental design. (Of course, after the observations having been made, for every response quantity its own set of parameters is estimated.) This approach is feasible if the linear models for the individual response quantities do not differ too much; in this case all methods described in Section 4.2 can be used.

2. If quite different linear models are related to the response quantities, and/or if partial derivatives of the quantities will be brought into account, it is recommended to use those methods described in Section 4.2, through which an experimental design can be optimized based on several simultaneous linear models.

- Next, the set of design points, which build a complete experimental design, can be formulated. Generally a complete design is infeasible because it contains far too many design points and, in the case of $2^n$-designs, a fractional $2^n$-design is developed. In the case of optimal experimental design the points of a complete design build a possible set of candidate points. So many candidate points, however, may be infeasible as well and we should reduce their number. A reasonable choice may be defining the set of candidate points as a fractional $2^n$-design. Sometimes there are practical reasons to eliminate certain points from the set of candidate points.
- Finally we estimate the number of design points needed in the ultimate fractional design, and we decide how many points will be used in the next experimental design. Choices depend on the method used in the experimental design. In fractional $2^n$-designs one may use blocks derived from related defining contrasts. In optimal experimental design methods one is freer to choose the number of design points. Generally the points already used in the preliminary design will be maintained and the next experimental design is formulated augmenting the current one.

We proceed with short explanations of the remaining steps of the scheme in Fig. 4.2.

**Experimental design.**
The major task in this step is to determine the set of design points. Possible choices are, for instance, a new block in a $2^n$-design, or an augmentation of the current set with a rather arbitrary number of new points in the case that the optimal design theory is used.

**Collecting the data.**
Usually the revision of the experimental design has no consequences for the collection of the data, and the same procedures as described in Subsection 4.3.1.3 will be used.

**Model fitting and testing.**
After the data has been collected, for every response quantity a linear model must be fitted. These models must subsequently be tested on their validity and predictive ability. The tests, which are relevant for our purpose, are mentioned in Subsection 4.2.5. Once more we emphasize not to rely solely on computer tests; one should also check the models with common sense and consult experts on the problem under study for their appraisal.

**Checks on project goals.**
Finally the goals and the resources of the project will be checked and we must find out whether the model building process can be successfully terminated or a next cycle will have to be initiated.

We will conclude this subsection with a typical use of the scheme in Fig. 4.2 in conjunction with optimal experimental design. If the experimental design is augmented with a certain set of design points, the optimality criterion does not always have to remain the same. In subsequent model building cycles a reasonable alternative may be chosen, for instance:

- Minimization of $\det((\boldsymbol{X}^T\boldsymbol{X})^{-1})$, resulting in good parameter estimates (D-criterion).
- Minimization of the maximum response variance, resulting in good predictive ability in limited areas (G-criterion).
- Minimization of the mean response variance, resulting in good predictive ability in a wide area (V-criterion).

The number of points added in the subsequent augmentations need not always be the same. If the model building process evolves, the model becomes of higher quality and the number of points in augmentations may be lowered.

#### 4.3.1.6 Operating Regression Models

Regression models based on numerical experiments can be used in several ways; in the next section we sum up the most important kinds of application. In these lecture notes we are most concerned with procedures for structured development of regression models. However, till now the use of the developed models happened merely according to "ad hoc" procedures. Such a use of regression models is not the most efficient one, and we will write down some ideas, which probably will enhance future use.

After a regression model has been developed, the following data should be properly recorded:

- The coding and physical meaning of the response quantities.
- A measure for the predictive ability, for instance the variance of the response quantity.
- The coding and physical meaning of the control variables.
- The ranges of the control variables and perhaps some additional constraints. The user of the model should respect these limits as a necessary demand to avoid extrapolation using the model, probably resulting in poor behaviour of the model. Some additional comments on this subject will be made later.
- The number and types of the model functions in the regression equations.
- The estimated model parameters.

Finally it is recommended to record the set of design points on which the regression model is based. This information can be used to check whether a future design point perhaps may specify a so-called hidden extrapolation (Montgomery & Peck, 1992). Although a point lies within the ranges of the control variables, it may lie in a region of the design space, which is not occupied by points used in the experimental design. Montgomery and Peck point out some procedures to check for hidden extrapolation. Such procedures should become tools in future software, which supports the use of regression models.

The means by which the data of the experimental design are collected should also be recorded; for instance one can think of the used finite element package and the file containing the element model. This ensures the possibility to check future results of the regression model and to update it if this might be necessary. For this purpose it will also be useful to establish procedures to gather user's experiences.

### 4.3.2 Applicability of Regression Models

The main reason to develop a regression model is the need for a fast analysis model for complex systems, where accurate analysis results can only be achieved by elaborate numerical analysis. Formulated in this way, regression models may be developed for a very wide class of applications. However, we are concerned with structural analysis and optimization, and we will restrict the following considerations to this case and some related topics.

#### 4.3.2.1 Fast Analysis Modules for Design Offices and Education

**Applicability.**
In mechanical engineering design offices a wide class of machine parts is used to build complex structures. In most cases these parts can be described by not too many (say less than 10) design variables. Textbooks on machine parts show many examples, usually together with rather simple analytical analysis models for the parts. Some typical examples, which are pretty standard in mechanical engineering, are illustrated in Fig. 4.3.

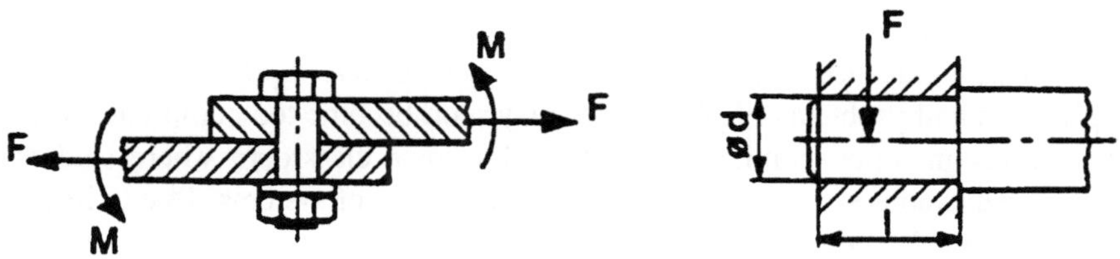


**Figure 4.3** Structural design components.

A second class of interesting parts may be less general, but quite common in certain production organizations or branches. Some examples are illustrated in Fig. 4.4.:

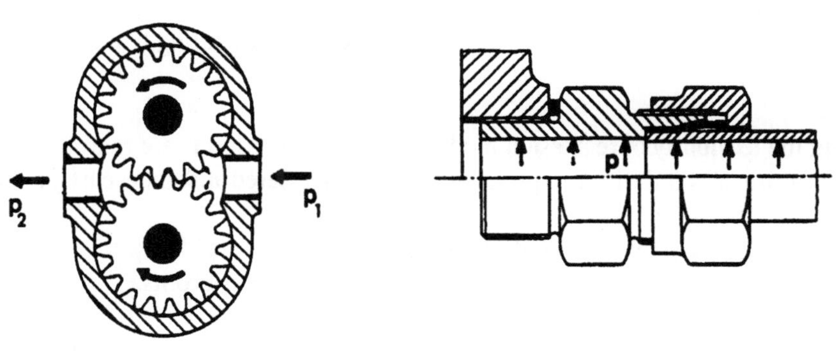


**Figure 4.4** Standardized special products.

A third typical class of problems is that of more or less isolated continuum mechanics problems, for instance stress concentration problems as illustrated in Fig. 4.5.

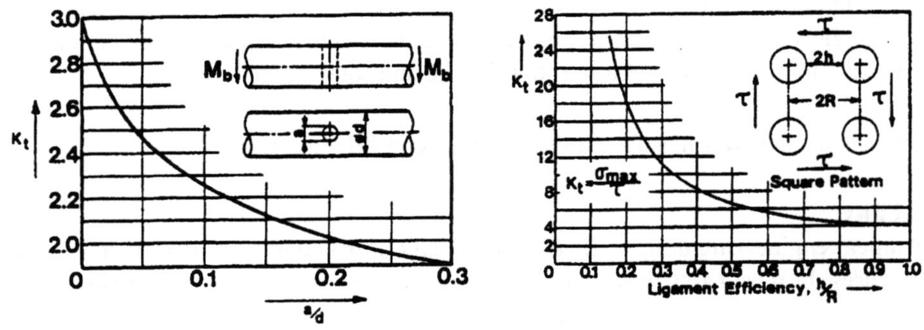

**Figure 4.5** Stress concentration problems

Nowadays the real mechanical behaviour of the machine parts mentioned before can be approximated much better using the finite element method. FEM-analysis of problems like stress concentrations demands also less effort than experimental investigations. However, even for such simple parts and problems FEM-analysis may soon become too comprehensive for ordinary use in a design office. Similar difficulties arise in a classroom where students attend computer-assisted exercises to learn about the real behaviour of systems. Due to limitations on hardware and time only rather simple problems can be treated.

In situations as described above the execution of direct FEM-analyses can be troublesome and the development and use of a number of well-chosen regression models may be useful. The regression models can be developed using standard software. Some other reasons whether to develop or to use regression models are:

- One has not available the hardware, software, knowledge or budget to perform comprehensive FEM-analyses.
- One tries to avoid repeated expensive FEM-analyses.
- Building a catalogue of standard regression models may be a vendible product for software houses. Collecting the FEM-data can be used to reduce idling of computer capacity.

**Feasibility**.

Concerning the feasibility of regression models the following aspects are relevant:

1. The measure of similarity of specimens of the considered machine part or mechanical problem.
2. The demanded accuracy for the response quantities.
3. The effort for development of the regression model versus its profits.

**Sub 1.** A necessary condition for the development of a regression model is a well-defined and stable set of design variables and their ranges, by means of which all specimens of the structure can adequately be described. If during the development changes occur in the set of design variables, one has to decide which part of the already collected data remains valuable.

The demand for similarity cannot always be fulfilled easily, and some additional modeling may be necessary. It can also result in dependencies between design variables; we will give an example. If the elements of a certain class of similar machine parts vary much in size, very often some material properties vary together with the size. For this reason one may decide to incorporate, besides variables concerning the size, those material properties in the set of design variables. It will be clear such a set of design variables is dependent.

**Sub 2.** Evaluation of a developed regression model in a certain design point will result in estimates of the response quantities, which in turn are estimates of the real behaviour of the structure. Since the regression model is fitted on FEM-data, these estimates obviously are less accurate than those obtained directly from a FEM-analysis. However, depending on the effort that is spent on the development of the regression model, approximation of direct FEM-results may be sufficiently close.

**Sub 3.** The effort for the development of a regression model is influenced by the following, certainly dependent factors:

- The number of control variables and their ranges.
- The number of parameters in the regression model and the types of the model functions.
- The number of levels for each control variable.
- The number of design points in the experimental design and the number of candidate points.
- The number of response quantities and the desired accuracy of response estimates.
- Modeling and computer costs for collecting the FEM-data.
- Computer costs for definition of the experimental design and model fitting and testing.

It is impossible to give general rules to derive the costs for the development of a regression model from the above factors. Our advice is to investigate a possible application carefully and to record the experiences for future learning. Following the procedures established in Subsection 4.3.1 one is guided to estimate the costs of development at an early stage and this estimate is subsequently updated.

The possible profits of a developed regression model are determined by:

- The expected number of future FEM-analyses which possibly can be replaced by an evaluation of the regression model, and the savings in human effort and computer costs per analysis.
- The smaller schedule to produce relevant results for projects.
- The possible savings in hardware and software facilities.

It will also be clear that estimating these profits is not an easy task. A problem is that substantial investments for the development of a regression model precede possible profits veiled in future. For the final decision whether or not to develop a particular regression model clearly no general rules can be given; good engineering judgment and some experience will be essential.

#### 4.3.2.2 Tool for Parameter Study and Model Building

Parameter studies and model building are often based on physical experimentation. The experimental design theory is often used to structure such investigations. Nowadays physical experimentation in many cases is combined with numerical analysis, thus resulting in better experiments and better analyses. With today's availability of very advanced numerical analysis packages, parameter studies in certain stages of a project may be solely based on numerical analyses. This situation is very similar to that in the previous subsection.

However, now the problem under study does not concern, for instance, a more or less standard machine part, but a unique physical problem. In such situations sometimes tremendous numerical investigations are performed, the results of which are analyzed with comparatively simple methods. Planning of such parameter studies using the experimental design techniques as described in Section 4.2 will not raise the project's budget, but on the contrary, the project will only gain by doing so.

#### 4.3.2.3 Use in Structural Optimization and Simulation

In Subsection 4.1.2 we already summed up some reasons for integration of experimental design into structural optimization. We repeat those reasons very shortly:

- The design variable concept appears in both disciplines.
- The need for minimizing the number of expensive elementary operations, whether it may be physical experiments or FEM-analyses.
- FEM-analyses can be regarded as numerical experiments.
- Structural optimization programs are suitable to collect data for experimental designs.
- A regression model can serve as a fast analysis model in a structural optimization program.
- Using such an optimization program an approximated global optimum can be found. It is possible to investigate several objective and constraint functions at low computing costs.
- The approximated global optimal design can be used as a starting point for a more accurate optimization, using direct FEM- analysis.

An additional, not yet mentioned reason is the use of gradients in both methods. Effective optimization algorithms use gradients of the objective function and the constraints; in finite element formulations such gradients can be computed effectively.

In developing regression models such gradients can also be used with advantage, because in a structure with $n$ design variables each FEM-analysis provides, per response quantity, $1 + n$ figures (response value + $n$ gradient values) which can be used to estimate the parameters in the regression model. In other words: if gradients are used the fractional experimental design can be reduced globally by a factor of $n$.

Having discussed the potential of regression models in structural optimization, we must find out in which situations it can be used. Globally we may say that the same arguments apply as in the two preceding subsections, because use of experimental design in structural optimization starts with the development of one or more regression models. However concerning optimization some additional remarks can be made.

One remark refers to the extent of the experimental design in relation to the number of direct FEM-analyses used in iterative optimizations. In a problem with n design variables a lower bound to produce a local optimum, using an efficient optimization algorithm, is about n FEM-analyses. If the problem is not too simple at least 5 to 10 different starting points must be used to allow a guess for the global optimum point. This counting applies for one optimization problem. If, for instance, several objective functions must be investigated, a multiple of the above-mentioned $5n$ to $10n$ FEM-analyses will be necessary. Such figures may give an idea about a reasonable number of FEM-analyses planned to develop a regression model.

In the weighting process whether to use direct FEM-analyses or a regression model in a certain optimization problem, the following comparative evaluation of the features and drawbacks of both approaches may be helpful. We consider the following two cases:

1. Iterative optimization based on direct FEM-analyses.
2. Development and use of regression models as an analysis module in iterative optimization processes.

We evaluate a number of aspects for these cases.

**Number of design variables.**

Sub 1. Number may be large (50 up to several hundreds). Nevertheless one should try to reduce the number; design variable linking can be used for that purpose.

Sub 2. Number is rather small; maximum is about 10 if the number of levels per variable is larger than 1. If the used model is a polynomial of first order and if gradients of the response can be used, then 1 design variable level will be sufficient, and a larger number of design variables can be used.

**Gain of insight into the problem**

Sub 1. Rather little insight is gained; this insight is restricted to a small area along the trajectory of the iteration steps.

Sub 2. Insight can rather easily be obtained over the full extent of the design variable space.

**Accuracy of solutions**

Sub 1. Possible high accuracy, which can be controlled by mesh refinement.

Sub 2. Lower accuracy, because regression models approximate a direct FEM-analyses; the element meshes are chosen in advance and usually are relatively coarse.

**Flexibility of problem solution**

Sub 1. With respect to solving the mechanical problem the flexibility is high; in the last extremity one can modify the FEM-model during every iteration step. Flexibility is low concerning the optimization problem, because modification of the objective and/or the constraint functions will require large additional computations.

Sub 2. With respect to solving the mechanical problem the flexibility is low, because the regression model is based on a certain fixed set of design variables and a fixed design variable space. Flexibility is high concerning the optimization problem, because several objective and constraint functions can be investigated at low computer costs.

**Value of the results of a FEM-analysis**

Sub 1. The FEM-analyses usually are only intermediate results in the iteration process.

Sub 2. The FEM-analyses have enduring value due to their incorporation in the regression model.

**Initial effort in problem solution**

Sub 1. Little initial effort is required, because iterative optimization runs can be controlled well and the number of FEM-analyses can be dosed very well.

Sub 2. Great initial effort is required, because the solution procedure starts with the development of the regression model, usually requiring a fairly large number of FEM-analyses.

Again, to make the right choices, one will require good engineering judgement and experience.

## References

Atkinson, A.C., and Donev, A.N. (1992). *Optimum Experimental Designs.* Oxford University Press, Oxford.

Bishop, C.M. (1995). *Neural Networks for Pattern Recognition.* Clarendon Press, Oxford.

Box, G.E.P., and Draper, N.R. (1987). *Emperical Model Building and Response Surfaces.* John Wiley, New York.

Box, G.E.P., Hunter, W.G., and Hunter, J.S. (1978). *Statistics for Experimenters.* John Wiley, New York.

Draper, N.R., and Smith, H. (1998). *Applied Regression Analysis.* 3rd ed., John Wiley, New York.

Fedorov, V.V. (1972). *Theory of Optimal Experiments.* Academic Press, New York.

Friedman, J.H. (1991). Multivariate Adaptive Regression Splines. *The Annals of Statistics*, 19: 1-141

Giunta, A., Watson, L.T., Koehler, J. (1998). A Comparison of Approximation Modeling Techniques: Polynomial versus Interpolating Models. 7th AIAA/USAF/NASA/ISSMO Symposium on Multidisciplinary Analysis and Optimization, paper AIAA-98-4758.

Haftka, R.T., and Gürdal, Z. (1992). *Elements of Structural Optimization.* Kluwer Academic Publishers, Dordrecht.

Jin, R., Chen, W., Simpson, T.W. (2000). Comparative Studies of Metamodeling Techniques under Multiple Modeling Criteria. 8th AIAA/USAF/NASA/ISSMO Symposium on Multidisciplinary Analysis and Optimization, paper AIAA-2000-4801.

Kiefer, J., and Wolfowitz, J. (1959). Optimum Designs in Regression Problems. *Canadian Journal of Mathematics*, 12:363-366.

Koehler, J.R., Owen, A.B. (1996). Computer Experiments. In: *Handbook of Statistics, vol.13* (eds. Ghosh, S., Rao, C.R.), Elsevier Science, Amsterdam.

Lancaster, P., Salkauskas, K. (1986). *Curve and Surface Fitting: an Introduction.* Academic Press, London.

Marquardt, D.W. (1970). Generalized Inverses, Ridge Regression, Biased Linear Estimation, and Nonlinear Estimation. *Technometrics*, 12:591-612.

Mitchell, T.J. (1974). An Algorithm for the Construction of D-optimal Experimental Designs. *Technometrics*, 16:203-210.

Montgomery, D.C., and Peck, E.A. (1992). *Introduction to Linear Regression Analysis.* 2nd ed., John Wiley, New York.

Montgomery, D.C. (1996). *Design and Analysis of Experiments*, 4th ed. John Wiley, New York.

Myers, R.H., Montgomery, D.C. (1995). *Response Surface Methodology: Process and Product Optimization using Designed Experiments.* J. Wiley, New York.

Nagtegaal, R. (1987). *Computer Aided Design of Experiments. A program for experimental design and model building.* (in Dutch). Report WFW 87.005, Eindhoven University Press.

Powell, M.J.D. (1987). Radial Basis Functions for Multivariable Interpolation: a Review. in: *Algorithms for Approximation* (eds. Mason, J.C., Cox, M.G.) Oxford University Press, London.

Sacks, J., Welch, W.J., Mitchell, T.J., Wynn, H.P. (1989). Design and Analysis of Computer Experiments. *Statistical Science*, 4: 409-435.

Schoofs, A.J.G., Klink, M.B.M., van Campen, D.H. (1992). Approximation of structural optimization problems by means of designed numerical experiments. *Structural Optimization*, 4: 206-212.

Schoofs, A.J.G., Houten, M.H. van, Etman, L.F.P., Campen, D.H. van (1997). Global and Mid-Range Function Approximation for Engineering Optimization. *Mathematical Methods of Operations Research*, 46:.335-359.

Smith, K. (1918). On the Standard Deviations of Adjusted and Interpolated Values of an Observed Polynomial Function and its Constants and the Guidance They Give Towards a Proper Choice of the Distribution of Observations. *Biometrica*, 12:1-85.

Welch, W.J. (1982). Branch and Bound Search for Experimental Designs Based on D-Optimality and Other Criteria. *Technometrics*, 24:41-48

Welch, W.J. (1983). A Mean Squared Error Criterion for the Design of Experiments. *Biometrica*, 70:205-223.

# CHAPTER 5

# Modelling and Approximation Strategies in Optimization - Global and Mid-Range Approximations, Response Surface Methods, Genetic Programming, Low / High Fidelity Models

V.V. Toropov

School of Engineering, University of Bradford, UK

**Abstract.** In this chapter global and mid-range approximations of the objective and constraint functions are introduced, discussed and illustrated by examples of real-life applications. Particular attention is paid to the development of techniques applicable to difficult design optimization problems in which the objective and constraint functions are computationally expensive, can be affected by numerical noise and at some combinations of design variables could be impossible to evaluate. Genetic programming methodology is introduced as a systematic way of selecting a structure of high quality global approximations. Mechanistic approximations and techniques based on the interaction of high and low fidelity numerical models are discussed and illustrated by examples.

## 5.1 Introduction

Many real-life design optimization problems have the following common features:

- the objective and constraint functions are evaluated as a result of expensive numerical computations, e.g. using a FEM or a CFD code;
- function values and/or their derivatives may contain some level of numerical noise;
- domain-dependent calculability of some of the response functions, i.e. situations when these functions cannot be evaluated at some points of the design variable space.

The presence of numerical noise means that function values, corresponding to two slightly different sets of design variables, may have a considerable random variation. Moreover, even function values, corresponding to the same set of design variables, may be different depending on the optimization search history and/or iteration history of an iterative response analysis. As an example, we mention FEM analysis in conjunction with Adaptive Mesh Refinement (AMR), see van Keulen et al. (1997), van Keulen and Toropov (1997) and Polynkine et al. (1996), where the mesh density information can be influenced by the optimization history. The level of noise can vary from negligible to significant (or even intolerable) depending on the formulation and the size of the optimization problem as well as the size and specific features of the response analysis model.

As an example of the noisy behavior of response functions, consider a simple optimization problem of a square plate with a circular central hole, van Keulen and Toropov (1998). The radius of the hole is treated as a single design variable. For a coarse (Fig. 5.1) and a fine mesh (Fig. 5.2), the normalized stress constraint and normalized stability constraint are shown in Fig. 5.3 and Fig. 5.4, respectively. In each case the mesh density was kept constant. It can be seen that the coarse mesh results possess a significantly higher level of noise. Another obvious observation is that the numerical solution is also characterized by an offset from the exact solution. The magnitude of the offset depends on the specified mesh density and tends to zero when the mesh density is increased.

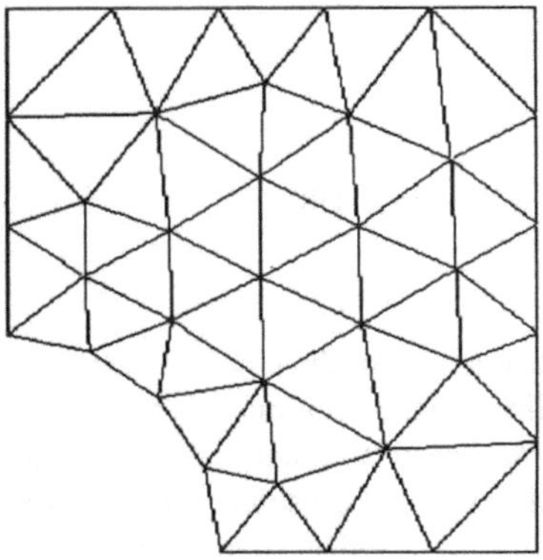

**Figure 5.1** Coarse mesh.

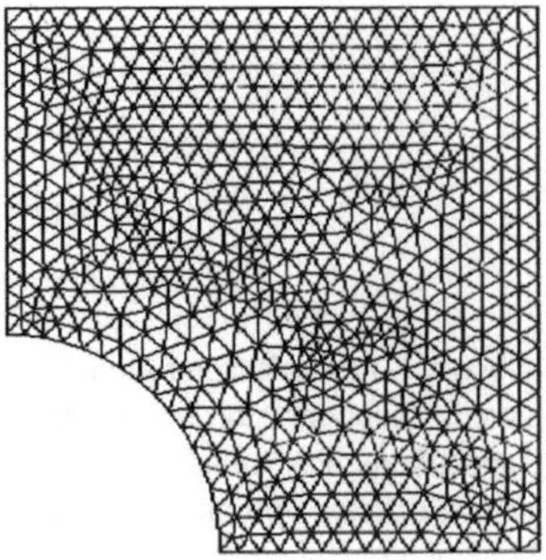

**Figure 5.2** Fine mesh.

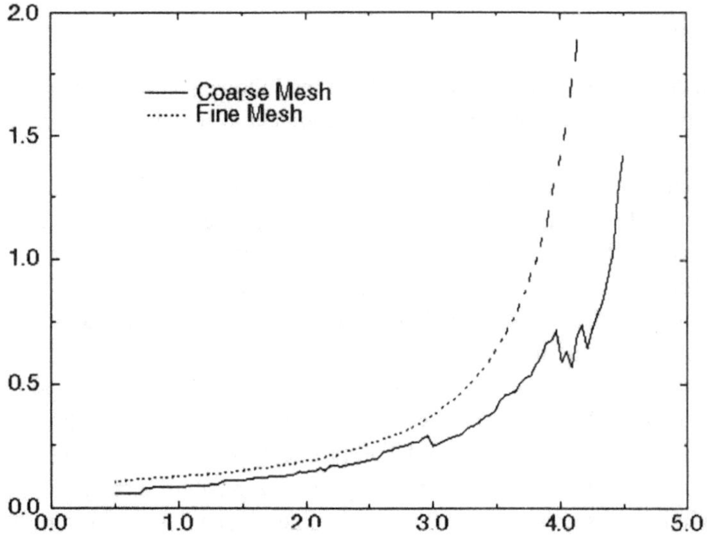


**Figure 5.3** Normalized stress constraint.

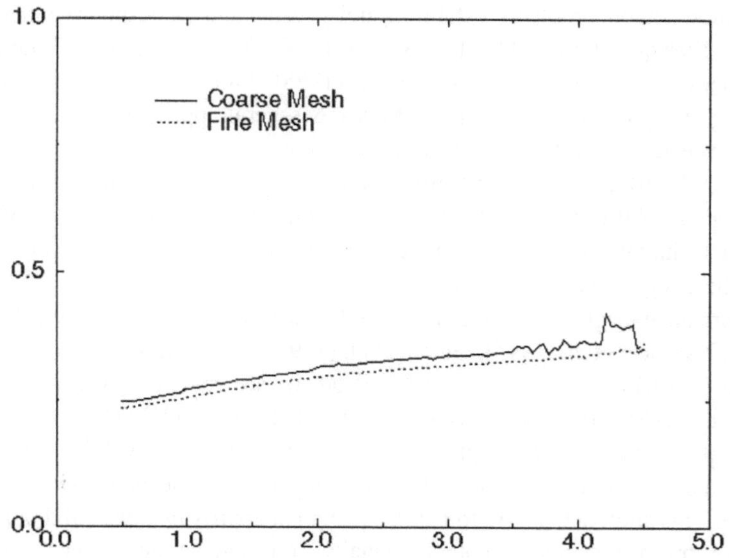


**Figure 5.4** Normalized stability constraint.

Another example (Polynkine et al., 1996) relates to the optimization of a conical shell made of a continuous fibre reinforced thermoplastic (CFRTP) material (Fig. 5.5). The response analysis in this case involves two modelling steps, namely, simulation of the thermoforming process and the FE analysis of the obtained product. The problem is formulated as maximization of the strain energy subject to the constant volume constraint. The design variables are the height of the cone and the base radius. Fig. 5.6 shows the noisy behavior of the objective function.

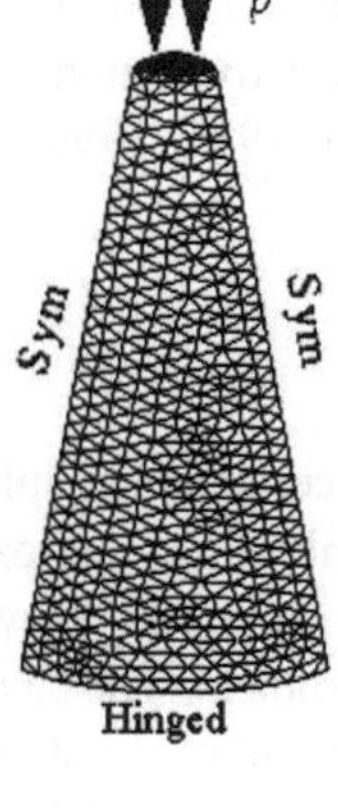


**Figure 5.5** CFRTP conical shell.

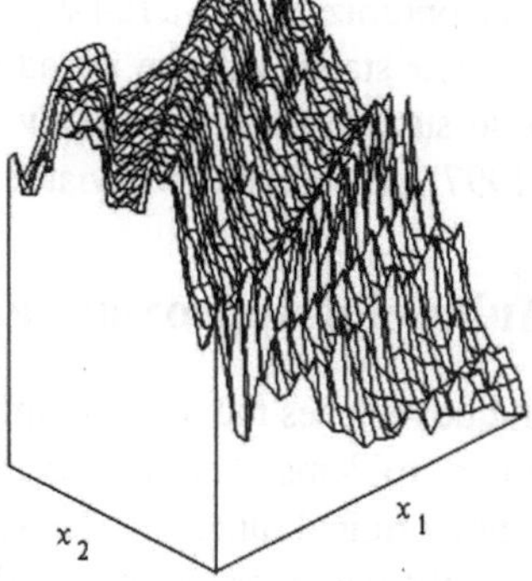


**Figure 5.6** Objective function for the CFRTP conical shell problem.

Design optimization problems exhibiting noisy behaviour of response functions have been considered by Free et al. (1987), Giunta et al. (1994), Narducci et al. (1995), Valorani and Dadone (1995), Toropov et al. (1996), Polynkine et.al. (1996), Etman et al. (1996), van Keulen et al. (2000), among many others. During the last decade, several approximation techniques were proposed which take account of the negative effect of noise.

Rasmussen (1990) investigated the influence of noise in the first order sensitivities on the performance of his ACAP technique, it was shown that even a considerable level of noise could be tolerated. The technique does not take into account the noise in the function values, i.e. deals with the noise in the design sensitivities only.

The performance of global approximation techniques was investigated by Schoofs (1987), Rikards (1993), Narducci et al. (1995), Giunta et al. (1997), Venter et al. (1998), Roux et al. (1998), Madsen et al. (2000), Sevant et al. (2000) and others, see reviews by Barthelemy and Haftka (1993), Haftka (1997) and Sobieszanski-Sobieski and Haftka (1997). These techniques are based on the least-squares surface fitting approach which was originally developed for problems where the function values are obtained as a result of physical measurements inherently containing some level of noise. This approach allows to construct explicit approximations valid in the entire design variable space, but it is becoming computationally expensive as the number of design variables grows.

The issue of domain-dependent calculability arises when at some points of the design variable space the response functions cannot be evaluated by the adopted simulation code, Toropov et al. (1999). A traditional measure to prevent the failure of the optimization process when such a problem is encountered is to try to reduce the step length in the optimization technique. However, this may considerably slow down the optimization process and, still, does not guarantee the trouble-free convergence. The non-existence of the response function often means that an unrealistic design has been considered which cannot be analysed by a response analysis procedure. Sometimes this can be a result of some numerical problems in a simulation techniques (loss of convergence, etc.). Such situations can lead to a premature termination or failure of the optimization method. Examples of such problems include optimization of dynamic behaviour of nonlinear mechanical systems by Markine (1999), optimization of aerodynamic characteristics of aerofoils by Toropov et al. (1999), optimization of a railway track structure by Markine et al. (2000) to name but a few.

This chapter starts with the introduction into the multipoint approximation method based on the response surface fitting (Toropov 1989, Toropov et al. 1993, 1995, 1996, van Keulen and Toropov 1997), sometimes abbreviated to MAM or MARS.

## 5.2 Multipoint Approximation Method

This technique replaces the original optimization problem by a succession of simpler mathematical programming problems. The functions in each iteration present mid-range approximations to the corresponding original functions. These functions are noise-free or, at least, the level of noise does not influence the convergence of an individual optimization sub-problem. The solution of an individual sub-problem becomes the starting point for the next step, the move limits are changed and the optimization is repeated iteratively until the optimum is reached. Each approximation function is defined as a function of design variables as well as a number of tuning parameters. The latter are determined by the weighted least squares surface fitting using the original function values (and their derivatives, when available) at several points of the design variable space. This selection of points will be referred to as a plan (design) of numerical experiments. Some of the plan points are generated in a current iteration, and the rest is taken from the pool of points considered in the previous iterations.

In order to deal with the issue of domain-dependent calculability, the points of a plan where a response function is impossible to compute (e.g. resulting in a crash of a simulation code) are to be identified in the process of planning of numerical experiments, and a different point allocated.

There are several important aspects in the MAM which considerably influence the rate of convergence, namely:

- choice of the structure of the approximate expressions;
- strategy for allocation of the weight coefficients;
- choice of the plan of numerical experiments;
- move limit strategy.

A general optimization problem can be formulated as

$$F_0(\boldsymbol{x}) \rightarrow \min, \quad F_j(\boldsymbol{x}) \leq 1 \ (j = 1,\ldots,M), \quad A_i \leq x_i \leq B_i \ (i = 1,\ldots,N) \tag{5.1}$$

where $\boldsymbol{x}$ is the vector of design variables. In order to reduce the number of calls for the response function evaluations and to lessen the influence of noise, the MAM replaces the optimization problem by a sequence of approximate optimization problems:

$$\tilde{F}_0^k(\boldsymbol{x}) \rightarrow \min, \quad \tilde{F}_j^k(\boldsymbol{x}) \leq 1 \ (j = 1,\ldots,M), \quad A_i^k \leq x_i \leq B_i^k, A_i^k \geq A_i, \ B_i^k \leq B_i \ (i = 1,\ldots,N) \tag{5.2}$$

where $k$ is the iteration number.

The selection of the noise-free approximate response functions $\tilde{F}_j^k(\boldsymbol{x})$ $(j = 0,\ldots, M)$ is such that their evaluation is inexpensive as compared to the evaluation of the response functions $F_j$, although they are not necessarily explicit functions of the design variables, see Section 5.5. The approximate response functions are intended to be adequate in a current search sub-domain. This is achieved by appropriate planning of numerical experiments and use of move limits defined by the side constraints $A_i^k$ and $B_i^k$. Moreover, it is attempted to achieve the best quality of the approximation functions in those regions of the design variable space where the solution of the approximate optimization problem can be expected, e.g. on the boundary of the feasible region.

The approximations are determined by means of weighted least squares:

$$G_j(\boldsymbol{a}_j) = \sum_{p=1}^{P} w_{pj} \left[ F_j(\boldsymbol{x}_p) - \tilde{F}_j(\boldsymbol{x}_p, \boldsymbol{a}_j) \right]^2 \rightarrow \min \tag{5.3}$$

Here minimization is carried out with respect to the tuning parameters $\boldsymbol{a}_j$ and $w_{pj}$ refers to the weight coefficients. Their selection will be discussed in a subsequent section.

If the first order derivatives $F_j(\boldsymbol{x})_{,i} = \partial F_j(\boldsymbol{x}) / \partial x_i$ at plant points $\boldsymbol{x}_p$ are known, the least-squares surface fitting is equivalent to the minimization of the function (Toropov et al., 1993)

$$G_j(\boldsymbol{a}_j) = \sum_{p=1}^{P} w_{pj} \left\{ \left[ F_j(\boldsymbol{x}_p) - \widetilde{F}_j(\boldsymbol{x}_p, \boldsymbol{a}_j) \right]^2 + \gamma \delta_{pj} \sum_{i=1}^{N} \left[ F_j(\boldsymbol{x}_p)_{,i} - \widetilde{F}_j(\boldsymbol{x}_p, \boldsymbol{a}_j)_{,i} \right]^2 \right\} \rightarrow \min \quad (5.4)$$

where $\delta_{pj} = 1 / \sum_{i=1}^{N} \left[ F_j(\boldsymbol{x}_p)_{,i} \right]^2$ is a normalizing coefficient and $0 < \gamma < 1$ reflects the weight of the information on the sensitivities as compared to the information on the function values.

### 5.2.1 Intrinsically Linear Approximations

To construct the simplified expressions $\widetilde{F}_j^k(\boldsymbol{x})$, it is necessary to define them as a function of tuning parameters $\boldsymbol{a}$. The approximate response functions must be flexible enough to mimic the behaviour of the response functions with sufficient accuracy. On the other hand, they must be easy to evaluate and should not possess any significant level of numerical noise. To simplify notation, we will suppress the indices on the functions $F_j(\boldsymbol{x})$ and $\widetilde{F}_j^k(\boldsymbol{x})$. The simplest case is a linear function of the tuning parameters $\boldsymbol{a}$:

$$\widetilde{F}(\boldsymbol{x}) = a_0 + \sum_{l=1}^{L} \varphi_l(\boldsymbol{x}) a_l \quad (5.5)$$

The optimization problem (5.4) can be rewritten using the matrix notation as follows:
Minimize

$$G(\boldsymbol{a}) = (\boldsymbol{f} - \widetilde{\boldsymbol{f}})^{\mathrm{T}} \boldsymbol{W} (\boldsymbol{f} - \widetilde{\boldsymbol{f}}) \quad (5.6)$$

where

$$\boldsymbol{f} = \begin{Bmatrix} F(\boldsymbol{x}_1) \\ F(\boldsymbol{x}_1)_{,1} \\ \ldots \\ F(\boldsymbol{x}_1)_{,N} \\ \ldots \\ F(\boldsymbol{x}_P) \\ F(\boldsymbol{x}_P)_{,1} \\ \ldots \\ F(\boldsymbol{x}_P)_{,N} \end{Bmatrix}, \quad \widetilde{\boldsymbol{f}} = \begin{Bmatrix} \widetilde{F}(\boldsymbol{x}_1) \\ \widetilde{F}(\boldsymbol{x}_1)_{,1} \\ \ldots \\ \widetilde{F}(\boldsymbol{x}_1)_{,N} \\ \ldots \\ \widetilde{F}(\boldsymbol{x}_P) \\ \widetilde{F}(\boldsymbol{x}_P)_{,1} \\ \ldots \\ \widetilde{F}(\boldsymbol{x}_P)_{,N} \end{Bmatrix}, \quad \boldsymbol{W} = diag\,[w_1^{(0)}, w_1^{(1)}, \ldots, w_1^{(N)}, \ldots, w_P^{(0)}, w_P^{(1)}, \ldots, w_P^{(N)}].$$

The vector $\boldsymbol{f}$ is the vector of $P(N+1)$ elements containing information from experiments obtained at $P$ plan points, i.e. values of implicit functions $F(\boldsymbol{x}_p)$ and their derivatives $F(\boldsymbol{x}_p)_{,i}$

$(p = 1,..., P$ and $i = 1, ..., N)$ ; the vector $\tilde{f}$ presents simplified function values and their derivatives with respect to the design variables at the same plan points; $\boldsymbol{W}$ is the diagonal matrix $P(N+1)\times P(N+1)$ consisting of weight coefficients $w_p^{(0)}$ and $w_p^{(i)}$.

If we assume that the simplified function $\tilde{F}(x)$ is a linear function (5.5) of tuning parameters $\boldsymbol{a} = (a_0, a_1, ..., a_L)^{\mathrm{T}}$, then the vector $\tilde{f}$ can be expressed in the following form:

$$\tilde{f} = \boldsymbol{\Phi a} \tag{5.7}$$

where

$$\boldsymbol{\Phi} = \begin{bmatrix} 1 & \varphi_{11} & \cdots & \varphi_{L1} \\ 0 & \varphi_{11,1} & \cdots & \varphi_{L1,1} \\ \cdots & \cdots & \cdots & \cdots \\ 0 & \varphi_{11,N} & \cdots & \varphi_{L1,N} \\ \cdots & \cdots & \cdots & \cdots \\ 1 & \varphi_{1P} & \cdots & \varphi_{LP} \\ 0 & \varphi_{1P,1} & \cdots & \varphi_{LP,1} \\ \cdots & \cdots & \cdots & \cdots \\ 0 & \varphi_{1P,N} & \cdots & \varphi_{LP,N} \end{bmatrix}, \quad \boldsymbol{a} = \begin{Bmatrix} a_0 \\ a_1 \\ \cdots \\ \cdots \\ \cdots \\ \cdots \\ \cdots \\ \cdots \\ a_L \end{Bmatrix}, \quad \varphi_{lp} = \varphi_l(x_p), \ (l = 0, ..., L \text{ and } p = 1,..., P).$$

The necessary condition for the minimum of the quadratic form (5.6) with respect to the tuning parameters $\boldsymbol{a}$ leads to the linear system of $L+1$ normal equations with $L+1$ unknowns $a_l$, $(l=0,1,...,L)$:

$$\boldsymbol{\Phi}^{\mathrm{T}} \boldsymbol{W} \boldsymbol{\Phi} \boldsymbol{a} = \boldsymbol{\Phi}^{\mathrm{T}} \boldsymbol{W} f, \tag{5.8}$$

Note that the structure of the simplified expression (5.5) is rather general because the individual regressors $\varphi_l$ can be arbitrary functions of design variables.

The procedure described above can be further generalized by the application of *intrinsically linear* functions (Draper and Smith, 1981). Such functions are nonlinear, but they can be led to linear ones by simple transformations. The most useful function among them is the multiplicative function

$$\tilde{F}(x) = a_0 \prod_{l=1}^{L} \varphi_l(x)^{a_l} \tag{5.9}$$

with the logarithmic transformation $\ln \tilde{F}(x) = \ln a_0 + \sum_{l=1}^{L} a_l \ln \varphi_l(x)$.

In this case, the system of linear equations (5.8) contains the following matrix and vectors:

$$
\boldsymbol{\Phi}=\begin{bmatrix}
1 & \ln\varphi_{11} & \dots & \ln\varphi_{L1} \\
0 & \dfrac{F(\boldsymbol{x}_1)}{\varphi_{11}}\varphi_{11,1} & \dots & \dfrac{F(\boldsymbol{x}_1)}{\varphi_{L1}}\varphi_{L1,1} \\
\dots & \dots & \dots & \dots \\
0 & \dfrac{F(\boldsymbol{x}_1)}{\varphi_{11}}\varphi_{11,N} & \dots & \dfrac{F(\boldsymbol{x}_1)}{\varphi_{L1}}\varphi_{L1,N} \\
\dots & \dots & \dots & \dots \\
1 & \ln\varphi_{1P} & \dots & \ln\varphi_{LP} \\
0 & \dfrac{F(\boldsymbol{x}_P)}{\varphi_{1P}}\varphi_{1P,1} & \dots & \dfrac{F(\boldsymbol{x}_P)}{\varphi_{LP}}\varphi_{LP,1} \\
\dots & \dots & \dots & \dots \\
0 & \dfrac{F(\boldsymbol{x}_P)}{\varphi_{1P}}\varphi_{1P,N} & \dots & \dfrac{F(\boldsymbol{x}_P)}{\varphi_{LP}}\varphi_{LP,N}
\end{bmatrix},\quad
\boldsymbol{f}=\begin{Bmatrix}
\ln F(\boldsymbol{x}_1) \\ F(\boldsymbol{x}_1)_{,1} \\ \dots \\ F(\boldsymbol{x}_1)_{,N} \\ \dots \\ \ln F(\boldsymbol{x}_P) \\ F(\boldsymbol{x}_P)_{,1} \\ \dots \\ F(\boldsymbol{x}_P)_{,N}
\end{Bmatrix},\quad
\boldsymbol{a}=\begin{Bmatrix}
\ln a_0 \\ a_1 \\ \dots \\ \dots \\ \dots \\ \dots \\ \dots \\ \dots \\ a_L
\end{Bmatrix}.
$$

Another useful intrinsically linear function is the power function

$$
\tilde{F}(\boldsymbol{x})=\left[(a_0+\sum_{l=1}^{L}\varphi_l(\boldsymbol{x})a_l)\right]^{\alpha} \tag{5.10}
$$

with the transformation $\tilde{F}(\boldsymbol{x})^{1/\alpha}=a_0+\sum_{i=1}^{L}\varphi_l(\boldsymbol{x})a_l$. In this case

$$
\boldsymbol{\Phi}=\begin{bmatrix}
1 & \varphi_{11} & \dots & \varphi_{L1} \\
0 & \alpha F(\boldsymbol{x}_1)^{(1-1/\alpha)}\varphi_{11,1} & \dots & \alpha F(\boldsymbol{x}_1)^{(1-1/\alpha)}\varphi_{L1,1} \\
\dots & \dots & \dots & \dots \\
0 & \alpha F(\boldsymbol{x}_1)^{(1-1/\alpha)}\varphi_{11,N} & \dots & \alpha F(\boldsymbol{x}_1)^{(1-1/\alpha)}\varphi_{L1,N} \\
\dots & \dots & \dots & \dots \\
1 & \varphi_{1P} & \dots & \varphi_{LP} \\
0 & \alpha F(\boldsymbol{x}_P)^{(1-1/\alpha)}\varphi_{1P,1} & \dots & \alpha F(\boldsymbol{x}_P)^{(1-1/\alpha)}\varphi_{LP,1} \\
\dots & \dots & \dots & \dots \\
0 & \alpha F(\boldsymbol{x}_P)^{(1-1/\alpha)}\varphi_{1P,N} & \dots & \alpha F(\boldsymbol{x}_P)^{(1-1/\alpha)}\varphi_{LP,N}
\end{bmatrix},\quad
\boldsymbol{f}=\begin{Bmatrix}
F(\boldsymbol{x}_1)^{1/a} \\ F(\boldsymbol{x}_1)_{,1} \\ \dots \\ F(\boldsymbol{x}_1)_{,N} \\ \dots \\ F(\boldsymbol{x}_P)^{1/a} \\ F(\boldsymbol{x}_P)_{,1} \\ \dots \\ F(\boldsymbol{x}_P)_{,N}
\end{Bmatrix},\quad
\boldsymbol{a}=\begin{Bmatrix}
a_0 \\ a_1 \\ \dots \\ \dots \\ \dots \\ \dots \\ \dots \\ \dots \\ a_L
\end{Bmatrix}
$$

When the sensitivity information is not available, i.e. the formulation (5.3) is used, then the matrix $\boldsymbol{\Phi}$ and the vector $\boldsymbol{f}$ do not contain the terms related to the derivatives of the functions *F(x)* and $\varphi_l(x)$.

Intrinsically linear functions have been successfully used for a variety of design optimization problems by Toropov (1989), Toropov et al. (1993, 1996), Toropov and Carlsen (1994), Polynkine et al. (1995, 1996a, 1996b), Markine et al. (1996a, 1996b), van Keulen et al. (1997), van Keulen and Toropov (1997), Markine (1999). The advantage of these approximation functions is that a relatively small number of tuning parameters $a_i$ is to be determined, and the corresponding least squares problem is solved easily. After application of an appropriate transformation only a set of linear equations has to be solved. It should be mentioned, though, that the transformation changes the definition of the weight coefficients $w_{pj}$ used in the weighted least squares fitting.

When the simplest case of the linear function (5.5) $\varphi_i(\boldsymbol{x}) = x_i$ ($i=1,...,N$) is used, the convergence at the end of an optimization process often deteriorates. Generally, the multiplicative approximations show better convergence characteristics (even when $\varphi_i(\boldsymbol{x}) = x_i$ ) but the functions of the optimization problem should be normalized, for otherwise a floating point overflows can be encountered. Higher order approximations, e.g. full quadratic polynomials, lead to a significantly better quality of approximation. However, these approximation functions require a considerably larger number of designs to be evaluated.

It should be noted that the requirements to the accuracy of the mid-range approximations can be more relaxed as compared to the case of the global ones but, obviously, more accurate approximations would produce a faster convergence. Several new approaches have been investigated in the attempt to produce new high quality approximations valid for a larger range of design variables, see subsequent sections of this chapter.

### 5.2.2 Weight Coefficients

The weight coefficients influence the relative contribution of information on function values (and their derivatives, if available). Their choice strongly influences the difference in the quality of the approximations in different regions of the design variable space. Since the optimum point usually belongs to the boundary of the feasible region, the approximation functions should be more accurate in that region. Thus, the information at the points located near the boundary of the feasible region is to be treated with greater weights. This can be achieved by allocating an appropriate weight to the value of a constraint function at a point, e.g. $w_{pj}^{C} = e^{-4\left|F_j(x_p)-1\right|}$. The weights will then depend on the values of the constraint functions. It should be noted that some other definitions of weights are possible, see, for example, Toropov et al. (1993).

In a similar manner a larger weight $w_{pj}^{O}$ can be allocated to a design with a better objective function, see Toropov et al. (1995), van Keulen et al. (1996, 1997), van Keulen and Toropov (1997) for more details. Numerical examples showed, however, that the effect of this additional weight coefficient is relatively small.

In certain cases, e.g. when adaptive mesh refinement techniques are applied, it is possible to specify the error requested from the numerical analysis procedure (error of the numerical experiment). The difference in the quality of data, produced by response analyses with different prescribed errors, can also be taken into account by means of an additional weight coefficient $w_{pj}^{E}$. This can be efficiently used for the solution of problems where the response analysis is obtained using models of variable complexity and accuracy. Van Keulen et al. (1997) and van Keulen and Toropov (1997) showed examples where an appropriate strategy for the choice of the prescribed error (i.e. complexity of the model) throughout the optimization process could significantly reduce the overall computational effort. The basic idea is not to keep a constant prescribed error (and model complexity) but to use a relatively rough model in the beginning of the optimization process and to refine it as the process progresses. Correspondingly, the quality of approximations can be low in the beginning and improve in the process of optimization.

The weight coefficients finally used for the weighted least squares fitting are now taken as $w_{pj}^E = w_{pj}^C w_{pj}^O w_{pj}^E$.

When no information on the accuracy of the response functions is available or a certain contribution is not relevant, the corresponding term is taken as unity.

When a few optimization steps have been carried out, a database with response function values becomes available. In addition to these design points available from the database, new designs are evaluated as part of a step in the optimization process. In order to achieve good quality approximations in a current search sub-domain, a selection of plan points must be made. All designs located in the current search sub-domain will be included in the weighted least squares fitting (if the corresponding response function has been successfully evaluated, of course). Generally, points located far from the current search sub-domain would not contribute to the improvement of the quality of the resulting approximation functions in the current search sub-domain. For this reason only points located in the neighbourhood of the current search sub-domain are taken into account, as depicted in Fig. 5.7. A box in the space of design variables, which is approximately 1.5 to 2.0 times larger than the box representing the current search domain, was found by numerical experimentation to be a reasonable choice for the size of the neighbourhood.

It should be noted that rejection of points not belonging to the neighbourhood could also be looked upon as if a zero weight was assigned to a particular design.

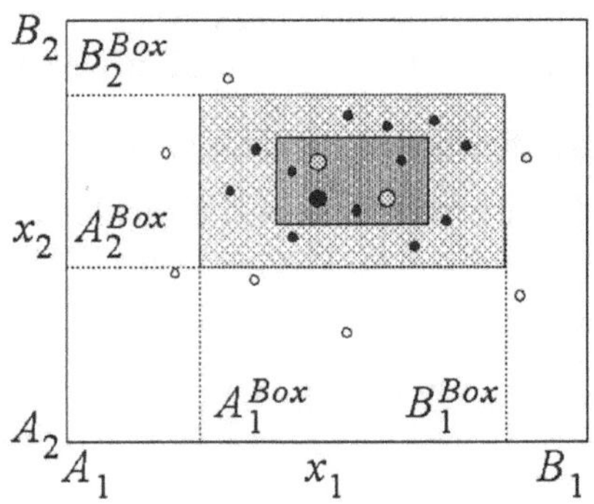


**Figure 5.7** Current search sub-domain and its neighbourhood (points marked ° are not included).

### 5.2.3 Plan of Experiments

In the course of the optimization process new points must be generated in specified domains of the design variable space. In the literature several schemes have been proposed for generating new plans of experiments, see Free et al. (1987) and Burgee et al. (1996) among others. A simple plan of experiments has been used (Toropov 1989, Toropov et al. 1993) which is based on

$N$ new points, where $N$ is the number of design variables. The locations of these points were obtained by a perturbation of each design variable by a fraction of the corresponding size of the current search sub-domain, as shown in Fig. 5.7 by larger circles. Generally, this scheme works well in conjunction with intrinsically linear approximations but, due to its simplicity, it may have some disadvantages. Firstly, it does not take into account design points which are already located in the present search sub-domain and, therefore, newly evaluated designs may be located too close to previously evaluated designs. Secondly, the scheme is inflexible with respect to the number of added points and, therefore, has to be modified if other than intrinsically linear approximation functions are used.

A different scheme has been suggested by van Keulen (see van Keulen and Toropov, 1999) which does not suffer from the disadvantages mentioned above. The scheme must work efficiently even if only a single new design point is generated. Hence if there is a need for a certain number of new design points, the algorithm will be used several times, each time funding the same number of unknowns, namely $N$. Another reason for the development of a scheme adding a single point per run comes from the need for parallel implementation, as discussed in Section 5.2.5. Finally, the scheme has to remain effective if different approximation functions are used within the same optimization task.

The simplest possible scheme is to add new points randomly. This is one of the ways of dealing with the issue of domain-dependent calculability of response functions (Toropov et al. 1999). Each added point is checked for calculability of the response function and, if the check fails, a new point is generated until a required number of plan points (all passing the check) are obtained. However, in this simple scheme the previously created points are not taken into account which can lead to poor plans.

It has been suggested (van Keulen and Toropov, 1999) to distribute points as homogeneously as possible in the sub-domains of interest. This is done by the introduction of a cost function

$$Q=\sum_{p=1}^{P}\frac{N^2}{\left\|\bar{\boldsymbol{x}}_p-\bar{\boldsymbol{d}}\right\|^2}+\sum_{p=1}^{P}\sum_{i=1}^{N}\frac{1}{([\bar{x}_i]_p-\bar{d}_i)^2}+\sum_{i=1}^{N}\frac{1}{(2\bar{d}_i)^2}+\sum_{i=1}^{N}\frac{1}{(2-2\bar{d}_i)^2}+ \sum_{i=1}^{N}\sum_{j=i+1}^{N}\frac{1}{(\bar{d}_i-\bar{d}_j)^2} \tag{5.11}$$

which is minimized with respect to the location of the new point $\boldsymbol{d}$. Symbols denoted $\overline{\ldots}$ refer to coordinates which are normalized in the sub-domain of interest. The first term in the expression attempts to maximize the distance between points, and the second term promotes a homogeneous distribution along the coordinate axes. The third and fourth terms ensure that points do not belong to the boundary of the sub-domain. The last term prevents points from aligning along the diagonal of the search sub-region when only a few points are available. It is reminded that if more points are to be allocated to a specific sub-domain, the minimization of $Q$ is repeated to obtain coordinates for each new point.

As already mentioned, in each step of the optimization a new set of designs, located in the current search sub-domain, will be evaluated. The number of additional points can be varied, e.g. the minimum number of new plan points equals the difference between the largest number of tuning parameters used in any of the approximation functions and the number of plan points already available in the current search sub-domain. As all points in the current search sub-domain and its neighborhood are involved in the weighted least squares fitting, more points are included in the plan of experiments than the minimum possible number. This has the advantage that the approximations become more reliable and it also reduces the effect of noise. Consequently, the convergence of the optimization process is improved. It may be advantageous to add even more points during a step of the optimization process. These additional points can be located in the current search sub-domain, but may also be placed somewhere in the entire search domain. In the latter case the algorithm should be accompanied by a scanning process of the continuously enhanced database. After solving the approximate optimization problem and evaluation of the corresponding design, the database is scanned for better points in terms of both objective and constraint functions. If a better point is found, this will serve as the starting point of the subsequent optimization step. If this happens, there will be a jump in the optimization trajectory. The resulting algorithm thus could be considered as a blend of a random search and a successive approximation technique. Disadvantages are peculiar convergence characteristics and a higher cost of computations. The main advantage is that the algorithm tends to be more robust in terms of lesser sensitivity to local optima. Also, such a strategy can become beneficial for some of parallel implementations, see Section 5.2.5.

It should be stated that when design sensitivities are available, i.e. formulation (5.4) is used in the least-squares surface fitting, a plan of new numerical experiments becomes trivial: it adds the point obtained as the solution of the approximated problem (5.2) in a previous iteration to the data base of previously examined points.

### 5.2.4 Move Limit Strategy

After having solved the approximate optimization problem, a new search subregion must be defined, i.e. its new size and its location have to be specified. This is done on the basis of a move limit strategy which is summarized here. The reader is referred for more details to van Keulen and Toropov (1997) and van Keulen et al. (1996).

The first step is to evaluate several indicators which are the basis for the move limit strategy. The first indicator is the quality index of the approximation functions. It is defined as the largest relative approximation error at the set of design variables corresponding to the solution of the approximate optimization problem. The quality of the approximation functions is then categorized as "poor", "reasonable" or "good". The second indicator is the location of the sub-optimum point in the current search subregion. When none of the current move limits is active, the solution is considered as "internal", otherwise it is denoted as "external". The third and fourth indicators are based on the movement history. For that purpose the angle between the last two move vectors is calculated. This will be the basis to identify the movement as "backward" or "forward". If the move vectors are nearly parallel, the convergence history will be labelled as "straight", otherwise it is marked as "curved". The fifth indicator, used in the termination criterion, is the size of the present search subregion. According to this indicator, the size can be "small" or "large". The sixth indicator is based on the value of the most active constraint. It is used to label the current solution as "close" or "far" from the boundary of the feasible and infeasible regions in the space of design variables. Once the indicators have been determined, the sub-domain is moved and resized. In the present implementation only an isotropic resizing is used. Both reduction and enlargement of the sub-domain can be applied. Several levels of resizing can be applied, ranging from a moderate resizing, typically changing the size in the order of 10%, up to a quite aggressive resizing when the changes can be of the order of 100% or more. A summary of the move limit strategy as well as termination criteria is presented in Table 5.1.

If the obtained point does not pass the check for calculability of the response functions, the search sub-domain is reduced and the approximated problem (5.2) is solved again. The only essential assumption here is that all functions of the optimization problem exist at the starting point.

**Table 5.1** Overview of move limit strategy.

<table>
<tr><td colspan="7">Approximations “poor”</td></tr>
<tr><td colspan="2">“Small”</td><td colspan="5">“Large”</td></tr>
<tr><td colspan="2" rowspan="2">Stop: No convergence found</td><td colspan="3">“Straight”</td><td colspan="2">“Curved”</td></tr>
<tr><td colspan="3">Reduce moderately or even enlarge</td><td colspan="2">Reduce</td></tr>
<tr><td colspan="7">Approximation “reasonable”</td></tr>
<tr><td colspan="4">“Internal”</td><td colspan="3">“External”</td></tr>
<tr><td>“Small”</td><td colspan="3">“Large”</td><td>“Backward”</td><td colspan="2">“Forward”</td></tr>
<tr><td rowspan="2">Stop: Convergence found</td><td colspan="2">“Close”</td><td>“Far”</td><td rowspan="2">Reduce</td><td>“Straight”</td><td>“Curved”</td></tr>
<tr><td colspan="2">Reduce</td><td>Reduce</td><td>Enlarge</td><td>Keep size</td></tr>
<tr><td colspan="7">Approximations “good”</td></tr>
<tr><td colspan="4">“Internal”</td><td colspan="3">“Boundary”</td></tr>
<tr><td>“Small”</td><td colspan="3">“Large”</td><td>“Backward”</td><td colspan="2">“Forward”</td></tr>
<tr><td rowspan="2">Stop: Convergence found</td><td colspan="2">“Close”</td><td>“Far”</td><td rowspan="2">Reduce</td><td>“Straight”</td><td>“Curved”</td></tr>
<tr><td colspan="2">Reduce</td><td>Reduce</td><td>Keep model & Enlarge</td><td>Keep model & Keep size</td></tr>
</table>

### 5.2.5 Parallel Implementation

In this section a specific implementation of the MAM in a heterogeneous parallel computing environment is discussed.

As outlined in the previous sections, the algorithm includes two stages at which design evaluations are carried out: firstly, when new design points have been generated and, secondly, when the approximate nonlinear programming problem has been solved and a sub-optimum point obtained. At the former stage a relatively large number of designs are to be evaluated, whereas at the latter stage only a single design is evaluated. This single design evaluation is required to determine whether the iterative process is to be terminated and, if not, provides information required by the move limit strategy. The tasks of the first stage can be easily carried out in parallel, although a careful planning may be required to achieve a proper load balancing. The second phase is more difficult to execute in parallel as it requires a single design evaluation.

Fig. 5.8 illustrates a straightforward implementation without any planning or scheduling. A block corresponds to the evaluation of a single design by a slave which can be a processor of a multi-processor computer or a computer in a network. Black indicates an active slave, grey indicates non-active slaves. It is seen that all available slaves are used to evaluate the new plan points. Thereafter the approximations are determined and the approximate optimization problem is solved. For the obtained design the response functions are evaluated by one of the slaves (the isolated block in the middle of Fig. 5.8). All remaining slaves are inactive at that time. It is seen that this simple implementation may lead to a relatively large idle time for individual nodes.

**Figure 5.8** Example of a load on slaves in a heterogeneous network with five nodes.

Apart from load balancing, the most obvious improvement would be the elimination of the single design evaluation which is carried out after identification and solution of the approximate optimization problem. However, this would make the move limit strategy much less effective as no assessment of the quality of approximations is then available. Instead, a different strategy for parallelization of the algorithm has been suggested by van Keulen (van Keulen and Toropov, 1999). The idea is to use idle nodes for the enhancement of the database by adding selected response function evaluations as explained below. At the design evaluation stages of the optimization process an additional design evaluation is submitted to an idle node when at least two nodes become idle. The remaining idle node shall be kept ready for the execution of the sequential step of the algorithm, i.e. the design evaluation at the obtained sub-optimum point. The selection of coordinates of the additional points is carried out on the basis of the strategy described in Section 5.2.3. The size of the sub-domain to which the new points are to be added is taken identical to that of the current search sub-domain. Before solving the approximate optimization problem, the centre of the sub-domain is taken as the best design evaluated so far. As soon as the approximate optimization problem has been solved, its solution is taken as the center. This rule leads to a good probability of putting the additional designs either in the next search sub-domain or sufficiently close to it. Thus, if an additional design belongs to the next search sub-domain, a smaller number of regular new design points have to be added to the plan corresponding to the next step of the optimization process. If an additional design falls outside the next search sub-domain, it may still contribute to the quality of the approximation functions, as all points belonging to the neighbourhood of a search sub-domain are involved in the weighted least-squares fitting process. It should be noted that the proposed algorithm becomes most efficient when it is combined with the scanning mentioned in Section 5.2.3.

Figure 5.9 illustrates the proposed algorithm in a heterogeneous environment. As before, black indicates an active slave, grey indicates non-active slaves. A white block corresponds to an active slave which is carrying out an additional design evaluation as a free node becomes available. Obviously, the proposed algorithm could be combined with an appropriate load balancing technique.

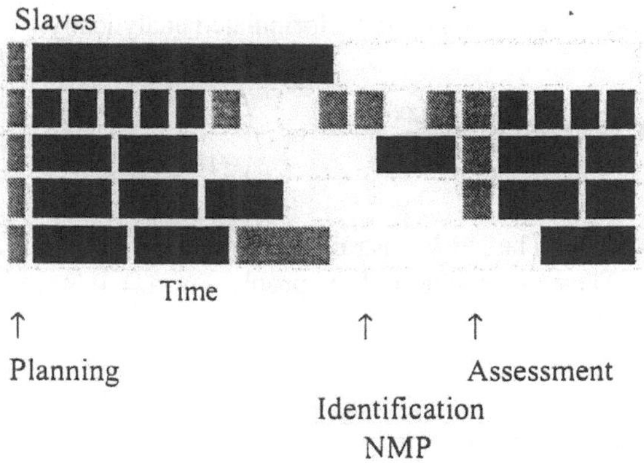


**Figure 5.9** Load on several slaves in a heterogeneous network with five nodes according to the modified algorithm.

### 5.2.6 Examples

**Testing on a two-bar truss problem.** The algorithm is controlled by several parameters that affect the efficiency of the method. They are all user-defined and can be adjusted for a particular problem if necessary. However, for most problems the default set of the parameters gives a good convergence within a reasonable number of iterations. To determine a "good" set of parameters, the method was intensively tested on simple problems with noisy responses (van Keulen et al., 1996 and Toropov et al., 1996).

A two-bar truss under an external force $\boldsymbol{P}$ ($P_y = 8\ P_x$, $|\boldsymbol{P}| = 2 \times 10^5\ N$) is shown in Fig. 5.10. The two design variables are the cross section areas of the bars and the half of the distance between the nodes 1 and 2. The vertical coordinate of the node 3 is fixed. The objective function is the mass of the structure and limitations are imposed on the stresses in the bars which must be less than $\sigma_a = 1000\ MPa$.

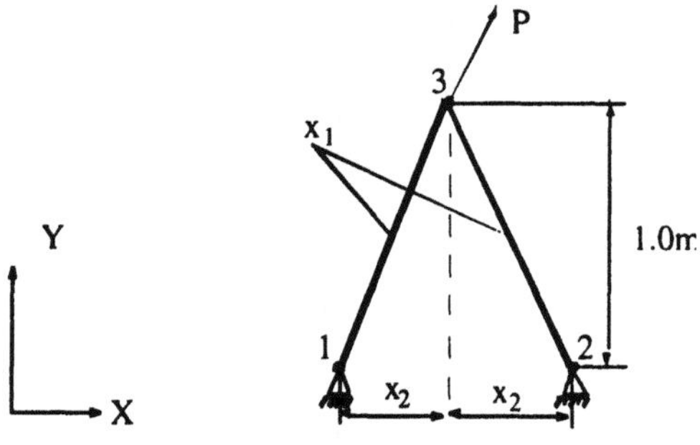


**Figure 5.10** Two-bar truss.

For this simple problem all functions can be formulated analytically:

$$F_0(\boldsymbol{x}) = C_1 x_1 \sqrt{(1+x_2^2)}\ ,\ F_1(\boldsymbol{x}) = \frac{\max \sigma(\boldsymbol{x})}{\sigma_a} = C_2\left(\frac{8}{x_1} + \frac{1}{x_1 x_2}\right)\sqrt{(1+x_2^2)} \le 1$$

where $C_1 = 1.0$, $C_2 = 0.124$. The side constraints are given by $A_1 = 0.2\ cm^2$, $B_1 = 4.0\ cm^2$ and $A_2 = 0.1\ m$, $B_2 = 1.6\ m$. The exact solution of the problem is $x_1 = 1.41\ cm^2$, $x_2 = 0.38\ m$, $F_0 = 1.51$ $kg$, $F_1 = 1.0$, see Fig. 5.11.

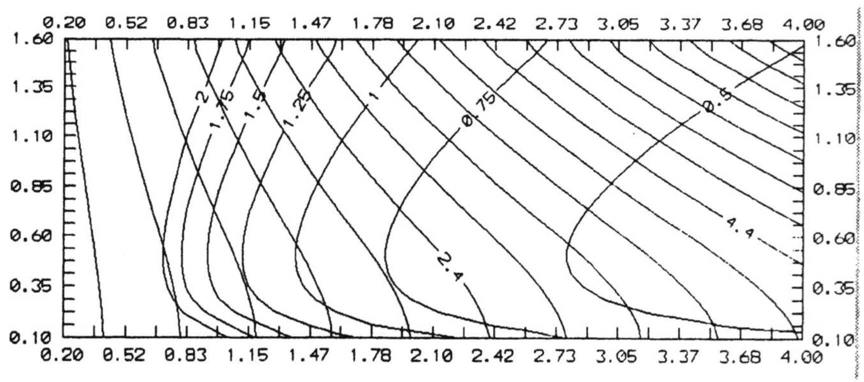


**Figure 5.11** Contours of the objective and constraint functions for the two-bar truss problem.

To simulate numerical noise, an error has been deliberately brought in the evaluation of the functions using a random number generator: $F_0^R = F_0\,[1 + A(2R-1)]$ and $F_1^R = F_1\,[1 + 2A(2R-1)]$ where $A$ is the amplitude of the random oscillations and $R$ is a random variable, $0 \le R \le 1$. Since in many practical optimization problems the evaluation of the objective function is performed with a higher accuracy compared to the evaluation of the constraints, the constraint function $F_1^R$ has a doubled value of the error. The level of noise can vary during the optimization process. It can be relatively high in the beginning of the optimization when a rough model is used, e.g. a coarse FE mesh, and it must be reduced as the optimum is approached, see, e.g. van Keulen et al. (1997) where the optimization procedure is linked to the adaptive mesh refinement technique with varying prescribed error. To simulate a gradually reducing level of noise, the amplitude of the random perturbation can be adapted using a simple rule $A^{(i+1)} = r\,A^{(i)}$ where $r < 1$ is a reduction parameter. This is done after every function evaluation. Note that the case of a constant level of noise is still included by $r = 1$.

Fig. 5.12 shows the objective function in the original form and spoiled by a noise level of 5% ($A$ = 0.05). A conventional method of mathematical programming could easily be trapped in the multiple "local minima" of the noisy objective function.

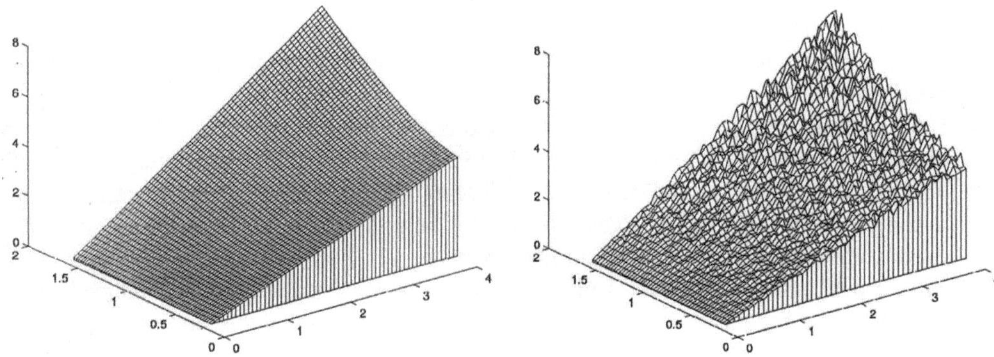


**Figure 5.12.** Objective function for the two-bar truss problem: original (left) and with random noise of 5 % (right).

Since the functions $F_0^{\mathrm{R}}$ and $F_1^{\mathrm{R}}$ are randomised, the optimization has been repeated several times with a different seed in the random number generator in order to adjust the set of parameters of the algorithm. To estimate the performance of the method, the following indicators have been introduced: the average number $\overline{n}$ of iteration steps; the average location $\overline{x}$ of the optimum points obtained in a series of runs, and the average distance $\overline{d}$ of the individual location of the optimum points from $\overline{x}$. For each set of parameters a series of 51 trials has been used.

For $A$ = 0.1 and a constant level of numerical noise, i.e. $r$ = 1, the following performance characteristics have been obtained: $\overline{n}$ = 24.9, $\overline{x}$ = (1.42, 0.46), and $\overline{d}$ = 0.104. For $A_{\text{initial}}$ = 0.1 and gradually reducing level of numerical noise ($r$ = 0.98), better performance of the algorithm has been observed: $\overline{n}$ = 22.9, $\overline{x}$ = (1.43, 0.40), and $\overline{d}$ = 0.061. When no noise has been added to the values of the functions, the solution (1.42, 0.37) has been found in 11 iterations.

**Optimization of a small Stirling engine.** The above described optimization technique was used for the performance optimization of small Stirling engines developed by Dr H. Carlsen of the Technical University of Denmark (Toropov and Carlsen, 1994). They are designed to be used as individual combined heat and power plants (co-generation plants). Such engines driven from natural gas or biomass can supply a single family house with heat and power with a high level of energy utilization. It is considered a one of possible ways of decentalization of energy supply in rural areas.

The basic Stirling engine working cycle is a closed cycle (Fig. 5.13), where the working gas is kept inside the cylinders, and heat is added and removed from the working space through heat exchangers. The Stirling engine differs therefore from conventional internal combustion engines by having external combustion, like in a boiler. All sorts of gas, liquid and solid fuels can be used depending on the proper external burner system only. The cycle is divided into four parts:

1 - 2 The working piston is moved to the left and the gas is compressed. The gas is cooled from the outside by cooling water in order to obtain compression at constant temperature.

2 - 3 The displacer piston is moved to the right forcing the gas through the connection channel to the hot volume, where the gas is heated by a burner. When the gas is heated, the gas pressure increases.

3 - 4 Both pistons are now moved to the right expanding the gas. The gas is heated from the outside by the burner in order to obtain the expansion at constant temperature.

4 - 1 The displacer piston is moved to the left forcing the gas through the connection channel to the cold volume, where the gas is cooled by the cooling water. When the gas is cooled, the gas pressure decreases to the starting pressure.

From the p-V and T-s diagrams (Fig. 5.13) it can be seen that power is produced on the working piston. If a regenerator is placed in the channel between the cold and the hot cylinder volumes, heat can be stored, when the gas flows from the hot to the cold cylinder volume and used for reheating the gas, when it flows back to the hot volume.

The ideal Stirling working cycle has the maximum obtainable efficiency defined by Carnot efficiency, and highly efficient Stirling engines can therefore be built, if designed properly. Heat exchangers, regenerators and cylinder volumes have to be optimized carefully for heat transfer, flow losses, dead volume, etc. To analyse the power output and the efficiency of a Stirling engine under consideration, a simulation program has been developed, which calculates power output, efficiency and heat flow from the thermodynamic equations. These differential equations are solved by means of the fourth order Runge-Kutta method. The initial pressure and temperatures for steady state operation are found by iterative procedure. To verify the simulation program, the numerical results have been compared to the results of laboratory experiments with a 10 kW natural gas driven Stirling engine. The comparison showed that the simulation program predicts the Stirling engine performance quite accurately.

To describe the engine design for the optimization purposes, it is necessary to introduce a number of design variables. The objective and constraint functions of the optimization problem present a considerable level of noise, i.e. can only be estimated with a finite accuracy due to the iterative nature of the simulation program. For that reason, it is impossible to estimate the gradients by finite differences accurately enough.

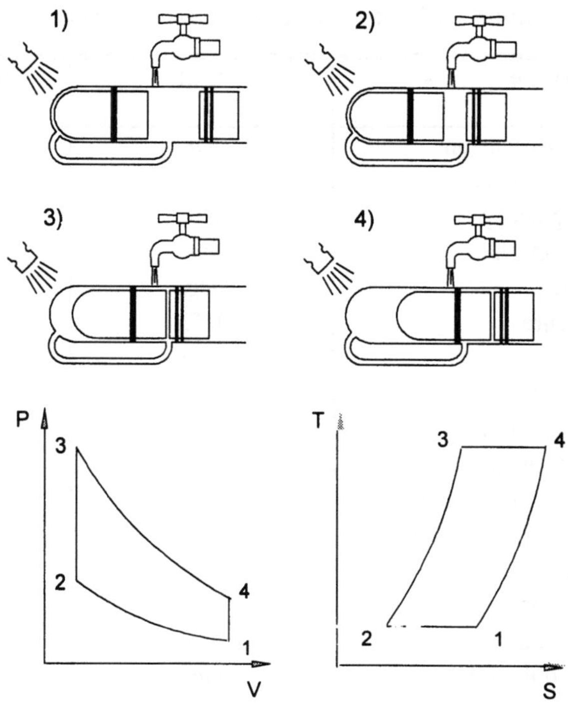


**Figure 5.13** Stirling engine working principle.

The efficiency has been maximized, while the power output has been controlled by two constraints. The first constraint forces the power output not to be less than 10 kW, and the second constraint keeps the specific power above a certain minimum. Other constraints ensure the strength characteristics of the engine components.

The objective and constraint functions have been approximated by multiplicative expressions (5.9) in the simplest form:

$$\tilde{F}(\boldsymbol{x},\boldsymbol{a}) = a_0 x_1^{a_1} \ldots x_{18}^{a_{18}} .$$

The optimization was carried out using 18 design parameters, the results (Table 5.2) showed that the efficiency increased from 30.1% to 41.2% with the same power output without changing the important design parameters like bore and stroke by more than 20%. Solutions were found which had never occurred by the traditional "trial and error" approach. The number of calls for the numerical simulation program in all runs has not been greater than 200.

**Table 5.2** Stirling engine optimization results.

| No | Design variable | Unit | Starting value | Optimum value |
|---|---|---|---|---|
| 1 | Bore, displacer piston | *mm* | 105 | 121 |
| 2 | Bore, power piston | *mm* | 95 | 125 |
| 3 | Stroke, displacer piston | *mm* | 51 | 63 |
| 4 | Stroke, power piston | *mm* | 60 | 63 |
| 5 | Phase angle | *deg.* | 70 | 66.8 |
| 6 | Mean pressure | *MPa* | 8 | 4.73 |
| 7 | Speed | *rpm* | 1550 | 1050 |
| 8 | Displacer gap | *mm* | 0.75 | 0.7 |
| 9 | Regenerator width | *mm* | 25.4 | 39.2 |
| 10 | Regenerator length | *mm* | 44 | 66 |
| 11 | Regenerator filler factor | | 0.22 | 0.224 |
| 12 | Regenerator thread diameter | *mm* | 0.04 | 0.71 |
| 13 | Number of heater tubes | | 24 | 45 |
| 14 | Number of cooler tubes | | 216 | 270 |
| 15 | Length of heater tubes | *mm* | 365 | 477 |
| 16 | Length of cooler tubes | *mm* | 180 | 146 |
| 17 | Internal diameter of heater tubes | *mm* | 8 | 5.1 |
| 18 | Internal diameter of cooler tubes | *mm* | 2.5 | 2.68 |
| Efficiency | | % | 30.1 | 40.2 |
| Power output | | *kW* | 10.1 | 10.0 |

**Optimization of a manipulator for a flight simulator.** This application concerns kinematic optimization of a Stewart platform manipulator for a flight simulator. The Stewart platform consists of six prismatic actuators and two platforms as shown in Fig. 5.14. Each actuator is connected by a universal joint (two orthogonal intersecting revolute joints) to the lower (base) platform whereas a spheric joint (three orthogonal intersecting revolute joints) is used to connect it to the upper (mobile) platform. To optimize the manipulator, a procedure based on the maximum workspace criterion proposed by Markine et al. (1998) and Markine (1999) has been used. The goal of optimization is to design a manipulator with maximum workspace whose characteristics are defined *a priori* according to the manoeuvres of a real aircraft to be simulated. The workspace of the manipulator assessed by the size of the inscribed hyperellipsoid similar to the one containing all required trajectories is then characterized by the corresponding ratio of similitude of the inscribed hyperellipsoid. Thus, for the optimal manipulator the inscribed hyperellipsoid should have maximum ratio of similitude (RS). The design variables comprised the geometry of upper and lower platforms as well as the lengths $L_{min(i)}$ ($i$=1, 2, 3) and attachment points of the actuators (Fig. 5.15).

**Figure 5.14** The Stewart platform.

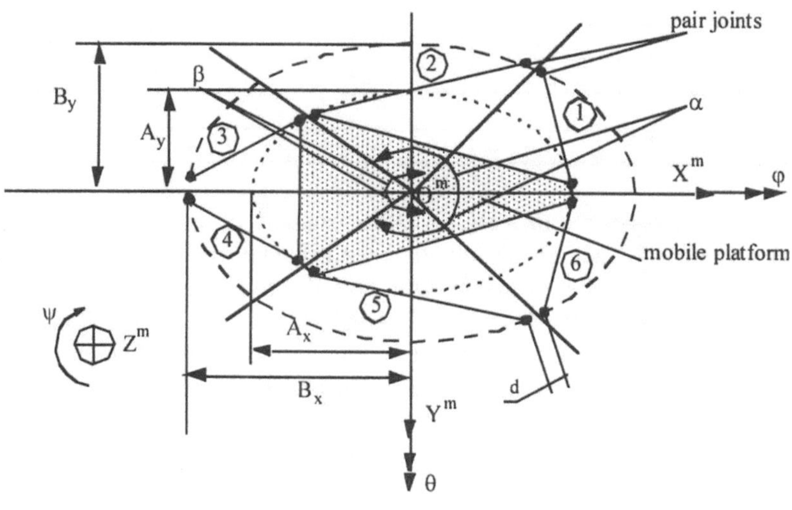


**Figure 5.15** Parameters of a modified Stewart platform.

The optimal design has to satisfy a dexterity (manipulability) requirement along with the constraints to avoid the interference between the actuators. The dexterity index (DI) describes the ability of the manipulator to arbitrarily change its position and orientation, or apply forces and torque in arbitrary directions. The position and rotation of the Stewart platform is completely defined by the lengths of its actuators. However, the control of the manipulator might become difficult if a singular configuration has been approached, Markine (1999). The problem has been solved for the workspace required for the simulation of a large transport aircraft.

A difficulty occurs when a singular configuration of the manipulator is approached. The constraints and objective function cannot be evaluated for such a design of the manipulator. To overcome the difficulty, the modified plan of experiments presented above has been used. An optimum design of the Stewart platform has been found in 15 iterations. The results are given in Table 5.3. It should be noted that for the optimal design the ratio of similitude has been slightly more than one that means that the obtained manipulator could perform all the required manoeuvres.

**Table 5.3** Optimization of a flight simulator.

| Design variable | Lower bound | Upper bound | Optimum design | Units |
|---|---|---|---|---|
| $A_x$ | 1.5 | 3.0 | 1.555 | m |
| $A_y$ | 1.5 | 3.0 | 1.533 | m |
| $B_x$ | 1.5 | 3.0 | 1.766 | m |
| $B_y$ | 1.5 | 3.0 | 2.281 | m |
| $\alpha$ | $0.5\ \pi$ | $0.95\ \pi$ | 2.131 | rad |
| $\beta$ | $0.5\ \pi$ | $0.95\ \pi$ | 1.944 | rad |
| $L_{min(1)}$ | 1.381 | 2.981 | 2.211 | m |
| $L_{min(2)}$ | 1.381 | 2.981 | 2.043 | m |
| $L_{min(3)}$ | 1.381 | 2.981 | 2.140 | m |
| RS | | | 1.010 | |
| DI | | | 0.2 | |

**Optimization of an assembly process.** Another example of application of the method to a system with geometrically and physically nonlinear behaviour is optimization of an automated assembly process performed by a manipulator. The determination of the position and elastic properties of the remote centre of the ARCC (Adjustable Remote Centre Compliance) mechanism (Fig. 5.16), which is used to avoid failure of error-corrupted assembly, has been formulated as an optimization problem. The production speed should be as high as possible while imposing restrictions on the values of contact forces and initial errors. The vector of the design variables comprises the parameters of the ARCC and the velocity of insertion (Fig. 5.17).

Again, for some set of the design variables the constraints (contact forces) cannot be evaluated because of the failure of a numerical integration procedure in the response analysis software. Physically, such a situation corresponds to a failure of the assembly process. The problem has been solved by Markine (1999) using a multilevel approach. Due to a difficulty of finding a feasible initial design for a manipulator with flexible links, the problem for a manipulator with rigid links is solved first. The optimal design is then used as a starting point for the optimization of a flexible link manipulator. Alternatively, a straightforward optimization of a flexible link manipulator has been performed (Toropov et al., 1999) using the modified plan of experiments suggested above. The optimum design has been obtained in 18 iterations. The results are collected in Table 5.4.

**Figure 5.16** Adjustable remote centre compliance mechanism.

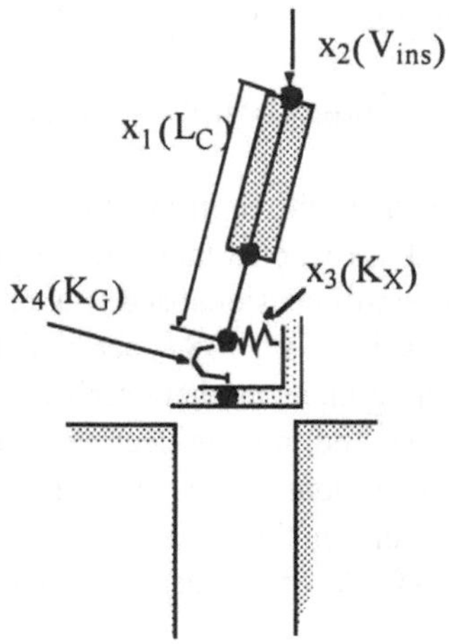


**Figure 5.17** Design variables in the optimization problem.

**Table 5.4** Optimization of an assembly process.

| Design variable | Lower bound | Upper bound | Optimum design | Units |
|---|---|---|---|---|
| $L_C$ | 0 | 250 | 34.01 | *mm* |
| $V_{ins}$ | 5 | 30 | 18.2 | *mm/s* |
| $K_X$ | 10 | 900 | 116 | *KN/m* |
| $K_G$ | 0.02 | 2 | 1.2 | *Nm/rad* |
| $F_1$ | | | 10 | *N* |
| $F_2$ | | | 10 | *N* |

**Optimization of an aerofoil.** The method has been applied to airfoil design by C. Holden, see Toropov et al. (1999). The airfoil was modelled with a separate Bezier curve for each of the upper and lower surfaces. This gives a convenient surface geometry representation with a relatively small number of defining parameters. The leading and trailing edge airfoil poles were held fixed to prevent the airfoil from drifting in space. The poles vary in both $x$ and $y$ directions to allow the shape of the airfoil to be adjusted smoothly. To ensure that the leading edge tangency condition at the leading edge was met, the $x$ values of the two poles adjacent to the leading edge on the upper and lower surfaces were held fixed. In total, 42 design variables were used in the problem. The bounds on the design variables were such that the poles could move up to half the distance between adjacent poles at the starting position. This ensured that they did not cross, which could cause inflexions, undesirable aerodynamically.

In this optimization problem, the objective function, formulated as a weighted sum of the values of aerofoil drag coefficient at Mach 0.73 and Mach 0.76

$$F_0(\boldsymbol{x}) = 2.0\ C_{d\ \mathrm{total}\ (M=0.73)} + 1.0\ C_{d\ \mathrm{total}\ (M=0.76)}\ ,$$

is minimised for a fixed lift coefficient $C_l = 0.7$, which was achieved by modifying the aerofoil angle of attack and a Reynolds number, based on aerofoil chord, $Re2d = 30.0 \times 10^6$. The objective function was non-dimensionalised to a value close to 1.0 after the first test optimization.

The BVGK (Ashill et al., 1987) aerodynamic prediction code used to calculate the section lift, drag and pitching moment characteristics is a field integral method, incorporating an inviscid full potential method, an engineering boundary layer method and a wave drag prediction method. The pitching moment was taken about the ¼ chord position and the transition position was fixed at 2 % chord on the upper and lower surfaces. Typical run times for the discretization of an airfoil section into 65 points on each surface was ~ 50 seconds on a SUN Sparc station 20.

The problem represents an airfoil redesign from a BAe Systems research section, in which new thickness requirements are specified. These were that the upper section spar depth at 5 % chord was 13 % thinner, the forward spar was 4 % thicker, the rear spar was 6 % thinner and the 80 % chord thickness was 26% thicker than the starting airfoil. The value of the pitching moment was restricted as a larger pitching moment would incur a trim drag and possibly a weight penalty. There was also constraint on form factor at the upper surface trailing edge in order to prevent the boundary layer being too close to separation.

Aerodynamic codes sometimes fail to converge unexpectedly when an input geometry is not a conventional aerodynamic shape. This could be considered as domain-dependent calculability of the functions in the optimization problem.

The method used 50 function evaluations (or experiments, slightly more than the number of design variables) per plan (or step) and it took 14 steps to convergence, i.e. 700 evaluations in total. The history of the optimization process shows that the optimum design was found in 10 iterations and the satisfaction of the termination criterion took the remaining 4 iterations. If the functions at a point in the plan was not calculated, i.e. BVGK failed to converge, then an additional point, randomly chosen from the current search subregion, was added to the plan until the requested number of evaluations (50 in this case) had run successfully. The obtained geometry with exaggerated vertical scale is presented in Fig 5.18, and the drag polars at $M = 0.73$ and $M = 0.76$ are shown in Fig. 5.19. Dashed curves correspond to the initial configuration of the aerofoil and the solid curves correspond to the obtained configurations. Fig. 5.18 shows the modifications made to the geometry are small. The leading edge is rounder. The increased trailing edge thickness is achieved. The drag polars in Fig. 5.19 show an improved polar above $C_l = 0.6$ at Mach 0.73 and above $C_l = 0.65$ at Mach 0.76.

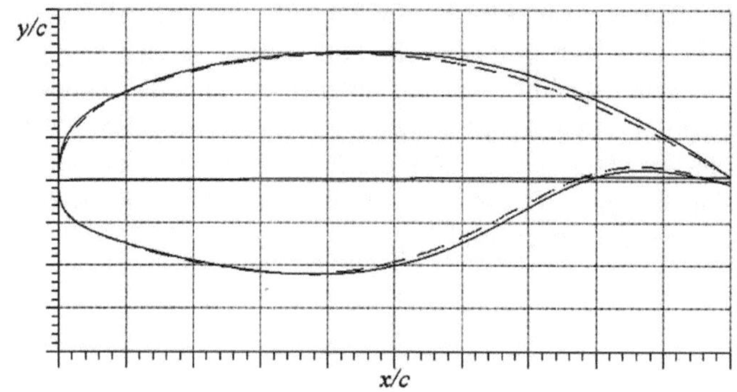


**Figure 5.18** Aerofoil optimization: geometry.

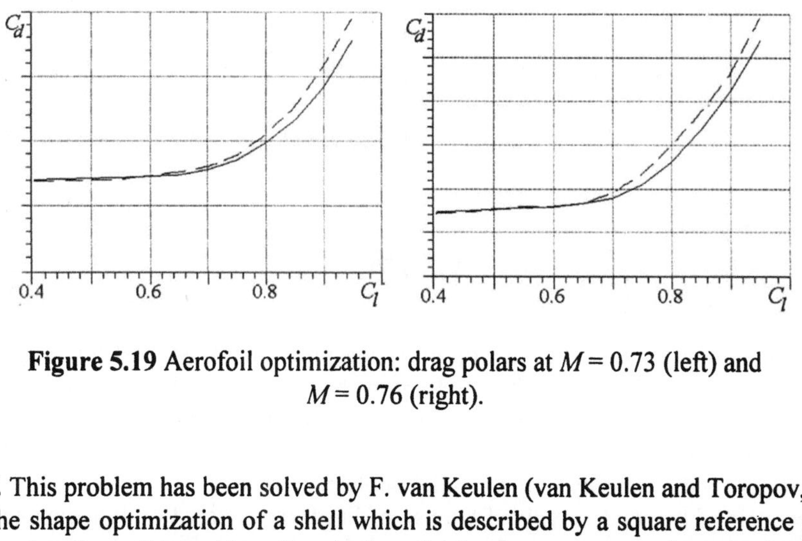


**Figure 5.19** Aerofoil optimization: drag polars at $M = 0.73$ (left) and $M = 0.76$ (right).

**Shell roof.** This problem has been solved by F. van Keulen (van Keulen and Toropov, 1998). It concerns the shape optimization of a shell which is described by a square reference plan. The out-of-plane location of its mid-surface is described using square patches. At the keypoints (corner nodes of the patches) both the out-of-plane coordinate and its derivatives with respect to the in-plane coordinates can be specified. The patches and the keypoints are depicted in Fig. 5.20. The geometry is assumed to be symmetric with respect to the diagonals and the lines A-A and B-B. The design variables are the out-of-plane coordinates of the keypoints and the corresponding derivatives. The out-of-plane coordinates of the corners are fixed. Furthermore, the thickness of the shell is taken as a design variable. The in-plane dimensions in millimetres are depicted in Fig. 5.20. The shell is supported at its corner nodes, for which all displacement components are prescribed. The shell is loaded by a uniform out-of-plane load with an intensity of 0.01 *MPa*. Homogeneous and isotropic linear elastic material behaviour is assumed with a Young's modulus $2.1 \cdot 10^5$ *MPa* and a Poisson's ratio 0.3. The optimization problem is formulated as minimization of the maximum displacement, whereas the volume should remain less than 40.103 $mm^3$. The thickness can vary from 0.1 *mm* to 2.0 *mm*. All lower bounds on the locations of the keypoints are given by 0 *mm*, whereas the upper bounds are selected as 50 *mm*, 80 *mm*, 100 *mm* or 150 *mm*, depending on their distance to the corner nodes. The derivatives should be in the range [-1, 1]. In total, 12 design variables are used. For all studies the entire search domain was selected as the search sub-domain of the first step of the optimization. The first plan of experiments has always been enhanced by means of 80 additional design evaluations. No design sensitivities have been used.

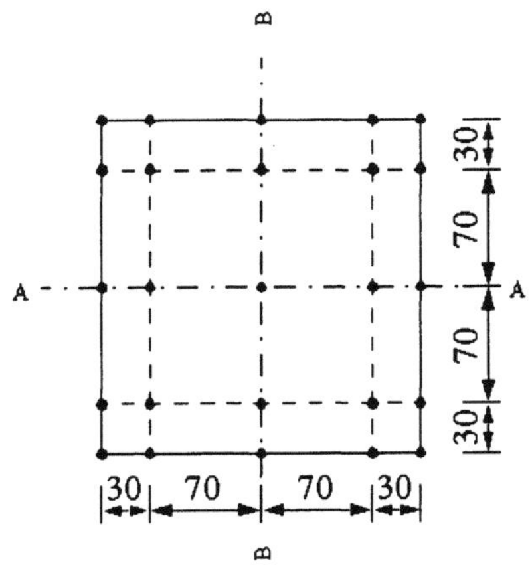


**Figure 5.20.** Rectangular patches and keypoints used for the description of the geometry of the shell.

Numerical studies showed that this optimization problem has several local optima. In the course of the study two almost equally good optima have been found. These design alternatives and corresponding values of the objective and constraint functions are given in Fig. 5.21 and Fig. 5.22. In contrast to the previous examples listed in this section, here the main focus will be on the robustness and efficiency of the parallel implementation.

The first part of the study involved the use of linear approximation functions and the proposed parallel implementation using four slaves with different performance characteristics. An extensive search which used 80 additional designs per step in the search domain and a minimum number of new points in the plan of experiments, resulted in the design depicted in Fig. 5.21. For this design the maximum displacement is $1.6556 \cdot 10^{-2}$ m, and the normalized constraint equals 0.993. As stated in Section 5.2.5, a minimum number of new points in the plan of experiments still leads to more points than the minimum to be taken into account in the weighted least squares fitting. A less extensive search with 12 additional design evaluations in the search domain and a minimum number of new points in each plan of experiments, gives the objective function equal to $3.73 \cdot 10^{-2}$ *mm* and the constraint equal to 0.997. However, when in addition always 12 new design points have been used in each plan of experiments, a significantly better result was obtained with the objective function equal to $1.77 \cdot 10^{-2}$ *mm* and the constraint equal to 0.999. It is concluded that accepting a greater number of designs to be evaluated, in combination with scanning of the database, produces an algorithm which is less sensitive to local optima.

**Figure 5.21** Shell roof: design found using linear approximation functions.

The effect of the proposed scheme for the use of idle slaves has also been studied using quadratic approximations. The study was carried out with a minimum number of points in each plan of experiments. The result obtained with the modified algorithm (Fig.5.9) is shown in Fig. 5.22. For this design the maximum displacement is $1.6578 \cdot 10^{-2}$ mm, and the normalized constraint equals 0.998. When the simpler scheme, which leaves idle nodes (Fig. 5.8), was used, the resulting design had an objective function equal to $4.27 \cdot 10^{-2}$ mm and the constraint equal to 0.999. It is seen that the modified scheme prevented the algorithm from getting trapped in a local minimum.

**Figure 5.22** Shell roof: design found using quadratic approximation functions.

## 5.3 Kriging

Kriging is an approach to constructing approximations using the original function (5.1) values from some plan of experiments. Kriging has been originally developed in the field of geostatistics (Matherton, 1963) and is attributed to the development of statistical techniques by mining engineer D.G. Krige. An approximation function $\widetilde{F}(x)$, built using kriging, has the property that it agrees with $F(x)$ at the plan points, i.e. this approach produces an interpolation model, see chapter on kriging in the book by Papalambros and Wilde (2000).

The main points of the algorithm are as follows. The approximation function is presented in the form:

$$y(\mathbf{x}) = f(\mathbf{x}) + Z(\mathbf{x}) \tag{5.12}$$

where $f(\mathbf{x})$ is some known function and $Z(\mathbf{x})$ is a Gaussian random function with zero mean. A typical choice of the function $f(\mathbf{x})$ is a regression model

$$f(\mathbf{x}) = \mathbf{g}^{\mathrm{T}}(\mathbf{x})\boldsymbol{\beta} \tag{5.13}$$

where $\mathbf{g}(\mathbf{x}) = [g_1(\mathbf{x}), \ldots, g_k(\mathbf{x})]^{\mathrm{T}}$ and $\boldsymbol{\beta} = [\beta_1, \ldots, \beta_k]^{\mathrm{T}}$.

The covariance matrix of $Z(\mathbf{x})$ is expressed as

$$Cov\,[Z(\mathbf{x}^{(i)}), Z(\mathbf{x}^{(j)})] = \sigma^2\, \mathbf{R}[R(\mathbf{x}^{(i)}, \mathbf{x}^{(j)})],\ i = 1, \ldots, N, j = 1, \ldots, N_{\mathbf{exp}} \tag{5.14}$$

where $\mathbf{R}$ is correlation matrix, and $R$ is the correlation function selected by the user, $\sigma^2$ is the variance of $Z(\mathbf{x})$.

As a typical example of $R$, consider the following function:

$$R(\mathbf{x}^{(i)}, \mathbf{x}^{(j)}) = \exp\left[-\sum_{k=1}^{N} \theta_k \left| x_k^{(i)} - x_k^{(j)} \right|^2\right] \tag{5.15}$$

The computed (estimated) model is $y^{(\mathrm{e})}(\mathbf{x})$. The minimization of the mean squared error is equivalent to the following maximization problem (with respect to $\theta_k$, $k = 1, \ldots, N$):

$$(-1/2)\ [(N_{\mathbf{exp}} \ln \sigma^{2(\mathrm{e})}) + \ln \det \mathbf{R}] \tag{5.16}$$

where $\sigma^{2(\mathrm{e})} = (\mathbf{y}_s - \mathbf{F}\boldsymbol{\beta}^{(\mathrm{e})})^{\mathrm{T}}\, \mathbf{R}^{-1}\, (\mathbf{y}_s - \mathbf{F}\boldsymbol{\beta}^{(\mathrm{e})}) / N_{\mathbf{exp}}$, $\quad \boldsymbol{\beta}^{(\mathrm{e})} = (\mathbf{F}^{\mathrm{T}}\, \mathbf{R}^{-1}\, \mathbf{F})^{-1}\, \mathbf{F}^{\mathrm{T}}\, \mathbf{R}^{-1}\, \mathbf{y}_s$,
and $\mathbf{F} = [\mathbf{g}(\mathbf{x}^{(1)}), \ldots, \mathbf{g}(\mathbf{x}^{(N\mathrm{exp})})]^{\mathrm{T}}$.

The prediction of the response is

$$y^{(\mathrm{e})}(\mathbf{x}) = \mathbf{g}^{\mathrm{T}}(\mathbf{x})\, \boldsymbol{\beta}^{(\mathrm{e})} + \mathbf{r}^{\mathrm{T}}(\mathbf{x})\, \mathbf{R}^{-1}\, (\mathbf{y}_s - \mathbf{F}\boldsymbol{\beta}^{(\mathrm{e})}) \tag{5.17}$$

where $\mathbf{y}_s$ is a vector of $N_{\mathbf{exp}}$ observed responses and $\mathbf{r}$ is the vector of correlations between the response at an arbitrary location $\mathbf{x}$ and sample points:

$$r(x) = [R(x, x^{(1)}), \ldots R(x, x^{(Nexp)})]. \tag{5.18}$$

Although kriging has been originally developed for situations when only the function values are available, the values of the derivatives of the response function can be also used, see Lewis (1998). Applications of kriging to various design optimization problems can be found in the papers by Booker (1998) and Booker et al. (1999), and it has been recently compared with other types of approximations by Chung and Alonso (2000) and Jin et al. (2000).

## 5.4 Use of Genetic Programming Methodology

Selection of the structure of an analytical approximation function is a problem of empirical model building (Box and Draper, 1987). Selection of individual regression components in a model results in solving a combinatorial optimization problem. Even if the bank of all regressors is established (which is a difficult problem on its own), the search through all possible combinations would result in prohibitive computational effort. Toropov and Alvarez (1998) attempted to develop and use a genetic programming (GP) methodology (Koza, 1992, Kinnear, 1994) for the creation of an approximation function structure of the best possible quality, and use it within a multipoint (or global) approximation technique.

GP is a branch of genetic algorithms (GA). While a GA uses a string of numbers to represent the solution, the GP creates a population of computer programs with a tree structure. In our case of design optimization, a program represents an empirical model to be used for approximation of a response function. A typical program, representing the expression $(x_1/x_2 + x_3)^2$, is shown in Figure 5.23.

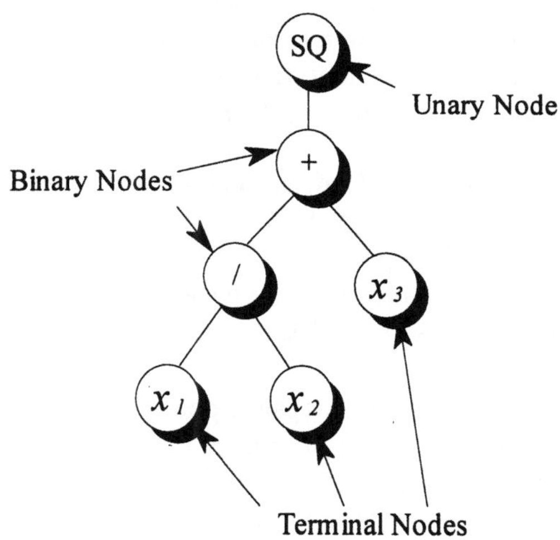


**Figure 5.23** Typical tree structure for $\left(\dfrac{x_1}{x_2} + x_3\right)^2$.

These randomly generated programs are general and hierarchical, varying in size and shape. GP's main goal is to solve a problem by searching highly fit computer programs in the space of all possible programs that solve the problem. This aspect is the key to finding near global solutions by keeping many solutions that may potentially be close to minima (local or global). The creation of the initial population is a blind random search of the space defined by the problem. In contrast to a GA, the output of the GP is a program (i.e. an empirical model used for approximation), whereas the output of a GA is a quantity.

The programs are composed of elements from a *terminal set* and a *functional set*, called nodes.

| Terminal Set | Design variables: $x_1$ , $x_2$ , ..., $x_N$ |
|---|---|
| Functional Set | Mathematical operators that generate the regression model: { +, *, /, $x^y$, etc. } |

The functional set can be subdivided into *binary nodes*, which take any two arguments (like addition), and *unary nodes*, which take one argument, e.g. a square root. All the functions and terminals must be compatible in order to faultlessly pass information between each other (closure property).

The evolution of the programs is performed through the action of the genetic operators and the evaluation of the fitness function.

### 5.4.1 Genetic Operators

Model structures evolve through the action of three basic genetic operators: reproduction, crossover and mutation. In the *reproduction* stage, a strategy must be adopted as to which programs should die. In this implementation, trees with fitness below the average are killed. The population is then filled with the surviving trees according to fitness proportionate selection.

*Crossover* (Fig. 5.24) combines good information from two trees (parents) in order to improve the fitness of the next generation. The basic algorithm is as follows:

- select two trees from the whole population;
- within each of these trees, randomly select one node;
- swap the subtrees under the selected nodes, thus generating two offspring belonging to the new population.

*Mutation* (Fig. 5.25) protects the model against premature convergence and improves the non-local properties of the search. The following algorithm is used:

- randomly select one node within a tree;
- replace this node with another one from the same set (a function replaces a function and a terminal replaces a terminal) except by itself.

An additional operator, *elite transfer*, is used to allow a relatively small number of the fittest programs, called the elite, to be transferred unchanged to a next generation, in order to keep the best solutions found so far. As a result, a new population of trees of the same size as the original one is created, but it has a higher average fitness value.

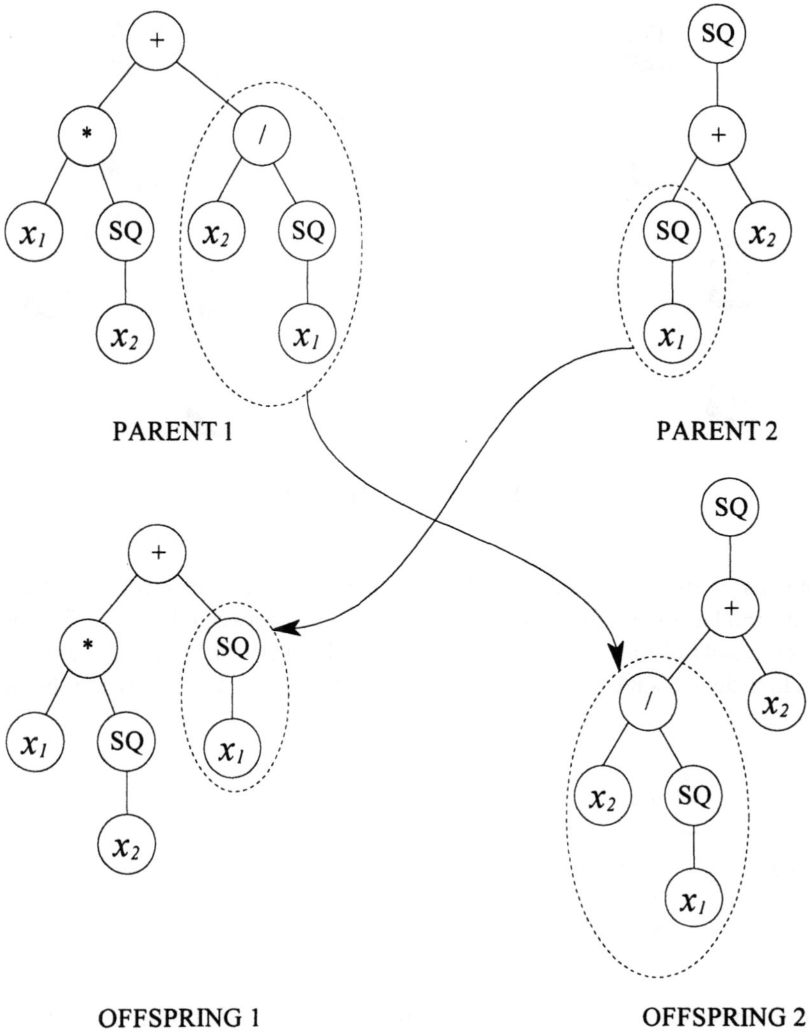


**Figure 5.24** Crossover.

### 5.4.2 Fitness Function

When selecting randomly a tree to perform any genetic operation, the so-called fitness proportionate method is used here. This method specifies the probability of selection on the basis of the fitness of the solution.

The fitness of a solution shall reflect the quality of approximation of the experimental data by a current expression represented by a tree, and the length of the tree in order to obtain more compact expressions.

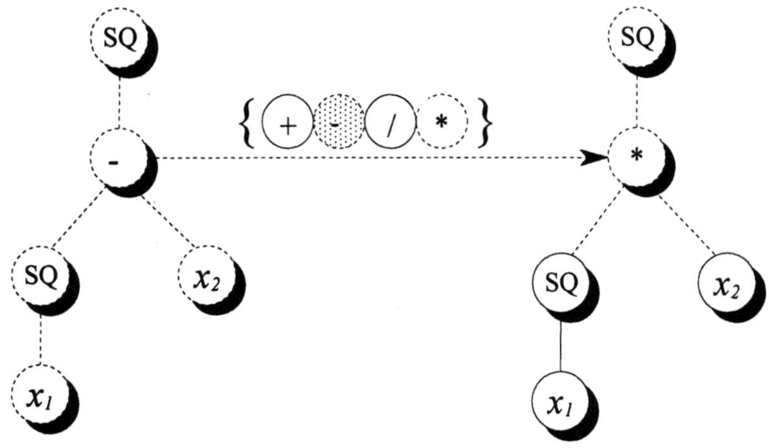


**Figure 5.25** Mutation.

In problems of empirical model building, the most obvious choice for the estimation of the quality of the model is the sum of squares of the difference between the simplified model output (5.2) and the results of runs of the original model (5.1) over some chosen plan (design) of experiments. In a dimensionless form this measure of quality of the solution can be presented as follows:

$$Q(S_i) = \frac{\sum_{p=1}^{P} \left(F_p - \tilde{F}_p\right)^2}{\sum_{p=1}^{P} F_p^2} \tag{5.19}$$

If in addition to the values of the original function $F_p$ their first order derivatives at point $p$ $F_{p,i} = \frac{\partial}{\partial x_i} F_p$, $(i = 1,\dots,N)(p = 1,\dots,P)$ are known, the numerator in (5.19) is replaced by the following expression:

$$\sum_{p=1}^{P} \left( \left(F_p - \tilde{F}_p\right)^2 + \gamma \frac{\sum_{i=1}^{N} \left(F_{p,i} - \tilde{F}_{p,i}\right)^2}{\sum_{i=1}^{N} F_{p,i}^{\;2}} \right) \tag{5.20}$$

where $\gamma > 0$ is the parameter characterizing a degree of inequality of the contribution of the response and the sensitivity data, taken here as 0.5.

If $Q(S_i)$ is the measure of quality of the solution $S_i$, $Q_{\max}$ is the maximum value of this quantity out of all $N_t$ members of the population, $ntp_{\max}$ is the maximum allowed number of tuning parameters, $ntp_i$ is the number of tuning parameters contained in the solution $S_i$, and $c$ is a coefficient penalizing the excessive length of the expression, the fitness function $\Phi(S_i)$ can be expressed in the following form:

$$\Phi(S_i) = Q_{\max} - Q(S_i) + c * (ntp_{\max} - ntp_i)^2 \tag{5.21}$$

The probability that the solution $S_i$ will be selected is

$$\frac{\Phi(S_i)}{\sum_{j=1}^{N} \Phi(S_j)} \tag{5.22}$$

Programs with greater fitness values $\Phi(S_i)$ have a greater chance of being selected in a subsequent genetic action. Highly fit programs live and reproduce, and less fit programs die.

The fitness function defined as the sum of squares of the error between model output and experimental data has been applied before to the model building in fluid flow problems (Gray et al., 1996a) and dynamic systems (Gray et al., 1996b). Another possible choice is the statistical concept of correlation, which determines and quantifies whether a relationship between two data sets exists. This definition was used for the identification of industrial processes (McKay et al. 1996).

### 5.4.3 Design of Experiments

The choice of the design of experiments can have a large influence on the accuracy of the approximation and the cost of constructing the response surface. In this work, the approach suggested by Audze and Eglais (1977) and later used by Rikards (1993) is adopted. It considers a non-traditional criterion for elaboration of plans of experiments which is not dependent on the mathematical model of the object or process under consideration. The input data for the elaboration of the plan only include the number of factors $N$ (number of design variables) and the number of experiments $K$. The main principles in this approach are as follows:

- the number of levels of factors (same for each factor) is equal to the number of experiments and for each level there is only one experiment;
- the points of experiments are distributed as uniformly as possible in the domain of variables. There is a physical analogy with the minimum of potential energy of repulsive forces for a set of points of unit mass, if the magnitude of these repulsive forces is inversely proportional to the distance squared between the points:

$$\sum_{p=1}^{P} \sum_{q=p+1}^{P} \frac{1}{L_{pq}^{\;2}} \rightarrow min \tag{5.23}$$

The plan of experiment is characterized by a matrix which contains the levels of factors for each of $P$ experiments. For example, for a number of factors (design variables) $N = 2$ and $P = 10$, the matrix is

$$\begin{vmatrix} 8 & 10 & 4 & 6 & 2 & 3 & 9 & 5 & 7 & 1 \\ 1 & 7 & 10 & 6 & 8 & 5 & 4 & 2 & 9 & 3 \end{vmatrix} \tag{5.24}$$

The corresponding plan of experiments is shown in Fig. 5.26.

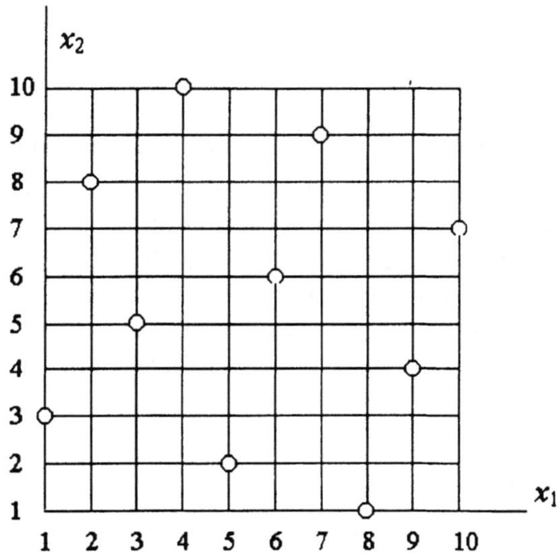


**Figure 5.26** Plan of experiments for $N = 2$ and $K = 10$.

### 5.4.4 Model Tuning

The approximation function is characterized not only by its structure (to be found by the GP) but also by a set of tuning parameters $\boldsymbol{a}$ to be found by the model tuning, i.e. the least squares fitting of the model into the set of values of the original response function:

$$G(\boldsymbol{a}) = \sum_{p=1}^{P} \left(F_p - \tilde{F}_p(\boldsymbol{a})\right)^2 \rightarrow min \tag{5.25}$$

The allocation of tuning parameters $\boldsymbol{a}$ to an individual tree (Fig. 5.27) follows the basic algebraic rules, see Fig. 5.28.

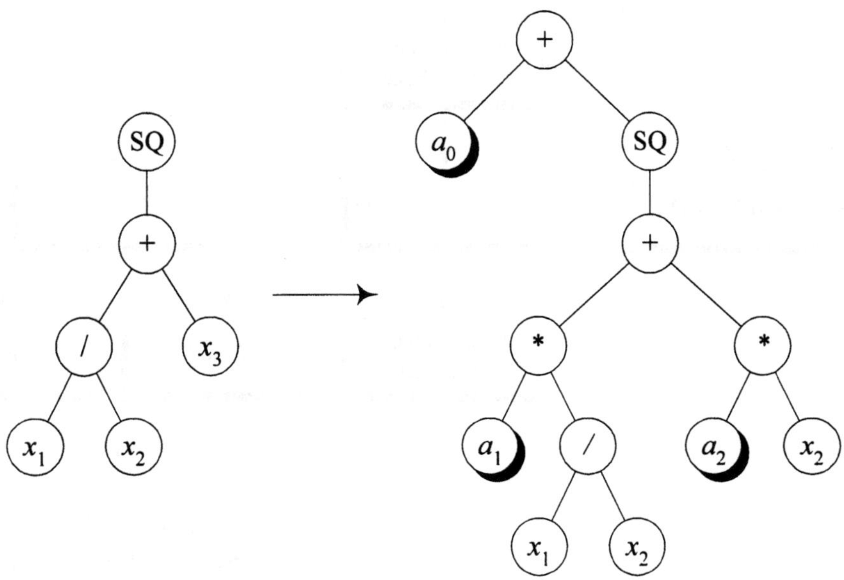


**Figure 5.27** Allocation of tuning parameters to a tree.

Going through the tree downwards, tuning parameters are allocated to a subtree depending on the type of the current node and the structure of the subtree according to the following algorithm:

1. Current node is of type *Binary*
   - multiplication and division operations only require one tuning parameter, e.g. $\tilde{F} = x_1 * x_2 \Rightarrow \tilde{F}(\boldsymbol{a}) = a_1 * x_1 * x_2$;
   - all other operations require two tuning parameters, e.g.

$$\begin{aligned} \tilde{F} = x_1 + x_2 \Rightarrow \quad & \tilde{F}(\boldsymbol{a}) = a_1 * x_1 + a_2 * x_2 \\ \tilde{F} = x_1 \wedge x_2 \Rightarrow \quad & \tilde{F}(\boldsymbol{a}) = (a_1 * x_1) \wedge (a_2 * x_2) \end{aligned};$$

   - when $\tilde{F}$ is a combination of the previous two approaches, tuning parameters are only applied to operations different from multiplication and division, e.g.

$$\tilde{F} = x_1 * (x_2 / x_3 + x_4) \quad \Rightarrow$$
$$\tilde{F}(\boldsymbol{a}) = x_1 * (a_1 * x_2 / x_3 + a_2 * x_4)$$

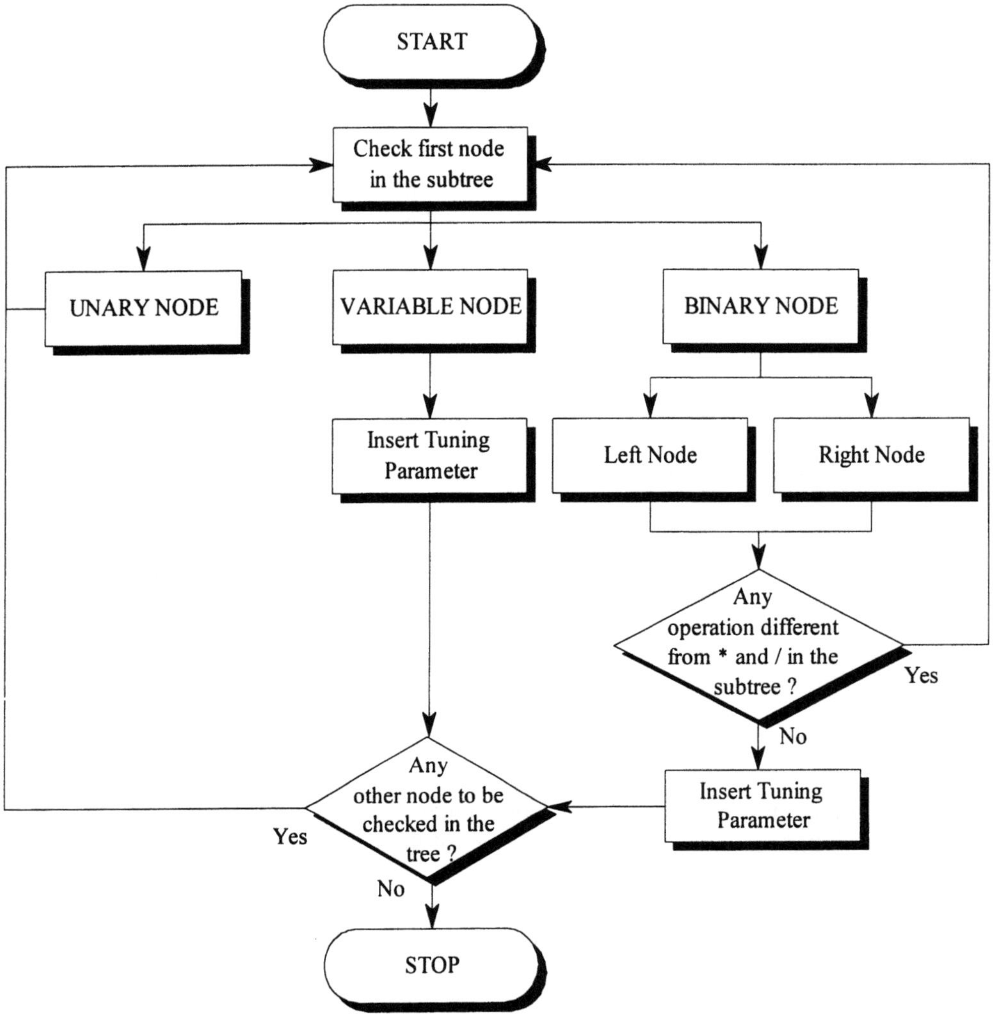


**Figure 5.28** Tuning parameter allocation diagram.

$$\widetilde{F} = (x_1 + x_2)^\wedge(x_3 * x_4) \Rightarrow$$
$$\widetilde{F}(\boldsymbol{a}) = (a_1 * x_1 + a_2 * x_2)^\wedge(x_3 * x_4)$$

2. Current node is of type *Unary*: ignore.
3. Current node is of type *Variable*
   - one tuning parameter is added, e.g.

$$\widetilde{F} = (x_1)^2 \Rightarrow \widetilde{F}(\boldsymbol{a}) = (a_1 * x_1)^2$$

4. Insert a free parameter, e.g.

$$\tilde{F}(\boldsymbol{a}) = a_1 * x_1 \quad \Rightarrow \quad \tilde{F}(\boldsymbol{a}) = a_1 * x_1 + a_0$$

To identify the parameters of the expression by the nonlinear least squares fitting, i.e. to solve the optimization problem (5.25), a combination of a genetic algorithm and a nonlinear optimization method by Madsen and Hegelund (1991) is used.

The output of the genetic algorithm is then used as the initial guess for the subsequent derivative-based optimization. The method by Madsen and Hegelund (1991) used here amounts to a variation of the Newton's method in which the Hessian matrix is approximated by the secant (quasi-Newton) updating method. Once the technique comes sufficiently close to a local solution, it normally converges quite rapidly. To speed up the convergence the algorithm uses the adaptive update of the Hessian and, consequently, the algorithm is reduced to either a Gauss-Newton or Levenberg-Marquardt method.

### 5.4.5 Applications

**Test example 1.** Rosenbrock's banana-shaped valley function is a classic optimization problem. The optimum point is inside a long, curved and narrow valley. The function is defined as follows:

$$F(x_1, x_2) \;=\; 100 * \left(x_2 - x_1^{\,2}\right)^2 \;+\; (1 - x_1)^2$$

Fig. 5.29 shows the contour plot and the surface plot of Rosenbrock's function.

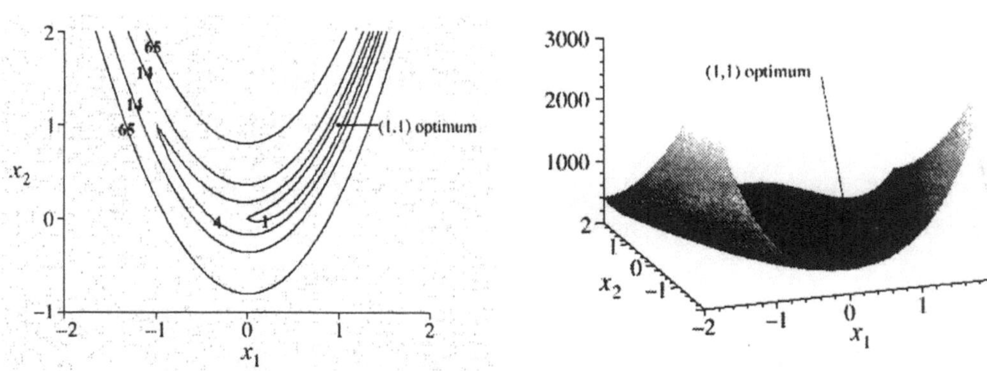


**Figure 5.29** Rosenbrock's function: contour plot (left) and surface plot (right).

With a population of 200 trees, the approximation of Rosenbrock's function has been tested with and without the use of sensitivity information. When no sensitivity information has been used, GP was run with a plan of experiments of 5 and 10 points. In the case of 5 points, a solution with good fitness has evolved, it had almost exact match at the plan points but very poor quality everywhere else. The reason is that insufficient information was passed to GP to represent an accurate solution, and the solution suffered from overfitting. When approximated with 10 plan points, the Rosenbrock's function emerged as the solution of the problem.

When the first order derivatives, were included in the approximation of Rosenbrock's function with a plan of experiments of 5 points, the algorithm exactly matched the original expression. This suggests that, if available, derivatives provide with more information, thus improving the convergence characteristics. If the derivatives are not available, the use of more points in the plan of experiments is necessary.

**Test example 2.** Generally, a large number of points in the plan of experiments is desirable in order to provide more information to the genetic programming algorithm. To illustrate these aspects, the following expression has been tested (see Fig. 5.30):

$$(30 + x_1 \sin x_1)(4 + e^{-x_2})$$

Two tests were performed with data generated with a plan of experiments of 20 and 10 points (Fig. 5.30 and Fig. 5.31, respectively). The sine and exponential functions were included in the functional set. Results show that the higher the number of experiments, the better the approximation.

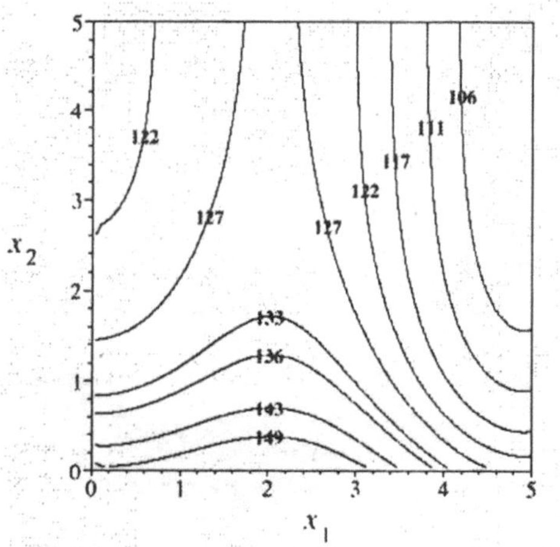


**Figure 5.30** The original function and the approximation with 20 point plan.

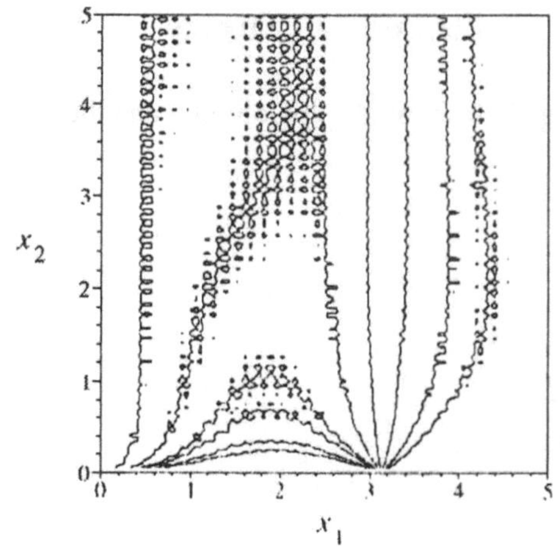


**Figure 5.31** Approximation with 10 point plan of experiments.

**Recognition of damage in steel structures.** The output error method of system identification has been used to assess the presence and extent of damage in steel structures. In the solution of the optimization problem the objective function and, alternatively, its individual terms corresponding to individual frequencies, have been approximated by analytical expressions using the genetic programming methodology.

Damage location in a typical steel portal frame is found by minimization of the difference between the measured and analysed structural response, namely, frequencies of vibration:

$$\sum_{i=1}^{F} w_i \left\{ \left[ \omega_i^M - \omega_i^A(x) \right] / \omega_i^M \right\}^2 \rightarrow \min \tag{5.26}$$

where $\omega_i^M$ are frequencies measured in the experiment, $\omega_i^A$ are frequencies obtained by the finite element analysis and $w_i$ are weights describing the relative importance of the match between frequencies of the $i$-th mode.

In the formulation of the optimization problem (5.26) the number of optimization variables $N = 3$, the number of used frequencies $M = 4$, and $x_1$, $x_2$, $x_3$ describe percentage of reduction of cross-sectional area in three locations at welded joints. The description of actual damage corresponds to the following set of optimization variables: $x_1 = 100$, $x_2 = 54$, $x_3 = 100$, i.e. damage in second location.

The approximation procedure using GP has been carried out following two different approaches: approximation of the original optimization problem (5.26), and approximation of the individual frequencies corresponding to the first four modes of vibration.

For the 3-dimensional graphical representation, the approximation functions have been plotted fixing one of the three optimization variables, corresponding to possible damage locations, i.e. $x_1 = 100$. Fig. 5.32a shows the original function in (5.26). Fig. 5.32b and 5.32c show the approximation functions obtained using the values of the function in (5.26) at $P = 50$ and $P = 20$ points of the optimization variable space. The following input parameters have been used:

- plans of experiments: 20 and 50 points
- population size: $N_t$=100
- proportion of the elite: $P_e$=0.2
- probability of mutation: $P_m$=0.001
- functional set:
  binary functions +, *, /, ^
  unary functions $(...)^2$, $\sqrt{(...)}$, -(...)
- terminal set: variables $x_1$, $x_2$, $x_3$.

The solution of the simplified optimization problem has been obtained in two steps of approximation building. In the first step the following values of lower and upper bounds have been selected: $A_j = 10$ and $B_j = 110$, $j = 1, 2, 3$. In the second step the size of the search domain of the optimization variable space, defined by $A_j$ and $B_j$, has been reduced by half and the new approximations have been constructed. When the approximation have been built using 50 points, the following solution has been obtained: $x_1 = 110.0$, $x_2 = 45.4$, $x_3 = 110.0$. Using 20 points, the following solution has been obtained: $x_1 = 74.0$, $x_2 = 50.7$, $x_3 = 110.0$.

When the approximation functions were obtained as a combination of approximations for the individual frequencies, as illustrated by Fig. 5.33, the following solutions have been obtained in one step: $x_1 = 92.6$, $x_2 = 50.1$, $x_3 = 110.0$ using 50 points, and $x_1 = 80.0$, $x_2 = 51.1$, $x_3 = 89.6$ using 20 points.

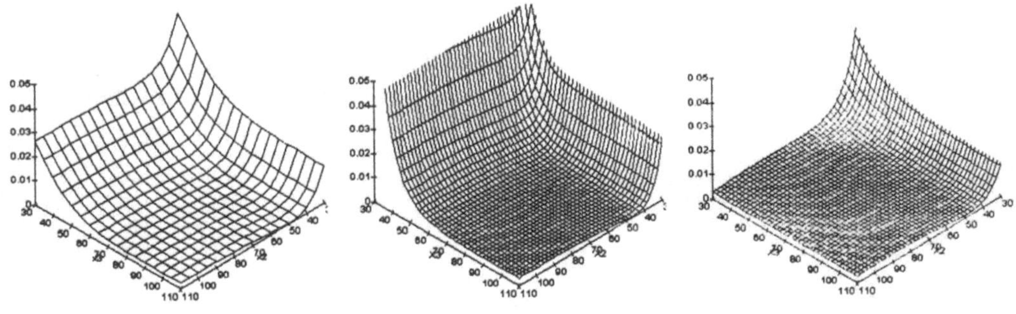

a) Original function b) Approximation (50 points) c) Approximation (20 points)

**Figure 5.32** Approximation of the overall expression, $x_1 = 100$.

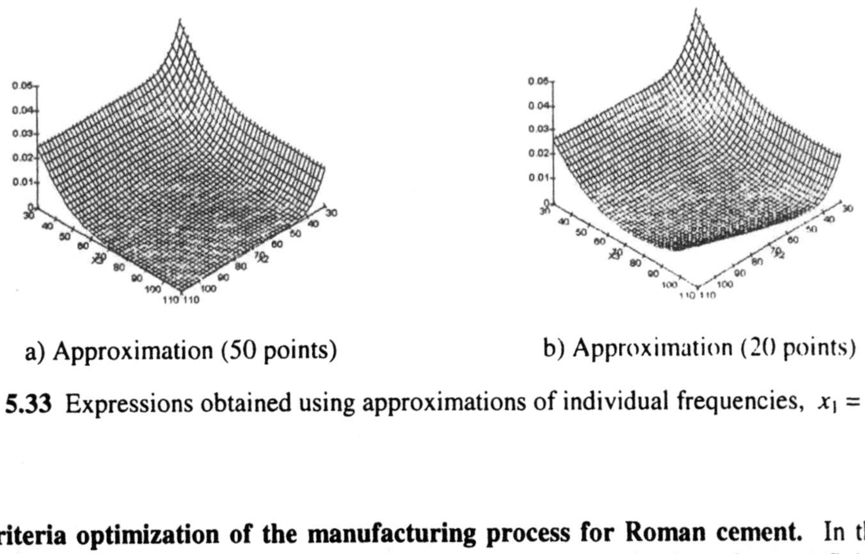

a) Approximation (50 points) b) Approximation (20 points)

**Figure 5.33** Expressions obtained using approximations of individual frequencies, $x_1 = 100$.

**Multicriteria optimization of the manufacturing process for Roman cement.** In the 19th century, Roman cement was used throughout Europe for the production of stucco finishes in architectural decoration. However, problems with the supply of raw materials and developments in Portland cements motivated the decline in its use and the corresponding loss of craft techniques.

The Charter of Venice (1964) states that the process of restoration should be based on respect for the original material. This is in contrast with the use of the current modern cement products, finishing layers and paints that do not match the original physical and mechanical properties of the stuccoes.

Consequently, for the re-introduction of Roman cement, there is a need to find a suitable range of similar materials among the re-emerging natural hydraulic binders, and to appreciate the technical knowledge and understanding of the original cement-makers. Experimental results on the calcination of cement-stones show that both setting time and strength development are functions of source and calcination temperature.

In this application, a single source of cement-stone was identified for experimentation within an optimization programme to relate mechanical and mineralogical characteristics to calcination conditions. Genetic programming has been used to illustrate the general trends of minerals and the strength development of the cement. The data will be useful for the selection of hydraulic binders and as an element in the re-introduction of Roman cement to the European market.

Experimental work. The cement-stones used in this research were collected from the Yorkshire coast at Whitby Long Bight. The calcination process is not well documented in the historic literature. Previous work at Bradford University has utilised temperatures of 940°C and 1100°C and found the lower temperature to offer better product. For the current research, the Audze-Eglais plan of experiments shown in Fig. 5.34 has been used with 10 initial and 2 supplementary points. Each cement is referred to using the nomenclature of temperature and residence time, e.g. 917, 276 taken as design variables $x_1$ and $x_2$ respectively. An electric kiln was used and no attempt made to either circulate air through it or to seal it during calcination.

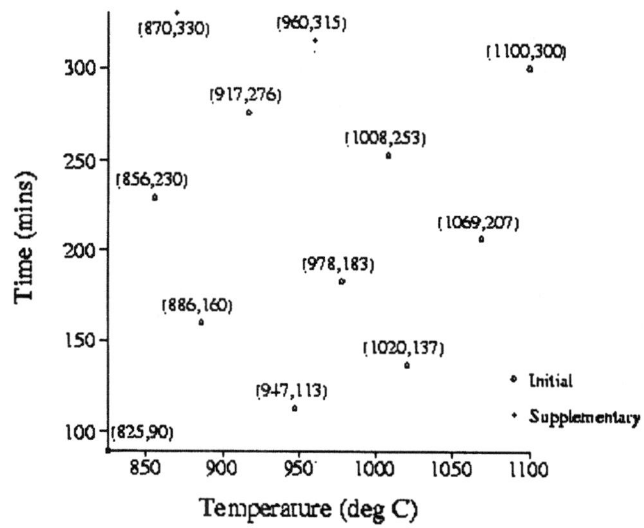


**Figure 5.34** Plan of experiments.

Results (Alvarez et al., 2000). At each combination of time and temperature given by the plan of experiments, strength at one week (*MPa*) and rate of strength enhancement between 8 and 17 weeks (*MPa*/week) have been obtained by a series of laboratory experiments. Using GP, these response quantities have been obtained as analytical expressions, their contour plots are shown in Fig. 5.35.

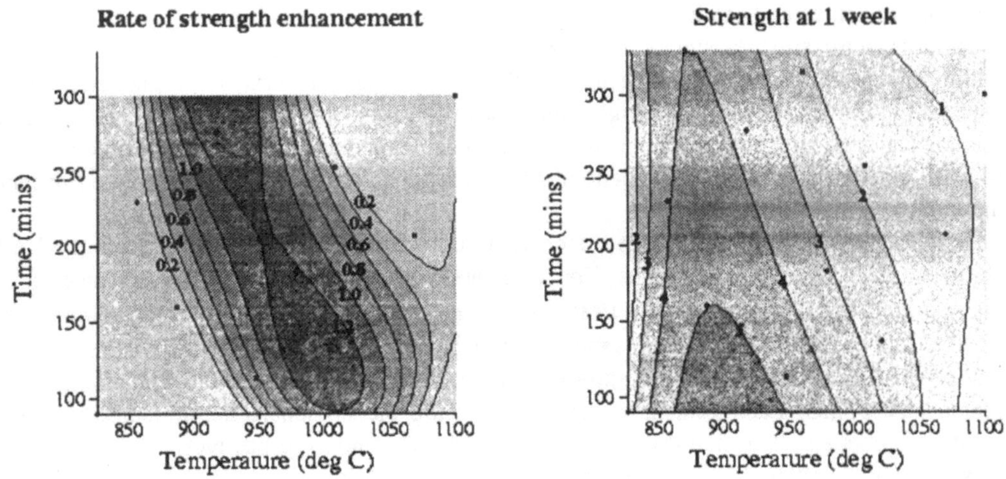


**Figure 5.35** Strength development for Roman cement.

The multicriteria optimization problem is stated as to maximize the early age strength and the rate of strength enhancement simultaneously. The constraints have been established following different technical aspects. Although the green discolouration of some pastes is associated with the presence of a mineral called calcium sulfite, its concentration is too low to quantify. Consequently this constraint has been represented by a maximum silica content (determined by X-Ray Difraction using boehmite as an internal standard) expressed as relative intensity 50. This correlates with pastes which do not turn green. Further work on the materials aspect of the investigation is required to refine this constraint and provide a direct measure of greenness. Another important quantity, the weight loss on calcination (LOC) of 28-30% has been selected to represent the confused historic statements of "calcine sufficient to decarbonate", "just enough to minimize the weight" and "underfire to economize on grounding" found in the literature, although not well documented.

The formulation of the optimization problem is as follows:

| | |
|---|---|
| Strength at one week | $\rightarrow$ *max* |
| Rate of strength enhancement | $\rightarrow$ *max* |
| Subject to: | Silica $\leq$ 50<br>28% $\leq$ LOC $\leq$ 30% |

To find the Pareto-optimal set, first the objective and constraint functions have been plotted to identify the feasible solution domains. Second, a series of optimization runs has been performed by the SQP algorithm at regular intervals. The final solution is represented in Fig. 5.36. The darker area defines the feasible solution domain, the solid contours show the function of the strength at one week, the lighter contours show the function of the rate of strength enhancement. The points define the discrete approximation of the Pareto-optimal set.

The analysis of the optimization results (Fig. 5.36) reveals that there are three main zones of study according to the obtained Pareto-optimal set: the upper, middle and lower zones.

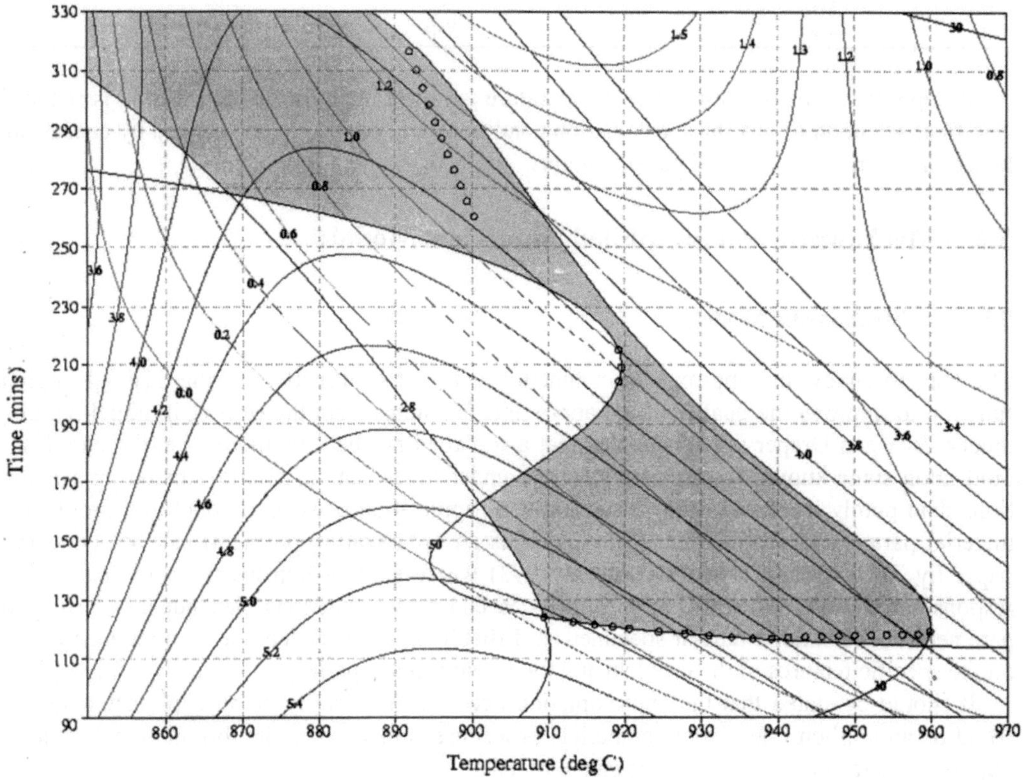


**Figure 5.36** Results of optimization.

The lower zone has a very narrow and essentially constant time band. As temperature increases, the strength at one week decreases but the rate of strength enhancement increases. The upper zone has a narrower temperature band than the lower zone. Here as the temperature reduces, the time increases. As this progression is followed; the same trend as before is noted in terms of the strength at 1 week and the rate of strength enhancement. The middle zone is located in a very narrow area of the feasible domain that suggests uncertainty about the validity of these solutions.

It is possible to identify three points which yield similar optimal performance but obtained under different calcination conditions as described in Table 5.5.

**Table 5.5** Optimal solutions.

| | Strength at 1 week ($MPa$) | Rate of strength enhancement ($MPa$/week) | Temperature (°$C$) | Time (min) |
|---|---|---|---|---|
| 1 | 4.4 | 1.0 | 957 | 120 |
| 2 | 4.3 | 1.0 | 920 | 205 |
| 3 | 4.2 | 1.1 | 900 | 260 |

It appears that as the calcination temperature is raised, then the residence time is reduced. The final selection of any these calcination conditions will be heavily influenced by energy and hence financial and ecological considerations.

## 5.5 Mechanistic and Simplified Numerical Models

### 5.5.1 Mechanistic Models

The idea to endow the approximation function with some properties of the original implicit function to improve the quality of the approximation stems from the empirical model-building theory. Box and Draper (1987) showed that a so-called mechanistic model, i.e. the one that is built upon some knowledge about the system under investigation, provides better approximations than purely empirical ones. An example of application of such a model to a problem of material parameter identification (formulated as an optimization problem) can be found in a paper by Toropov and van der Giessen (1993) where explicit non-linear approximations of response quantities for a solid bar specimen descend from a functional dependence on material parameters for a simpler tubular specimen. A difficulty in using such models is that they depend on the specific features of the problem and the researcher's experience.

It should be stated that the approximation need not necessarily be an explicit function. It could be an implicit one if some numerical procedure is involved in its formulation. The basic requirements to such a model can be summarized as:

- it must depend on the same design variables as the original function;
- it has to contain some tuning parameters to be defined using the general (non-linear) least-squares method;
- it must be simple enough to be used in numerous repeated calculations;
- it should not contain any considerable level of numerical noise in order not to cause convergence problems in the optimization process.

### 5.5.2 Simplified Numerical Models

A more general way of constructing high quality approximations is to obtain a simplified numerical model by simplifying the analysis model (e.g. by using a coarser finite element mesh discretization, a reduced number of the natural modes of the model in dynamic analysis, simpler geometry, boundary conditions, etc.). Such a model should inherit the most prominent features of the original one and the response analysis with this model should be computationally much less expensive than with the original model. In such a case a simplified numerical model provides a good basis for development of high quality approximations (Toropov and van der Giessen 1993, Toropov and Markine 1996, Toropov et al. 1997).

A simplified numerical model can be then used to build the approximation model:

$$\tilde{F}(\boldsymbol{x},\boldsymbol{a}) \equiv \tilde{F}(f(\boldsymbol{x}),\boldsymbol{a}) \tag{5.27}$$

where *f(x)* is the function presenting a response quantity using the simplified model. The approximation based on the simplified model satisfies all the above mentioned requirements. It reflects the main properties of the original complex model, it is computationally less expensive and relatively noiseless. Depending on the way of introduction and the number of the tuning parameters in the simplified expression (5.12), the following three types of the approximations have been proposed (Toropov and Markine, 1996):

1. As a simplest case, the approximation function is a linear or multiplicative function of two tuning parameters:

$$\tilde{F}(\boldsymbol{x},\boldsymbol{a}) = a_0 + a_1 f(\boldsymbol{x}) \quad \text{or} \quad \tilde{F}(\boldsymbol{x},\boldsymbol{a}) = a_0 f(\boldsymbol{x})^{a_1} \tag{5.28}$$

The vector of the tuning parameters then consists of two elements: $\boldsymbol{a} = [a_0, a_1]^T$.

2. Alternatively, the tuning parameters can be introduced in an explicit correction function $C(\boldsymbol{x},\boldsymbol{a})$, which also depends on the design variables and can be developed in exactly the same way as analytical approximation functions described above. For example, using the linear and multiplicative models as the correction function, the following approximations can be built:

$$\tilde{F}(\boldsymbol{x},\boldsymbol{a}) = f(\boldsymbol{x}) + C(\boldsymbol{x},\boldsymbol{a}), \quad C(\boldsymbol{x},\boldsymbol{a}) = a_0 + \sum_{i=1}^{N} x_i a_i \quad \text{or} \tag{5.29}$$

$$\tilde{F}(\boldsymbol{x},\boldsymbol{a}) = f(\boldsymbol{x}) C(\boldsymbol{x},\boldsymbol{a}), \quad C(\boldsymbol{x},\boldsymbol{a}) = a_0 \prod_{i=1}^{N} x_i^{a_i} \; . \tag{5.30}$$

The same approach has been used by Venkataraman *et al.* (1998) who referred to $C(\boldsymbol{x},\boldsymbol{a})$ as correction response surface approximations.

3. The simplified model, providing the basis for the approximation, depends in fact not only on the design variables $\boldsymbol{x}$ but also on the other parameters of the simulation model, e.g. material or geometry properties. To fit the simplified model into the data obtained with the complex model at the plan points, some of the parameters can be considered as the tuning parameters. The choice of the tuning parameters a should then be based on physical grounds. The approximation model can then be written in the form

$$\tilde{F}(\boldsymbol{x},\boldsymbol{a}) \equiv f(\boldsymbol{x},\boldsymbol{a}) \, . \tag{5.31}$$

### 5.5.3 Numerical Examples

The mechanism presented in Fig. 5.37 consists of three links of circular cross-section and one fixed link (link four in Fig. 5.37). A constant angular velocity is applied at the point $A_0$ . Each link is modelled by one planar beam finite element, which allows for the bending deformations. The lenghts of the links are: $l_1$ = 0.3048 *m*, $l_2$ = 0.9144 *m*, $l_3$ = 0.7620 *m*, $l_4$ = 0.9144 *m*. The material of the links has the mass density $\rho = 2757.24\ kg/m^3$ and modulus of elasticity $E = 6.895 \cdot 10^{10}\ Pa$. The cross-sectional areas of the three links $x_i$ ($i$ = 1, 2, 3) are chosen as the design variables. The optimization problem is to minimize the total mass of the system

$$F_0(\mathbf{x}) = \rho(l_1x_1 + l_2xl_2 + l_3x_3) \tag{5.32}$$

subject to constraints on the maximum values of the bending stresses in the links over the period of time of 0.2 *s*:

$$F_i(\mathbf{x},t) = \frac{\sigma_i(\mathbf{x},t)}{\sigma_{al}} \leq 1,\ i = 1, 2, 3,\ \ 0 \leq t \leq 0.2\ s. \tag{5.33}$$

The time dependence in the constraints is eliminated by using the critical points equivalent constraints and the time discretization of 100 points. The bending stresses are evaluated using the values of the bending moment

$$\sigma_i(\mathbf{x},t) = \frac{4\sqrt{\pi}}{x_i^{3/2}} M_i \tag{5.34}$$

where $M_i$ is the bending moment evaluated in the $i$-th link.

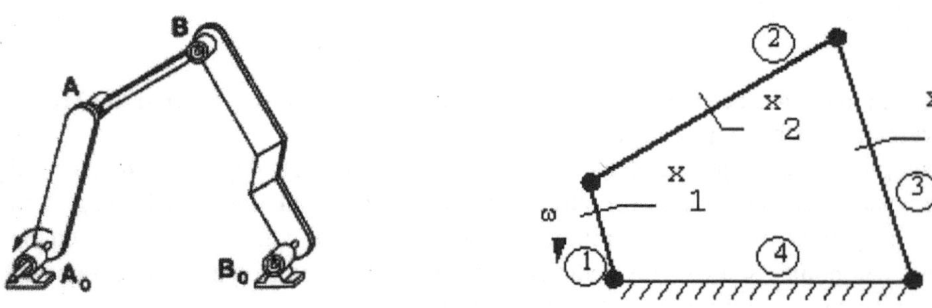


**Figure 5.37** Four-bar mechanism.

Using the definition of the problem and expressions (5.34), the following explicit mechanistic approximations of the constraint functions have been proposed (Toropov et al., 1997):

$$\widetilde{F}_1(x) = a_1 (4x_2 + x_3)\, x_1^{-3/2},\ \widetilde{F}_2(x) = a_2 x_2^{-1/2},\ \widetilde{F}_3(x) = a_3 x_3^{-1/2}. \tag{5.35}$$

Since the driving link is relatively short (almost rigid in bending) the bending moment $M$ in (5.34) weakly depends on the mass of the link defined by $x_1$ and, therefore, from (5.34) the approximation of the bending stress in the first link $\widetilde{F}_1(x)$ should depend on $x_1^{-3/2}$. The factor four in the formulation of $\widetilde{F}_1(x)$ representing the relative contribution of the masses of the second and third link to the bending moment of the first link can be replaced by a tuning parameter. Since the bending moments in the second and third links in (5.34) mostly depend on the masses of the elements which are proportional to $x_2$ and $x_3$, respectively, the approximations of the bending stresses $\widetilde{F}_2(x)$ and $\widetilde{F}_3(x)$ are assumed to be proportional to $x_2^{1/2}$ and $x_3^{1/2}$, respectively, see (5.35).

The dynamic analysis of a flexible mechanism (high fidelity model) is a time consuming procedure. It requires integration of a system of nonlinear differential equations of motion. Moreover, to find the periodic solution, the system has to be integrated over the period several times (Markine, 1999). As a first step in the analysis of a mechanism with flexible links, it is useful to investigate the simple one with rigid links. A dynamic analysis for such a model is two orders of magnitude faster. Moreover, if the input velocity is relatively low, the results of the analysis of the mechanism with flexible and rigid links are close. Therefore, the mechanism with the rigid links has been chosen as the simplified (low fidelity) model, i.e. $f(x)$ in (5.27) is the stress constraint function evaluated for the rigid link mechanism. Since the stress distribution along the link depends on its inertia properties, the mass densities of the links have been chosen as the tuning parameters to construct the approximations of the third type (5.31), namely $a_i = \rho_i$, $i = 1, 2, 3$.

The results of the optimization are collected in Table 5.6 where $N_{it}$ is the total number of iterations when the same starting point $x_i = 0.1\ m^2$, $i = 1, 2, 3$ was taken in all optimization runs. For comparison, the same problem (both with rigid and flexible links) was solved using the simple explicit multiplicative approximation (5.9) with $\varphi_i(x) = x_i$, $i = 1, 2, 3$.

As expected, the use of the knowledge about the original objective and constraint functions in the definition of the approximations (e.g. explicit mechanistic approximations (5.34)) results in better convergence of the optimization process (compare the results with the explicit multiplicative and explicit mechanistic approximations). From Table 5.6 it can also be observed that the use of the simplified numerical models in the approximation formulation significantly improves the quality of the approximation functions and speeds up the convergence.

From the results given in Table 5.6 it can be concluded that the use of the multiplicative correction in approximations of the first and the second type leads to better approximations than the use of the linear correction for the same type of the approximation. It reflects the fact that the linear correction functions of type 1 and type 2 do not allow sufficient reduction of the discrepancy between the functions representing complex and simplified models. The poor quality of the approximations of the type one (5.28) with the linear correction function explains the poor convergence characteristics of the corresponding optimization process (13 iterations). Since the initial point was far from the optimum, the search subregion has been considerably reduced during the first iteration steps because of the poor quality of the approximations.

**Table 5.6** Comparison of approximations.

| | Explicit (mult.) | | Mecha-nistic | Simplified numerical model | | | | | Unit |
|---|---|---|---|---|---|---|---|---|---|
| | Rigid links | Flex. links | | Type 1 | | Type 2 | | Type 3 | |
| | | | | Linear | Mult. | Linear | Mult. | $a_i = \rho_i$ | |
| $x_1$ | 0.94e-3 | 0.89e-3 | 0.98e-3 | 0.87e-3 | 0.99e-3 | 1.00e-6 | 0.98e-3 | 0.85e-3 | $m^2$ |
| $x_2$ | 0.42e-3 | 0.35e-3 | 0.41e-3 | 0.37e-3 | 0.43e-3 | 0.43e-6 | 0.42e-3 | 0.42e-3 | $m^2$ |
| $x_3$ | 0.26e-3 | 0.24e-3 | 0.28e-3 | 0.18e-3 | 0.26e-3 | 0.27e-3 | 0.27e-3 | 0.26e-3 | $m^2$ |
| $F_0$ | 2.39 | 2.14 | 2.44 | 2.03 | 2.46 | 2.49 | 2.45 | 2.40 | $kg$ |
| $F_1$ | 1.00 | 0.99 | 1.00 | 1.01 | 1.00 | 1.00 | 1.01 | 1.03 | - |
| $F_2$ | 1.00 | 1.01 | 1.00 | 1.12 | 1.00 | 0.99 | 1.02 | 1.02 | - |
| $F_3$ | 1.00 | 1.04 | 1.00 | 1.18 | 1.01 | 0.98 | 0.99 | 1.01 | - |
| $N_{it}$ | 4 | 12 | 4 | 13 | 7 | 6 | 4 | 5 | |

The best results presented in Table 5.6 were obtained using the explicit mechanistic approximations (5.35) and the approximation of type 2 with the multiplicative correction function (5.30). However, as mentioned above, the expressions (5.35) were obtained for approximation of the bending stresses of this particular four-bar mechanism and they cannot be used for other problems while the approximations (5.30) are general.

The use of some input parameters of a simplified model as the tuning parameters (type 3) seems to be promising because it leads to a more natural form of approximation. However, due to the limited freedom in the number of parameters of the simplified model (the mechanical system is quite simple) which could be taken as the tuning parameters in the particular example, the result was not as good as the one using the approximations of the second type with a multiplicative correction function. It can be expected that more complex problems would give greater freedom in the selection of tuning parameters from the simplified models.

In conclusion, the results showed that the use of simplified numerical models can improve the quality of the approximations and, ultimately, the convergence characteristics of the optimization process. Due to their generality, such approximations can be used for various design optimization problems.

## 5.6 References

Alvarez, L.F., Hughes, D.C., and Toropov,V.V. (2000). Optimization of the manufacturing process for Roman cement. In Sienz, J., ed., *Engineering Design Optimization. Process and Product Improvement. Proceedings of 2nd ASMO UK / ISSMO Conference*, 27-32.

Ashill, P.R., Wood, R.F., and Weeks, D.J. (1987) *An improved, semi-inverse version of the Viscous, Garabedian and Korn Method (VGK)*, RAE Technical Report 87002.

Audze, P., and Eglais, V. (1977). New approach for planing out of experiments. *Problems of Dynamics and Strengths* 35:104-107 (in Russian).

Barthelemy, J.-F.M., and Haftka R.T. (1993). Approximation concepts for optimum structural design - a review. *Structural Optimization* 5:129-144.

Booker, A.J. (1998). Design and analysis of computer experiments. AIAA-98-4757. *Proceedings of 7th AIAA/USAF/NASA/ISSMO Symposium on Multidisciplinary Analysis and Optimization*, St. Louis, AIAA. Part 1:118-128.

Booker, A.J., Dennis, J.E., Jr., Frank, P.D., Serafini, D.B., and Torzson, V. (1999). A rigorous framework for optimization of expensive functions by surrogates. *Structural Optimization* 17:1-13.

Box, G.E.P., and Draper, N.R. (1987). *Empirical model-building and response surfaces*. New York: John Wiley and Sons.

Burgee, S., Giunta, A.A., Balabanov, V., Grossman, B., Mason, W.H., Narducci, R., Haftka, R.T., and Watson, L.T. (1996). A coarse-grained parallel variable-complexity multidisciplinary optimization paradigm. *International Journal of Supercomputer Applications and High Performance Computing* 4:269-299.

Chung, H.-S. and Alonso, J.J. (2000). Comparison of approximation models with merit functions for design optimization. AIAA 2000-4754. *Proceedings of 8th AIAA/USAF/NASA/ISSMO Symposium on Multidisciplinary Analysis and Optimization*, Long Beach, CA: AIAA.

Etman, L.F.P., Adriaens, J.M.T.A., van Slagmaat, M.T.P., and Schoofs, A.J.G. (1996). Crash worthiness design optimization using multipoint sequential linear programming. *Structural Optimization* 12:222-228.

Free, J.W., Parkinson, A.R., Bryce, G.R., and Balling, R.J. (1987). Approximations of computationally expensive and noisy functions for constrained nonlinear optimization. *Journal of Mechanisms, Transmissions, and Automation in Design* 109:528-532.

Giunta, A.A., Dudley, J.M., Narducci, R., Grossman, B., Haftka, R.T., Mason, W.H., and Watson, L.T. (1994). Noisy aerodynamic response and smooth approximations in HSCT design. AIAA-94-4376-CP. *Proceedings of 5th AIAA/USAF/NASA/ISSMO Symposium on Multidisciplinary and Structural Optimization*, Panama City, Florida: AIAA. 1117-1128.

Giunta, A.A., Balabanov, V., Haim, D., Grossman, B., Mason, W.H., Watson, L.T., and Haftka, R.T. (1997). Multidisciplinary optimization of a supersonic transport using design of experiments theory and response surface modelling. *Aeronautical Journal* 101:347-356.

Gray, G., Li, Y., Murray-Smith, D., and Sharman, K. (1996). Structural system identification using genetic programming and a block diagram oriented simulation tool. *Electronics Letters* 32:1422-1424.

Gray, G., Murray-Smith, D., Sharman, K., and Li, Y. (1996). Nonlinear model structure identification using genetic programming. In *Late-breaking papers of Genetic Programming '96*, Stanford, CA.

Haftka, R.T. (1997). Optimization and experiments – a survey. In Tatsumi, T., Watanabe, E., and Kambe, T., eds., *Theoretical and Applied Mechanics 1996, Proceedings of XIX International Congress of Theoretical and Applied Mechanics*, Elsevier. 303-321.

Jin, R., Chen, W. and Simpson, T.W. (2000). Comparative studies of metamodeling techniques under multiple modeling criteria. AIAA 2000-4801. *Proceedings of 8th AIAA/USAF/NASA/ISSMO Symposium on Multidisciplinary Analysis and Optimization*, Long Beach, CA: AIAA.

van Keulen, F., Toropov, V.V., and Markine, V.L. (1996). Recent refinements in the multipoint approximation method in conjunction with adaptive mesh refinement. In McCarthy,

J.M., ed., *Proceedings of ASME Design Engineering Technical Conferences and Computers in Engineering Conference*, Irvine, CA: 96-DETC/DAC-1451. 1-12.

van Keulen, F., Polynkine, A.A., and Toropov, V.V. (1997). Shape optimization with adaptive mesh refinement: Target error selection strategies. *Engineering Optimization* 28:95-125.

van Keulen, F., and Toropov, V.V. (1997). New developments in structural optimization using adaptive mesh refinement and multi-point approximations. *Engineering Optimization* 29:217-234.

van Keulen, F., and Toropov, V.V. (1998). Multipoint approximations for structural optimization problems with noisy response functions. *Proceedings of 1st ISSMO/NASA Internet Conference on Approximations and Fast Reanalysis in Engineering Optimization.* Published on a CD ROM by ISSMO/NASA/AIAA.

van Keulen, F., and Toropov, V.V. (1999). The multipoint approximation method in a parallel computing environment. *Zeitschrift fuer Angewandte Mathematik und Mechanik* 79:567-570.

van Keulen, F., Haftka, R.T., and Qu, X.-Y. (2000). Noise and discontinuity issues in response surfaces based on functions and derivatives. AIAA-00-1363. *Proceedings of 41st AIAA/ASME/ASCE/AHS/ASC Structures, Structural Dynamics and Materials Conference.*

Kinnear, K.E. (1994). *Advances in genetic programming*, MIT Press.

Koza, J.R. (1992). *Genetic Programming: on the programming of computers by means of natural selection.* MIT Press.

Lewis, R.M. (1998). Using sensitivity information in the construction of kriging models for design optimization. AIAA-98-4799. *Proceedings of 7th AIAA/USAF/NASA/ISSMO Symposium on Multidisciplinary Analysis and Optimization,* St. Louis: AIAA. Part 2:730-737.

Madsen, J.I., Shyy, W., and Haftka, R.T. (2000). Response surface techniques for diffuser shape optimization. *AIAA Journal* 38:1512-1518.

Madsen, K., and Hegelund, P. (1991). *Non-gradient subroutines for non-linear optimization. Institute for Numerical Analysis*, Technical University of Denmark, Report NI-91-05.

McKay, B., Willis, M.J., Hidden, H.G., Montague, G.A., and Barton, G.W. (1996). Identification of industrial processes using genetic programming. In Friswell, M.I., and Mottershead, J.E., eds., *Proceedings of International Conference on Identification in Engineering Systems*, Swansea, The Cromwell Press Ltd. 328-337.

Markine, V.L., Meijers, P., Meijaard, J.P., and Toropov, V.V. (1996). Optimization of the dynamic response of linear mechanical systems using a multipoint approximation technique. In Bestle, D., and Schiehlen, W., eds., *Proceedings of IUTAM Symposium on Optimization of Mechanical Systems*, Stuttgart, Kluwer. 189-196.

Markine, V.L., Meijers, P., Meijaard, J.P., and Toropov, V.V. (1996). Multilevel optimization of dynamic behaviour of a linear mechanical system with multipoint approximation. *Engineering Optimization* 25: 295-307.

Markine, V.L., Meijers, P., Meijaard, J.P., and Toropov, V.V. (1998). Optimization of mechanisms using direct differentiation and a multipoint approximation method. *Engineering Optimization* 31:141-160.

Markine, V.L., Esveld, C., Kok, A.W.M., de Man, A.P., and Toropov, V.V. (1998). Optimization of a high-speed railway track using multipoint approximation method. AIAA-98-4980. *Proceedings of 7th AIAA/USAF/NASA/ISSMO Symposium on Multidisciplinary Analysis and Optimization,* St. Louis: AIAA. Part 2:1328-1331.

Markine, V.L., Meijers, P., and Toropov, V.V. (1998). Optimization of workspace of a manipulator for a flight simulator using multipoint approximation method. AIAA-98-4955.

*Proceedings of 7th AIAA/USAF/NASA/ISSMO Symposium on Multidisciplinary Analysis and Optimization,* St. Louis: AIAA. Part 2:1922-1932.

Markine V.L. (1999). *Optimization of the dynamic behaviour of mechanical systems.* PhD thesis, Delft University of Technology, The Netherlands. Shaker Publishing B.V.

Markine, V.L., Toropov, V.V., and Esveld, C. (2000). Optimization of ballastless railway track using multipoint approximations: issue of domain-dependent calculability. AIAA-00-4714. *Proceedings of 8th AIAA/USAF/NASA/ISSMO Symposium on Multidisciplinary Analysis and Optimization,* Long Beach, CA. Published on a CD ROM by AIAA.

Matherton, G. (1963). Principles of geostatistics. *Economic Geology* 58:1246-1266.

Narducci, R., Grossman, B., Valorani, M., Dadone, A., and Haftka, R.T. (1995). Optimization methods for non-smooth or noisy objective functions in fluid design problems. AIAA 95-1648 CP. *Proceedings of 12th AIAA Computational Fluid Dynamics Conference.*

Papalambros, P.Y. and Wilde, D.J. (2000). Principles of optimal design. Modeling and computation. 2nd ed. Cambridge University Press.

Polynkine, A.A., van Keulen, F., and Toropov, V.V. (1995). Optimization of geometrically nonlinear thin-walled structures using the multipoint approximation method. *Structural Optimization* 9:105-116.

Polynkine, A.A., van Keulen, F., and Toropov, V.V. (1996). Optimization of geometrically non-linear structures based on a multi-point approximation method and adaptivity. *Engineering Computations* 13:76-97.

Polynkine, A.A., van Keulen, F., de Boer, H., Bergsma, O.K., and Beukers, A. (1996). Shape optimization of thermoformed continuous fibre reinforced thermoplastic products. *Structural Optimization* 11:228-234.

Rasmussen, J. (1990). Accumulated approximations - a new method for structural optimization by iterative improvements. *Proceedings of 3rd USAF/NASA Symposium on Recent Advances in Multidisciplinary Analysis and Optimization,* San Francisco, CA. 253-258.

Rikards, R. (1993). Elaboration of optimal design models for objects from data of experiments". In Pedersen, P., ed., *Optimal design with advanced materials, The Frithiof Niordson volume. Proceedings of the IUTAM Symposium,* Lyngby, Denmark, Elsevier. 149-162.

Roux, W.J., Stander, N., and Haftka, R.T. (1998). Response surface approximations for structural optimization, *International Journal for Numerical Methods in Engineering* 42:517-534.

Schoofs, A.J.G. (1987). *Experimental design and structural optimization*: Ph.D. thesis, Eindhoven, The Netherlands.

Sevant, N.E., Bloor, M.I.G., and Wilson, M.J. (2000). Aerodynamic design of a flying wing using response surface methodology. *Journal of Aircraft* 37:562-569.

Sobieszanski-Sobieski, J., and Haftka, R.T. (1997). Multidisciplinary aerospace design optimization: Survey of recent developments. *Structural Optimization* 14:1-23.

Toropov V.V. (1989). Simulation approach to structural optimization. *Structural Optimization* 1:37-46.

Toropov, V.V., Filatov, A.A., and Polynkin, A.A. (1993). Multiparameter structural optimization using FEM and multipoint explicit approximations. *Structural Optimization* 6:7-14.

Toropov, V.V., and van der Giessen, E. (1993). Parameter identification for nonlinear constitutive models: Finite element simulation - optimization - nontrivial experiments. In Pedersen, P., ed., *Optimal design with advanced materials, The Frithiof Niordson volume. Proceedings of the IUTAM Symposium,* Lyngby, Denmark, Elsevier. 113-130.

Toropov, V.V., and Carlsen, H. (1994). Optimization of Stirling engine performance based on multipoint approximation technique. In Gilmore, B.J. et al., eds., *Advances in Design Automation 1994, Robust Design Applications, Decomposition and Design Optimization, Optimization Tools and Applications. 20th Design Automation Conference,* Minneapolis. Vol. 2: 531-536.

Toropov, V.V., Van Keulen, F., and Markine, V.L. (1995). Structural optimization in the presence of numerical noise. *1st World Congress of Structural and Multidisciplinary Optimization*, Extended Abstracts - Posters, Goslar. 24-25.

Toropov, V.V., van Keulen, F., Markine, V.L., and de Boer, H. (1996). Refinements in the multi-point approximation method to reduce the effects of noisy responses. *Proceedings of 6th AIAA/USAF/NASA/ISSMO Symposium on Multidisciplinary Analysis and Optimization*, Bellevue, WA: AIAA. 941-951.

Toropov, V.V., and Markine, V.L. (1996). The use of simplified numerical models as mid-range approximations. *Proceedings of 6th AIAA/USAF/NASA/ISSMO Symposium on Multidisciplinary Analysis and Optimization*, Bellevue, WA: AIAA. 952-958.

Toropov, V.V., Markine, V.L., Meijers, P., and Meijaard, J.P. (1997). Optimization of a dynamic system using multipoint approximations and simplified numerical model. In Gutkowski, W., and Mroz., Z., eds., *Proceedings of 2nd World Congress of Structural and Multidisciplinary Optimization*, 613-618.

Toropov, V.V., and Alvarez, L.F. (1998). Approximation model building for design optimization using genetic programming methodology. AIAA-98-4769. *Proceedings of 7th AIAA/USAF/NASA/ISSMO Symposium on Multidisciplinary Analysis and Optimization*, St. Louis: AIAA. Part 1:490-498.

Toropov, V.V., and Markine, V.L. (1998). Use of simplified numerical models as approximations: Application to a dynamic optimal design problem. *Proceedings of 1st ISSMO/NASA Internet Conference on Approximations and Fast Reanalysis in Engineering Optimization.* Published on a CD ROM by ISSMO/NASA/AIAA.

Toropov, V.V., Markine, V.L., and Holden, C.M.E. (1999). Use of mid-range approximations for optimization problems with functions of domain-dependent calculability, *Proceedings of 3rd World Congress of Structural and Multidisciplinary Optimization,* Buffalo, NY.

Valorani, M., and Dadone, A. (1995). Sensitivity derivatives for non-smooth or noisy objective functions in fluid design problems. *Proceedings of ICFD Conference on Numerical Methods for Fluid Dynamics,* Oxford University.

Venkataraman, S., Haftka, R., and Johnson, T. (1998). Design of shell structures for buckling using correction response surface approximations. AIAA-98-4855. *Proceedings of 7th AIAA/USAF/ NASA/ISSMO Symposium on Multidisciplinary Analysis and Optimization*, St. Louis: AIAA. 1131-1144.

Venter, G., Haftka, R.T., and Starnes, J.H. Jr. (1998). Construction of response surface approximations for design optimization. *AIAA Journal* 36:2242-2249.

# Chapter 6

## Strategies for Modeling, Approximation, and Decomposition in Genetic Algorithms Based Multidisciplinary Design

P. Hajela
Mechanical Engineering, Aeronautical Engineering & Mechanics
Rensselaer Polytechnic Institute, Troy, NY, USA

**Abstract:** This chapter discusses the applicability of new computational paradigms motivated by biological processes, in the realm of multidisciplinary engineering design, and particularly, in the context of using formal methods of design optimization. The computational models considered in this discussion include genetic algorithms, neural networks, and a modeling of the biological immune system. The focus of the chapter is two-fold. First, it introduces the reader to the implementation of these newly emergent methods. Second, it describes how the use of these methods alleviates some of the difficulties associated with the application of formal optimization methods in practical design problems. Such problems are typically characterized by the presence of a large number of design variables and constraints, the need to consider multiple objective criterion, and, in some cases, a fuzzy description of design specifications. The analysis associated with the multidisciplinary design problem is both complex and computationally expensive. The discussion focuses on methods to reduce the computational effort through development of efficient optimal search algorithms, and in the efficient management of couplings in the analysis problem.

### 6.1 Introduction

The role of formal optimization methods in multidisciplinary systems has received considerable recent attention (Sobieszczanski-Sobieski, 1993, Tolson and Sobieszczanski-Sobieski, 1985, Abdi et. al. 1988). This increased study has been largely motivated by a recognition that the design and development of complex engineering system can no longer be conducted by handling its different components in isolation. Both an increased level of complexity in the component design process, and a desire to take full advantage of any synergism that may exist among the different disciplines dictate the need for a more comprehensive strategy. Inspite of an increased awareness of the potential of formal optimization methods in the design process, their application in real-life multidisciplinary systems continues to be a challenge. These challenges stem from the fact that practical scale multidisciplinary design optimization (MDO) problems are characterized by the presence of a large number of design variables and constraints, and analysis from various contributing disciplines which are not completely independent, involving one- or two-way couplings between the disciplines. In a number of situations, these analysis may be either very expensive, or in a worse case, simply unavailable to a design engineer in a form other than physical experimentation. MDO can be defined as a multifaceted entity with many conceptual components, and comprising of many established disciplines and technologies. Sobieszczanski-Sobieski (1993) defines the principal building blocks of the MDO problem in Fig. 6.1. This taxonomy chart clearly shows the different areas in which research must be focussed if multidisciplinary optimization is to embraced as a standard procedure in industry design practice. These areas can be identified as follows.

- Approximations are needed to replace expensive exact analysis, where the latter can come from either sophisticated numerical codes or through physical experiments. Towards this end, the focus has resided in developing derivative-based extrapolations, or derivative-free approximations such as response surfaces, fuzzy-logic, and neural networks. In case of the latter, design of experiment approaches such as Taguchi methods have an important role in determining the data that must be used to develop such approximations.
- Mathematical modeling is critical to the success of the MDO approach. Such a model is required to predict the system behavior and its sensitivity to design variable changes. Models can be considered to belong to the broad categories of either physical or non-physical models. These models are necessary to answer 'what-if' questions.
- Design-oriented analysis basically deals with the issue of cost vs. accuracy in analysis for optimization. Inexpensive re-analysis is critical and is often based on a linear representation of the problem or on the basis of response-surface like approximations. In some cases, mathematical models can be constructed which only predict the gross system behavior, and provide crude models for preliminary design. Design oriented analysis must also consider how different components of the MDO problem must be executed so as to minimize the effort required in computing their interactions.
- Decomposition is a major concern in MDO, and entails the break-up of a large-scale problem into a sequence of smaller, more tractable subproblems. Depending upon the problem under consideration, the interactions between the decoupled subproblems may be either hierarchic or non-hierarchic.
- Design space search is a major component in MDO and is carried out by using traditional mathematical programming algorithms or newly emergent search techniques based on random sampling such as genetic algorithms and simulated annealing.
- Human intervention continues to be an important issue in MDO as the solution process is not a push-button operation. Operations such as algorithm termination, problem reformulation, interpretation of results, overriding the design process, require human intervention. Methods of artificial intelligence and expert systems have been explored in this context.
- Optimization strategies must continue to be studied in the context of the MDO problem. This includes optimization within a subproblem or a parallel optimization in all subproblems including ways to coordinate the solution towards a converged point.

A detailed discussion of these issue is clearly beyond the scope of this chapter. Instead, the approach taken here is to focus on some specific adaptations of soft-computing methods such as genetic algorithms, neural networks, immune network modeling, and fuzzy logic in problems of multidisciplinary design. Subsequent sections of this chapter describe the adaptation of genetic algorithms and neural networks in multidisciplinary design optimization; issues of problem decomposition, coordination of solution in decomposed subsystems, design space search, and design oriented analysis, are forefront in this discussion. The chapter also discusses the implementation of GA based modeling of immune networks. Immune network modeling is shown to have a role in enhancing the GA based search process, in providing a mechanism for solution coordination in decomposition based design, and in the solution of multicriterion design problems. The immune network modeling approach is also extended in combination with fuzzy-logic concepts to solve design problems with non-crisp problem statements. Finally, the chapter will introduce some new concepts in computational intelligence, and of the role of genetic algorithms in developing useful information from the computational domain for enhancing the efficiency of the optimization process. In this context, the approach of classifier systems will be discussed with special focus on its application in facilitating the creation of approximate analysis models.

## 6.2 Evolutionary Optimization

Under the banner of evolutionary methods for design optimization, this section will primarily focus on the genetic algorithm approach; a brief review of the related approaches of evolutionary programming and evolutionary strategies is also included for completeness. The discussion will focus on the basic approach, the status of new developments, and potential new areas of research.

### 6.2.1 Genetic Algorithms

In Holland (1974), genetic algorithms were characterized by bit string representations of possible solutions to a given problem, and by transformations used to vary and improve these coded solutions. An analogy may be drawn with the natural process of reproduction in biological populations, where genetic information stored in chromosomal strings evolves over generations to adapt favorably to a static or changing environment. This chromosomal structure represents the generational memory, and is altered when members of the population reproduce. The three basic processes which affect the chromosomal makeup in natural evolution are inversions of chromosome strings, an occasional mutation of genetic information, and a crossover of genetic information between the reproducing parents. The last process is an exchange of genetic characteristics between the parents, and allows for beneficial genes to be represented in the progeny. Genetic algorithms, in a manner similar to their natural counterpart, use chromosome type representations of possible solutions of the problem to search for improved solutions.

Genetic algorithms are based on an elitist reproduction strategy, where members of the population that are deemed most fit are selected for reproduction, and are given the opportunity to strengthen the chromosomal makeup of the progeny generation. This approach is facilitated by defining a fitness function or a measure indicating the "goodness" of a member of the population in a given generation during the evolution process. To see how GA's actually work, we first examine those strategies wherein designs may be represented by chromosome–like strings. Consider a design problem with two design variables, and prescribed lower and upper bounds. In order to represent these design variables as a chromosome-like string, different schemes may be adopted. As an example, the design variable values can be in the decimal form, with a fixed number of significant digits. The decimal numbers can be stacked head-to-tail to create a string that represents the total design. Alternatively, each design variable can be converted into its binary equivalent, and thereby mapped into a fixed length string of 0's and 1's. Clearly, the numerical precision with which the design variable is represented is determined by the string length. In fact, it is easy to see that the precision $A_c$ can be computed from the relation

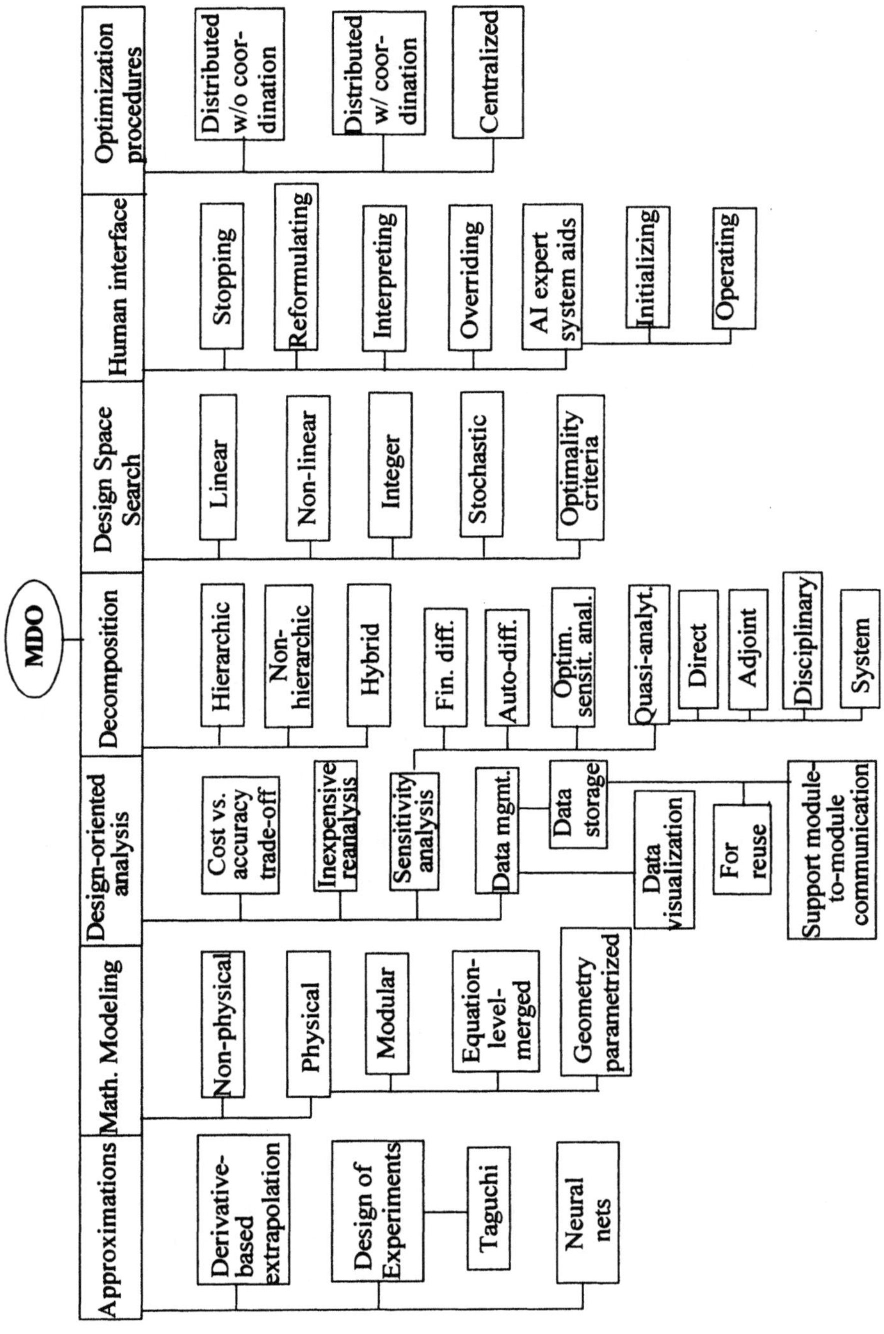


**Figure 6.1** Taxonomy of multidisciplinary design research.

$$A_c = \frac{(X_1^U - X_1^L)}{2^L - 1} \tag{6.1}$$

where L is the number of bits in the string. Assuming that each design variable is represented by a 10 digit bit string, the first step would be to map the maximum and minimum values of the design variables into binary number as follows:

$$X_1^L = 0000000000$$

$$X_1^U = 1111111111$$

A linear scaling can be introduced to convert intermediate values of the binary number into physical design variable values. Two such binary numbers, each corresponding to one design variable would then be stacked head-to-tail to create a 20 digit string of 0's and 1's. This 20 digit string represents one design; a sequence of such strings can be introduced to construct a population of designs, with each design having a corresponding fitness value $F_v$. This fitness value could be the objective function $F(X)$ for a function maximization problem. In the case of function minimization, at least two different alternatives can be considered as follows,

$$F_v = \frac{1}{F(X)}$$
$$F_v = F^* - F(X) \tag{6.2}$$

where $F^*$ is an arbitrary large value that is greater than the largest $F(X)$ in the population. We will defer the discussion of handling constraints to a later section. Once a chromosome–like representation of designs is available, simulations of genetic evolution and adaptation can be invoked. An artificial gene–transformation mechanism comprises of three principal components, namely reproduction, crossover, and mutation.

Reproduction: The simulation of genetic evolution here is contrived in that the population size is forced to remain unchanged during the evolution process; two mating parents create only two progenies but are themselves eliminated. This requirement is not essential, and simulations have been performed where a population increase is actually implemented. The reproduction process is one that biases the search toward producing more fit members in the population and eliminating the less fit ones. One simplistic approach to selecting members from an initial population to participate in the reproduction is to assign each member a probability of selection on the basis of its fitness. If $F_v(i)$ is the fitness measure of the i–th member, it can be assigned a probability of selection,

$$p_i = \frac{F_v(i)}{\sum_{NPOP} F_v(i)} \tag{6.3}$$

where NPOP is the population size. A new population pool of the same size as the original is created, but has a higher average fitness value. Note that no new designs are created in this process; the less fit ones are simply eliminated and additional copies of the more fit designs are brought into the population.

Crossover: The process of reproduction simply results in more copies of the dominant or fit designs to be present in the population. The crossover process allows for an exchange of design characteristics among members of the population pool with the intent of improving the fitness of the next generation. This is similar to exchange of genetic information in biological reproduction processes facilitated by DNA and RNA strings. While there are a number of different ways in which the crossover operation can be implemented, a widely practiced approach is the two-point crossover. In this approach, two mating parents are selected at random; the random number generator is invoked to identify two sites on the strings, and the strings of 0's and 1's enclosed between the chosen sites are swapped between the mating strings. An illustration of the crossover transformation for two mating strings, each with 20 bits is as follows:

Parent1 = 11001<u>0</u>01001<u>0</u>011010
Parent2 = 010111<u>1</u>00011<u>1</u>001000

Child1 = 11001100011001 1010
Child2 = 01011001001100 1000

The crossover sites on the parent strings are identified by the underlined bits. A probability of crossover $p_c$ is defined to determine if crossover should be implemented for the chosen pair of mating strings.

Mutation: Mutation safeguards the genetic search process from a premature loss of valuable genetic information during reproduction and crossover. As the population of designs begins to converge, it is entirely possible that bits at a few locations are identical in the entire population. Under these circumstances, there is no mechanism which would allow these bits to switch from a 0 to a 1 or vice versa. The process of mutation is added to correct for this deficiency, as computational expediency forces us to work with a limited size population and it is quite likely that converged sub-strings may occur prematurely in a population. The idea behind mutation is simply to choose few members from the population pool according to a probability of mutation $p_m$, and to switch a 0 to 1 or vice versa at a randomly selected mutation site on the chosen string.

The mechanics of genetic search are simple; however, there are some key differences from traditional methods to which a few of the strengths may be attributed. These are summarized as follows:

- Genetic algorithms work on function evaluations alone and do not require function derivatives. While derivatives contribute to a faster convergence towards the optimum, they may also direct the search towards a local optimum.
- Genetic search proceeds from several points in the design space to another such set of design points. Since the information for this update is drawn from several points distributed through the design space, the method has a better probability of locating global minima as opposed to those schemes that proceed from one point to another.
- Genetic algorithms work on a coding of the design variables rather than the variables themselves. This allows for an extension of these algorithms to design spaces consisting of a mix of continuous, discrete, and integer variables.

### 6.2.2 Other Evolutionary Algorithms

Evolutionary programming (EP) and evolutionary strategies (ES) are two other procedures that have been motivated by natural evolution (Fogel et. al. 1966, Rechenberg, 1973). These methods, like the genetic algorithm, use transformation operators such as selection, recombination (crossover), and mutation. However, the degree to which each operator is emphasized in the algorithm is where the principal differences can be found.

In EP, evolution of populations is done primarily through selection and mutation; recombination is not directly implemented although the contention is that the mutation operation is flexible enough to achieve the characteristics of recombination. All N members of the population are considered as parents, and mutated to produce N children. Using a probabilistic survival scheme, the best N of the 2N members then go over to the next generation. Real valued numbers are often used to represent the designs, and the mutation is normally distributed with a zero expectation.

The ES approach was both developed and more rigorously practiced in Europe. In this approach a population of designs is evaluated and a uniform random selection done to select the parents. In a standard recombinative ES, pairs of parents produce children via recombination, which are then further perturbed through mutation. The number of children may be greater than N, and either the N best children are selected to replace all parents or the best N out of the children and parents retained as the next generation of parents. The emphasis is again given to mutation operation, by adaptively selecting mutation rate for each variable that is normally distributed with a zero expectation. Note, however, that recombination is included, and this is in contrast to the EP. Design representations, however, are based on real numbers like in the EP approach, and averaging recombination is introduced to produce children with different characteristics than the parents.

For the industry design user, a careful selection of design variable representation strategy is even more important as it can significantly ease the computational requirements of a GA based search, and also improve the quality of results. A nonbinary representation of design variables has also been introduced in the genetic algorithm approach (Hajela and Lin, 2000, LeRiche and Haftka 1993). Two specific advantages have been put forward in support of going to a larger alphabet in the representation scheme. First, is the ease of implementation that is particularly attractive to an end-user. Second, and perhaps more importantly, is the enhanced efficiency of the search that is available in some instances. As an example, consider representing a composite laminate of fixed thickness with only ply orientations of 0, +45, -45, and 90 deg. In a binary representation scheme, a 2-digit binary number is needed to specify the ply orientation of each ply. For example, the strings 00, 01, 10, and 11 can be used to represent each of the above ply orientations. Alternatively, the above angles can be represented by the numbers 1,2,3, and 4, respectively. If these schemes are used to represent an n-ply composite laminate, each at a ply orientation selected from the four choices above, then the binary string representation would be of length 2n, and the number of designs represented by such a string would be $2^{2n}$. In the case where integers from 1 to 4 are used to represent the different angles, the length of the string would be n, and the number of design alternatives represented by this scheme be $4^n=2^{2n}$, or exactly the same as before. However, similarity in schema would be more difficult to detect in the second option given the larger alphabet. On the other hand, if the choice of angles were limited to just 0, 45, and 90 deg, then a 2 digit binary string would still be required, with one combination of this string either assigned to an inadmissible variable, or one ply orientation being represented by two different string combinations. Here, the integer representation scheme is at an advantage in that a n-ply laminate would simply result in a total of $3^n$ design alternatives, while the binary scheme would still represent $2^{2n}$ design alternatives. As an example, for a 5-ply laminate, the integer scheme would give rise to $3^5$=243 design alternatives, while a binary scheme would contain $2^{10}$=1024 design alternatives, many of which will be infeasible.

The use of the actual decimal values of the variables has also been explored with some success; however, this requires modifications to the crossover and mutation operations. It is easy to see that if real numbers are used, the crossover operation will simply produce progeny values that are present in the initial population. As an example consider a 4-ply laminate with ply thicknesses as the design variables. Two mating strings with crossover sites indicated by the underscore will produce progeny strings as follows:

Parent 1: $0.012\ 0.\underline{0}14\ 0.0\underline{1}1\ 0.002$ Child 1: 0.012 0.012 0.011 0.002
Parent 2: $0.008\ 0.\underline{0}12\ 0.0\underline{1}4\ 0.012$ Child 2: 0.008 0.014 0.014 0.012

Although the child designs have changed, they are simply a recombination of the original design variable values in the parent strings. To allow the crossover operation to be more exploratory, an averaging crossover has been proposed. For details of this and other operations designed for use of GA's in composite design, the reader is referred to Haftka and Gurdal (1993).

### 6.2.3 Enhancements to the GA Approach

It is most critical to bear in mind that GA's (binary representation) essentially search for an optimal design among a discrete set of design alternatives. A quick examination of the formula relating precision of representation $A_c$ to the string length (Eqn. 6.1), shows that string lengths increase when Ac is reduced or when the difference between the upper and lower bounds is increased. Hence, longer string lengths result by concatenating the string representations of a large number of design variables and/or considering a very fine precision in design variable representation as may become necessary when solving continuous variable problems. As stated earlier, longer string lengths result in an increase in the number of possible alternatives in the design space, and require that the population size which must be evaluated at every generation be proportionately increased.

A multistage search, wherein the granularity of the genetic algorithm based search is varied through a successive increase in the precision with which a design space is represented, provides one alternative to working with large population sizes. In this approach, a relatively smaller population is first used to identify promising regions of the design space; the design space itself is represented in a coarse granular manner in this early stage. Once the promising regions of the search space are identified, a biased search with higher precision of design space representation is conducted without requiring an attendant increase in population size. This search with higher precision can be implemented by reducing the upper and lower bounds on design variables on the basis of the previous level of search. This strategy, referred to as dynamic parameter encoding, was proposed by Schraudolph and Belew (1992). The approach, however, suffers from an inherent weakness in that it can restrict the GA search from a promising region that was not examined in the previous stage.

A better approach to increase the precision of search is to use longer string lengths in successive steps. Even though the design space is expanded by increasing the precision of representation, a proportional attendant increase in population size may not be required, as the previous stage solutions provide good seeds in regions that offer most promise. A slight increase in population may be included (random distribution of designs in regions not identified in previous stage) to account for the possibility that the previous stage search was unsuccessful in locating the most promising regions. It is entirely possible that a coarser representation of the design space in earlier stages of the search make it difficult to satisfy design constraints. An even more undesirable situation would be one where the coarser representation would bias the genetic search towards less promising regions of the design space. To counter the possibility of genetic search drifting away from the most likely optimal solutions, a relaxation of design constraints at earlier stages of the search must be introduced in conjunction with the multistage approach. This relaxation can be implemented by increasing the constraint allowables, or alternatively, by decreasing the constraint penalty appended to the objective function. The mathematical representation of this procedure can be summarized as follows.

Consider two stage in the search process for which the precisions of representation of the variables are $A_{c1}$ and $A_{c2}$, respectively.

$$\begin{array}{llll} \textit{Stage I} & X_i \in A^m & A^m \equiv A_1 \rightarrow A_m & A_{c1} = A_{j+1} - A_j \\ \textit{Stage II} & X_i \in B^p & B^p \equiv B_1 \rightarrow B_p & B_{c2} = B_{j+1} - B_j \end{array} \tag{6.4}$$

In the above $A_{c2}$=c*$A_{c1}$ and c < 1. An integral part of the multistage approach is the initialization of the population of designs at each stage of the search process. If designs from a previous stage are simply carried over as a starting population for the next stage, then there would be no seed designs for the new admissible points created by increasing the string lengths. Furthermore, crossover operations in the expanded design space are unlikely to distribute designs to the newly created points. To circumvent this problem, a probabilistic redistribution of designs to newly created points in the design space is required.

As an example, consider a design variable representation scheme with a precision level of $A_{c1}$=1.0, with admissible variable values of 10.0, 11.0, 12.0,...etc. In this case, a design variable value of 11.0 can be assumed to represent a region between 10.5 and 11.5. If in the next stage, the precision of representation is changed to $A_{c2}$=0.2, such that variable values 10.0, 10.2, 10.4,...12.0 are admissible, in this enhanced precision space, the variable value of 11.0 can be expected to represent the region 10.6, 10.8, 11.0, 11.2, and 11.4. If then, the value of the variable is selected as 11.0 in Stage I, when moving to Stage II, all design points 10.6, 10.8, 11.0, 11.2, and 11.4 must be given the chance to be represented in the starting population for that stage. A uniform probability distribution can be used for this purpose.

A probabilistic transformation scheme can be implemented for this purpose, wherein designs from a previous stage are first rank-ordered in terms of their fitness values. Since the central idea behind this approach is to bias the search in regions identified as most promising, a percentage of best designs from a stage is used to seed the permissible design points in the next stage. As an example, if the best 25% of final design are used as seeds for the next stage, each design in the top quartile of the population at Stage I is used four times to create a new design for Stage II according to the probabilistic transformation strategy discussed above. This would be equivalent to increasing the population density in the more promising regions by a factor of four, and would circumvent the need for large increase in population sizes when the string lengths become large.

Another approach which assigns significance to the previous generations of evolution in genetic search is referred to as directed crossover. In theory, if a binary string of length L is used for representing the design, a population size that is in proportion to the string length would have to be selected. If however, only a smaller fraction, *isig**L of the bits in the string were really significant to the search process, the population size could be reduced accordingly. The primary motivation behind the directed crossover strategy, therefore, is to identify significant bit position on the string, and to constrain the crossover operation to these bit locations. While many different options for this approach can be formulated, the strategy presented here develops a generational memory of the crossover gain available when different bit string locations are used as sites for crossover. This crossover gain curve is then used to bias the random selection of crossover sites.

The process is initiated in the usual manner with random selection of crossover sites on the mating strings, and with no preference allocated to any particular site. After the crossover operation, the fitness change of each mating pair is recorded. Use combinations of these fitness changes (either maximum, positive change only, average, etc.) to assign a crossover gain to each bit involved in the crossover. As an example, consider two 10-digit strings A and B, for which the fitness values before crossover are 10 and 20, respectively. Assume further that during crossover, the bits between locations 3 and 7 are exchanged among the mating strings, and that as a result of this crossover, the fitness of A and B are obtained as 14 and 19, respectively. If average fitness change is used as a criterion in directed crossover, an average gain of 1.5 would be assigned to each bit enclosed between locations 3 and 7. This step is repeated for all possible mating pairs, accumulating the crossover gains for each bitwise location of the string. This yields a crossover-gain curve similar to one shown in Fig. 6.2. The crossover gains are accumulated over a few initial cycles of evolution. The gain curve can then be used to perform two-point or multi-point crossover operations. Note that the crossover-gain curve is a summation of several curves from previous generation to provide a more stable generational gradient information. To remove the strong bias introduced in this strategy towards certain bits, and to ensure sufficient mixing, a portion of the more routine random crossover operation is retained; however, directed crossover is used as the major crossover operation.

Rather than using the difference of objective function values after crossover to establish the crossover gain curves, it is sometimes beneficial to scale this difference by dividing the crossover gain by the smaller of two objective function values (in a function minimization problem) before crossover. The effect of this scaling is to reference the gain during a crossover to the fitness value of the of two members involved in the crossover; the gain obtained from mating strings that were more fit before crossover is assigned a higher priority in building the gain curves. Furthermore, it eliminates the bias in the gain crossover curves that would otherwise result from initially poor strings making rapid improvements in fitness value. Details of both the multistage and the directed crossover strategies are presented in Lin (1990).

We next illustrate the applications of these strategies in a structural design problem. A ten bar truss shown in Fig. 6.3 was used a test problem. A load P=100 kips was applied at nodes 3 and 5 and the density of the truss material was selected as 0.1 lbm/in$^3$. The design variables were assumed to vary discretely between lower and upper bounds of 0.01 in$^2$ and 12.01 in$^2$, respectively; a spacing of 0.01 in$^2$ was assumed in this variation. The performance of directed crossover, multi-stage search, and a combination of these strategies was evaluated against a traditional genetic search. Optimal solution for this problem, for a lower bound on the design variables of 0.10 in$^2$, has been obtained in (Lin and Hajela, 1993) as {X}= {7.94, 0.10, 8.06, 3.94, 0.10, 0.10, 5.74, 5.57, 5.57}$^T$ in$^2$, for truss members identified as 1 through 10 in Fig. 6.3. The optimal weight corresponding to this design is 1593.2 lbs.

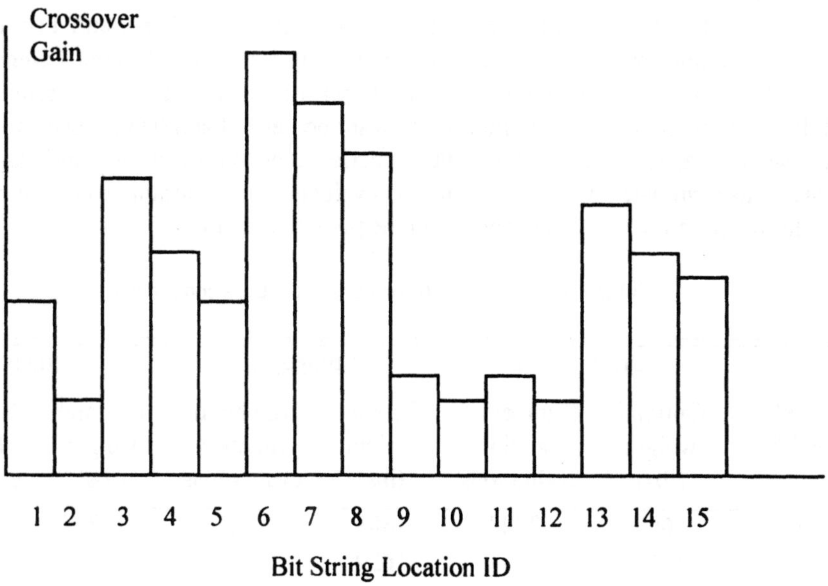


**Figure 6.2** Typical crossover gain curve.

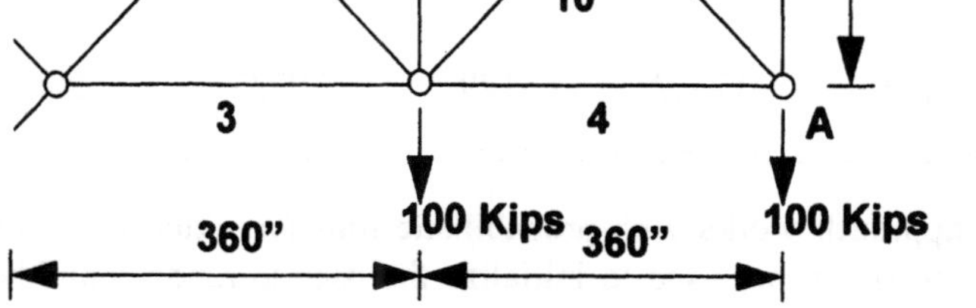


**Figure 6.3** A 10 bar planar truss.

A genetic search with 100% regular two-point crossover, a search with 70% directed crossover combined with 30% regular crossover, a multi-staged genetic search, and a combined directed crossover and multi stage strategy were implemented on identical initial populations. The first two searches with different crossover plans started with identical populations, and used identical design variable representations; i.e., both problems had the same lower and upper bounds for each variable, and the same string lengths for each design. The multistage problem was divided into four stages. While lower and upper bounds on the design variables were unchanged, precisions with which the variables were represented in stages 1 to 4 were set to 1.0, 0.2, 0.05, and 0.01 in$^2$, and required binary string lengths 40, 60, 80, and 110, respectively. Constraint relaxations associate with varying granularity for stage 1 through 4 were set as 10%, 5%, 2%, 0%, respectively. A population size of 100 was selected for all experiments, and the probabilities of crossover and mutation were prescribed as 0.8 and 0.005, respectively. In each case, the search was terminated after 20,000 function evaluations. In the multi stage approach, each stage was terminated after 5,000 function evaluations. Constraints were imposed through the penalty function formulation discussed in the previous section.

**Table 6.1** Results for 10 bar truss with stress constraints.

| Trial Number | Original GA | | Strategy 1 | | Strategy 2 | |
|---|---|---|---|---|---|---|
| | Optimal weight (lbs) | Number of function evaluations | Optimal weight (lbs) | Number of function evaluations | Optimal weight (lbs) | Number of function evaluations |
| 1 | 1685 | 32229 | 1688 | 6351 | 1678 | 23721 |
| 2 | 1713 | 32278 | 1639 | 27083 | 1677 | 34971 |
| 3 | 1821 | 30337 | 1639 | 25028 | 1639 | 14803 |
| 4 | 1901 | 21240 | 1685 | 18982 | 1635 | 17906 |
| 5 | 1717 | 32707 | 1724 | 6642 | 1688 | 16862 |
| 6 | 1773 | 16602 | 1740 | 11347 | 1666 | 26401 |
| 7 | 1776 | 26364 | 1639 | 25415 | 1692 | 23399 |
| 8 | 1914 | 33964 | 1768 | 9160 | 1725 | 33871 |
| Best results | 1685 | 16602 | 1639 | 6351 | 1635 | 14803 |
| Average of all runs | 1788 | 28215 | 1690 | 16251 | 1675 | 23992 |
| Deviation from best | 229 | 17362 | 129 | 20732 | 90 | 20168 |

Since the GA approach works with probabilistic transformation operators, the results are determined by the seed number used to initialize the pseudo-random number generator. It is therefore advisable to look at a statistical distribution of the results rather than the best performing case. For the 10-bar truss problem, each of the four search plans was executed five times, using five different seeds for random number generator. The results obtained were averaged over the five simulations using a normal crossover, directed crossover, a multi stage search, and a combined directed crossover and multi stage search. The best normalized solutions (using the continuous solutions from mathematical programming to normalize) obtained at 20,000 function evaluations were 1.058 (0.8%) for the normal crossover plan, and 1.031 (0.0%) for the directed crossover. Values in the parentheses indicate a percentage violation of design constraints defined in terms of a cumulative constraint measure (Hajela, 1981). In the multi stage and combined directed crossover-multi stage approaches, average objectives in first stage were significantly higher than that in two single-stage approaches. Nevertheless, good regions could be identified relatively quicker, and the best solutions obtained were 1.024 (0.0%) and 1.015 (0.0%) for multistage and the combined plan, respectively. A more direct comparison of the efficiency of each strategy is available in Table 6.1, which shows the number of function evaluations required by each strategy to obtain the minimum weight achieved at the end of 20,000 function evaluations by a regular genetic search. Use of these strategies can result in a 50% decrease in the required number of function evaluations. Further details of these numerical implementations, and results for larger dimensionality design problems are presented in Lin (1990).

## 6.3 Approximate Models of Analysis

The genetic algorithm approach is computationally expensive, requiring a large number of function evaluations during the search process. In complex multidisciplinary systems, the computational effort may be even more significant, given that the analyses must cycle between many different analysis modules. As genetic algorithms have continued to find applications in realistic multidisciplinary design problems, the focus on generating surrogate models of analysis that are computationally expedient, has sharpened considerably. One such approximate model that is well-suited to the global nature of genetic search is based on polynomial response surfaces. Here, a set of input-output patterns obtained from either a computational model or from a physical experiment are used to generate a polynomial response surface, the order of which is selected by the user. The neural network based function approximation may also be regarded in the general category of response surface models. Here, however, no polynomial function needs to be specified by the user. Rather, as seen in the following section, effective use is made of a nonlinear activation function in each neuron to develop approximations for fairly complex input-output relationships.

### 6.3.1 Neural Networks

Artificial neural networks were inspired by the impressive cognitive and data processing characteristics that are typical of the brain. While the functioning of the 'biological neural network' and its intricate system of sensors is not completely understood, there is some agreement that this biological machine comprises of about 100 billion threshold logic processing elements also referred to as neurons or brain cells. The number of interconnections between these neurons is estimated to be about one quadrillion. With such a high degree of complexity, it is not surprising that the similarities between the biological and artificial systems, are, at best, superficial. Nevertheless, models describing the cognitive process that have originated in neurobiology and psychology, have been embraced by computer scientists in the development of different neural network architectures.

It is perhaps more productive to view the ANN technology in much simpler terms. A computer science perspective of this field is provided in a two-volume compendium (Rumelhart and McClelland,1988, Vols. 1 and 2), and has been referred to as Parallel Distributed Processing (PDP). The artificial entity corresponding to a biological neuron is shown in Fig. 6.4. The processing element receives a set of signals $X_i$, i=1,2,...n, similar to the electro-chemical signal received by a neuron in the biological model. In the simplest implementation, the modeling of synaptic connections is achieved by multiplying the input signals by connection weights $w_{ji}$ (both positive and negative). The effective input to each processing element is therefore obtained as follows:

$$Z_j = \sum_n w_{ji} X_i \qquad j = 1, J \tag{6.5}$$

In a neurobiological system, the neuron fires or produces an output signal only if the combined input stimuli to the cell builds to a threshold value. In the artificial counterpart, this effect is simulated by processing the weighted sum of the inputs by an activation function F to obtain an output signal $Y_j=F(Z_j)$.

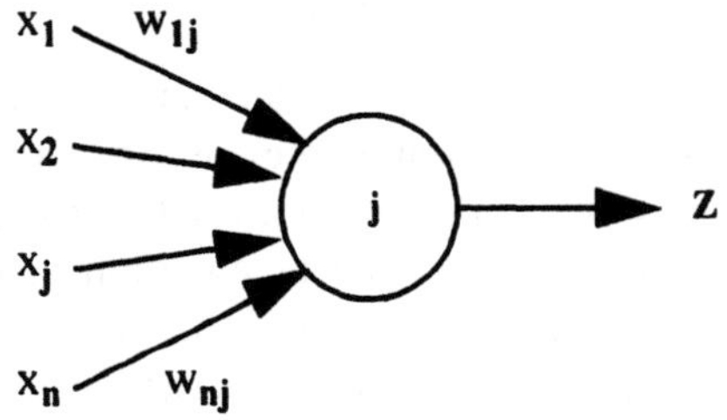


**Figure 6.4** The artificial neuron.

Although various forms of activation functions have been proposed, the most commonly used sigmoid function is given by the expression,

$$F(Z_j) = \frac{1}{\exp(-Z_j + T_j)} = Y_j \ 6. \tag{6.6}$$

where, $T_j$ is a bias parameter used to modulate the element output. A sketch of this function and two other activation functions are shown in Fig. 6.5. The principal advantage of this function is its ability to handle both large and small input signals. The slope of the function is representative of the available gain. For both large positive and negative values of the input signal, the gain is vanishingly small; at intermediate values of the input signal, the gain is finite. Hence, an appropriate level of gain is obtained for a wide range of input signals. The output obtained from the activation function may be treated as an input to other neurons in the network.

In principle, artificial neurons can be interconnected in any arbitrary manner. One classification is based on whether the flow of the stimuli is from the input to the output nodes only, or if neurons can also relay stimuli backwards to activate or inhibit neuron firings. Feedforward networks, characterized by the flow of stimuli in one direction only, will be discussed under the category of networks where the training is supervised. The multilayer perceptron model and the counterpropagation networks are specific architectures to be explored.

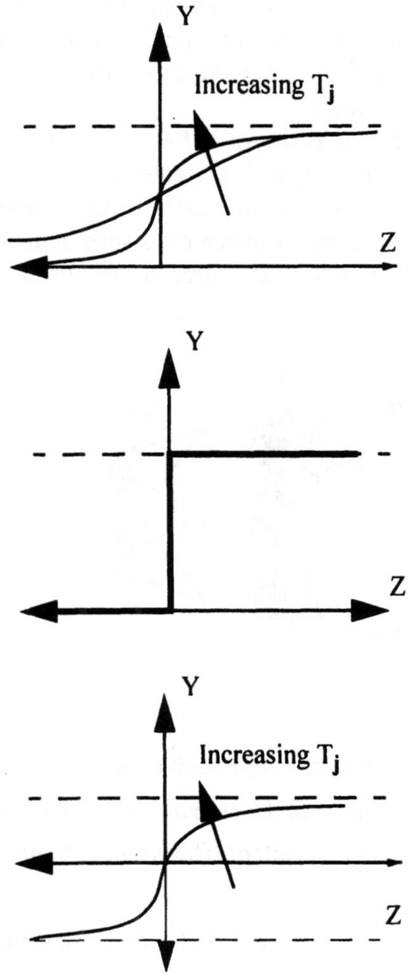


**Figure 6.5** The sigmoid, threshold and hyperbolic tangent activation functions.

**Backpropagation Neural Network.** The simple feedforward networks have a layer of neurons to which the external stimuli are presented, a series of hidden layers, and a layer of neurons at which the output is available. The input neurons do not process the input stimulus; they simply serve as "fan-out" points for connections to neurons in successive layers. The presence of the hidden layer, and the nonlinear activation functions, enhance the ability of the networks to learn nonlinear relationships between the presented input and output quantities. This "learning" or "training" in feedforward nets simply requires the determination of all interconnection weights $w_{ji}$ and bias parameters $T_j$ in the network. Once such a trained network is established, it responds to a new input within the domain of its training by producing an estimate of the output response. Variations of the generalized delta error back propagation algorithm have been used for this training; this scheme is essentially a special purpose steepest descent algorithm, and indeed, any optimization method can be used towards this end. The only concern would be the computational effort necessary for network training when the network size (number of independent networks parameters) increases. An abbreviated description of the training process is summarized here for completeness.

Consider a three-layer network as shown in Fig. 6.5. The interconnection weights are first initialized randomly and the input pattern is presented to the network from which the network output is determined. This output is compared to the expected output, and a sum of the squares of all errors is determined. The network parameters such as the interconnection weights and threshold levels are adjusted to minimize this error to some level of desired accuracy.

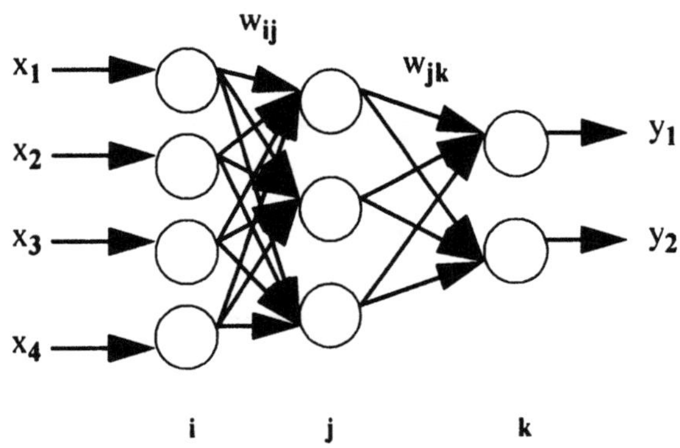


**Figure 6.5** A three-layer feedforward neural network.

Let $Y_j$ be the output at the j-th neuron of the output layer, and $Y_j^T$ be the expected or target value of this output from the chosen training pattern. The error at training step 't' is then determined as

$$e_j(t) = Y_j^T - Y_j \tag{6.7}$$

and the sum of squared errors over all neurons in the output layer are determined as follows:

$$E(t) = \frac{1}{2}\sum_{j \in J} (e_j(t))^2 \tag{6.8}$$

In the above, J is the total number of neurons in the output layer. The backpropagation (BP) algorithm provides an incremental change to the weights $w_{ji}$ in proportion to the instantaneous gradient Using the chain rule of derivatives, one can write the correction to the interconnection weights as follows:

$$\Delta w_{ji} = -\eta \frac{\partial E(t)}{\partial w_{ji}} \tag{6.9}$$

In the above, $\eta$ is the learning rate parameter and is indicative of the rate with which the weights are changed at any given learning cycle. Note that this can be computed for any neuron for which the error $e_j(t)$ can be explicitly computed, and this is possible only for the output layer where the error can be specifically known as the difference between the network predicted and known (target) output. In order to apply this weight change algorithm to neurons in the hidden layer (for which there is no known output), the error signal must first be back-tracked to that hidden layer. For such hidden layers, the expressions for weight correction (Hajela, 1997, Barron, 1992) is obtained as follows,

$$\Delta w_{ji} = -\eta\left(\frac{\partial F_j}{\partial Z_j(t)}\sum_{k \in J} e_k(t)\frac{\partial e_k(t)}{\partial Z_k(t)}w_{kj}(t)\right)Y_j(t) \tag{6.10}$$

and requires knowledge of the error signals for all neurons that lie to the immediate right of the hidden layer neuron 'j', and are connected to 'j'.

The training of the multilayer perceptron model using the BP algorithm can be improved by selecting an activation function which is asymmetric, as opposed to the nonsymmetric sigmoidal activation function. A popular form of an asymmetric activation function is a modified sigmoid given by the hyperbolic tangent function,

$$F(Z_j) = a\tanh(bZ_j) \tag{6.11}$$

where $a$ and $b$ are constants and typical recommended values for these constants are a=1.716 and b=0.6667. It is important to note that the hyperbolic tangent function is simply a biased and rescaled logistic function,

$$a\tanh(bZ_j) = \frac{2a}{1+\exp(-bZ_j)} - a \tag{6.12}$$

and hence, the modifications in the backpropagation algorithm required in implementing this function in place of the sigmoid are relatively minor.

There are essentially two modes of training that can be implemented. The first, referred to as the pattern mode updates the weights after each training pattern is presented. So each cycle of training involves weights to be changed a total of NPAT times, where NPAT is the total number of training patterns. The order in which these training patterns are presented to the network can be changed from one training cycle to another at random, giving the process some mechanism to avoid converging to a relative optimum in the weight space. A second approach, referred to as the batch mode, computes a cumulative error computed over all of the training patterns, and makes weight changes in proportion to the rate of decrease of this cumulative error. The relative effectiveness of the two approaches is problem dependent although the pattern mode generally performs better. It is worthwhile to add that it is sometimes possible to avoid being trapped into a relative minima of the error function by adding to the weight increment a term that is proportional to the weight change in the previous cycle of training. This term is referred to as the momentum term, and the constant of proportionality used to implement this addition is defined as the momentum coefficient.

Other helpful insights into the training of such networks include the following.

- The desired response of the trained network available at the output neurons should be offset by some small amount away from the limiting values (positive or negative) of the activation functions. This prevents neuron saturation and a considerable improvement in training efficiency.
- All interconnection weights and bias constants should be set at initially small values, randomly distributed within some narrow range. This is to prevent saturation of neurons early in the training process, resulting in slow learning.
- In some cases, it is beneficial to set the learning rate of each neuron individually, the weight adjustments are in proportion to the error gradients, and these gradients can vary significantly in magnitude. While, in general, an approach that uses lower learning rates for

the layers closer to the output layer, with increasing rates of learning towards the input layer has the desired effect, these learning rates can be derived using a one-dimensional search. The latter comes at some computational cost, and the user must weigh the associated benefits of this additional computation.

The reader also needs to be aware of a potential problem in network training that is referred to as overtraining. This is a phenomenon wherein the training error is forced to such low levels that the network essentially memorizes all the training patterns, with a very serious loss in generalization capabilities.

A process known as cross-validation (derived from statistics) is recommended during training, wherein the available training set is split into a training and a validation set. As the training proceeds, the validation samples are used to ensure that the generalization error continues to decline with the training error. When the trend reverses, and if the network training error is within acceptable bounds, the network is considered to be trained. Typically, 10% to 20% of the training samples should be set aside for cross validation purposes. Cross validation can also be used to study the effect of varying the network architecture.

When the BP neural network is used to map input-output functions, it is representing a special form of response surface where the response function is a nested squashing function; the interconnection weights correspond to the regression parameters in a typical regression model. The polynomial regression models are good for linear mappings but difficult, if not impractical for large dimensionality, nonlinear mappings. If for example the matrix inversion approach is used to solve for the regression parameters, then with NPAR parameters, the modeling capacity would be approximately proportional to the square root of the memory size. The neural network approach has no such restriction and the modeling capacity would be proportional to the actual memory size. In other words, if a polynomial model using inverse methods is used to derive a mapping, it could afford at most NPAR parameters while the BP neural network would allow the use of NPARxNPAR parameters.

In using a neural network for function response approximation, a number of associated issues must be resolved. First, a set of input-output training patterns must be obtained from the real process that one is attempting to simulate. Determination of the number of such training pairs, and how they should span the intended domain of training, requires experimentation and experience. The same statement is applicable to the selection of the network architecture, i.e., the number of hidden layers and the number of neurons in such layers. One approach to this problem is to assign a network architecture in accordance with the physical complexity of the problem, and to then determine the size and distribution of the training set. The second approach would be to fix the number of training patterns and to determine an optimal network architecture. The objective in either of these approaches is to both facilitate the network training and to ensure that the network generalizes effectively, i.e., the input-output relationship computed by the network is "correct" to within some tolerance levels. In practice, a bound on the size of the training set NPAT for good generalization is stipulated as $NPAT > \frac{W}{\varepsilon}$, where $\varepsilon$ is the bound on the generalization error, and $W$ is the number of interconnection weights in the network. So, with an error of 10%, the number of training samples should be ten times the number of interconnection weights. This is based on a distribution-free, worst case formula for a single-layer network, and as will be shown in subsequent examples, for many problems in engineering, the number of patterns required for good generalization is well below this bound. Recent developments in machine learning and computational intelligence also indicate that there may be formal methods of discovering the distribution of training patterns and/or network architecture.

**Counterpropagation Neural Network.** The counterpropagation (CP) neural network was first introduced by Hecht-Nielsen (1987) as a combination of two basic architectures - the Kohonen's self-organizing neural network and Grossberg's outstars neurons. This architecture required less computational effort to train than the multilayer perceptron architecture described in the previous section. Training times are of considerable importance when one considers modeling of extremely large structural systems. However, the original version of the network did not receive widespread attention due to its unimpressive generalization performance, particularly in comparison to the multilayer perceptron model. As shown in Fig. 6.6, this network contains three layers - a fan-out layer as in the BP network, a layer of Kohonen or feature sensitive neurons, and an interpolating or Grossberg layer. The inputs to the network are directed to the Kohonen layer, which acts like a clustering device. In other words, neurons in this layer classify all input vectors based on some identifiable features in these vectors. Each neuron in the Kohonen layer represents one such cluster, and the interconnection weights between the input nodes and this neuron are representative of an average of all input patterns of that cluster.

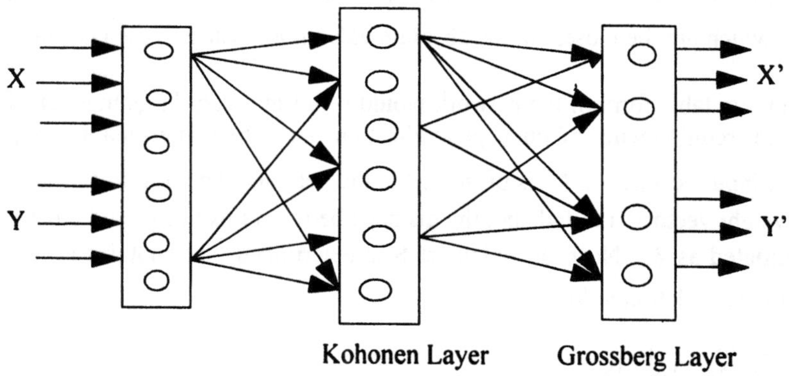


**Figure 6.6** Schematic of a counterpropagation neural network.

Similarly, the interconnection weights between each Kohonen neuron and the output or Grossberg layer neurons are representative of an averaged output of all patterns belonging to the cluster. If the radius of each cluster is infinitesimally small, then each Kohonen neuron will only represent one pattern, and the network would be a memory record of all input-output patterns. On the other hand, if the radius of the cluster is large, then several patterns would be classified in the same cluster. A larger radius generally results in significant errors during generalization, as an input pattern presented to the network is classified as belonging to a cluster on the basis of its similarity to the stored cluster weights for each Kohonen neuron. Even if the input pattern were classified into the right cluster, the output would only be an average of the outputs of all input patterns belonging to that cluster, and that were used in network training.

The unsupervised training of the Kohonen neuron is based on a minimum disturbance principle in which the weight vector of only one neuron, the one closest to the current input, is modified. When a new input vector activates one of the Kohonen neurons, the weight vector representing the connection between this neuron and the outstars is modified (in the training mode) or simply returned as an output (in the generalization mode). In the modified CP network described here, the output is actually a nonlinear blend of outputs from several Kohonen neurons which a given input may activate, albeit to different degrees. This nonlinear averaging allows for better generalization characteristics from the network.

As for the BP network, this network may be used to develop the mapping,

$$\phi|((X \in R^n) \rightarrow (Y \in R^m)) \tag{6.13}$$

and requires that weights of two layers be determined. As a consequence, there are two sets of weights to be considered - a set of Kohonen layer weights for the i-th Kohonen neuron,

$$w_i \equiv [w_{1i}, w_{2i}, \cdot w_{ni}] \in R^n, \ i = 1, k \tag{6.14}$$

and,

$$z_i \equiv [z_{i1}, z_{i2}, \cdot z_{im}] \in R^m, \ i = 1, k \tag{6.15}$$

for the Grossberg layer. Szewczyk and Hajela (1992) present a training algorithm for this network. It suffices for the purpose of this discussion to recognize that the training algorithm increases the size of the network, i.e., the number of Kohonen neurons k, if the new input pattern is more than $\delta_r$ away from the already defined neurons, or, it modifies the i-th neurons of both layers which are the closest in the absolute value norm sense to the currently presented input pattern.

Since a look-up table form of the network would result in a very large number of Kohonen neurons to meet requirements of good generalization, an output interpolation is performed. Consider some input vector, $\hat{x} \in R^n$ is presented to the network. The best match for this vector among all k weight vectors is found, and the distance between this vector and closest Kohonen neuron is computed as $\delta_o$. Next, an index set S is determined that contains all indices s for which the following relation holds:

$$\delta_s = \sum_{l=1}^{n} |\hat{x}_1 - w_{ls}| \le \delta_o + \delta \tag{6.16}$$

Here, the arbitrary parameter $\delta$ defines a neighborhood of the winning neuron, and is defined as the network interpolation parameter. Assuming that $\tau$ such neurons are found to be in the neighborhood, the output of each is used to develop the network output. To compute an individual contribution, $h_s$ of each such neuron to the output, a membership function $f(d_s)$ is used. This function is defined as a power function,

$$f(d_s) = 1 - d_s^r \tag{6.17}$$

where $d_s$ is a normalized distance calculated as follows:

$$d_s = |\delta_o - \delta_s| / \delta \tag{6.18}$$

The contributions $h_s$ of all $\tau$ neurons are normalized as,

$$h_s = \frac{1 - d_s^r}{\sum_s 1 - d_s^r} \tag{6.19}$$

so that they add up to unity, and the output is simply computed as,

$$out = \sum_{s=1}^{\tau} h_s z_s \tag{6.20}$$

Such network response, developed through interpolation, tends to yield better results as the sharp boundaries between clusters are removed. Clearly, the number of contributing neurons, $\tau$ depends on the value of $\delta$, set during testing to minimize the approximation error, and may vary for different input vectors. In a case where the input neuron is very close to an input vector, it becomes the sole contributor to the output response. The shape of the membership function is controlled by the exponent $r$, and its value must be adjusted for the problem at hand so that the approximation error is minimized.

### 6.3.2 Illustrative Examples

This example illustrates the use of backpropagation and counterpropagation neural networks in providing surrogate models for use in a genetic algorithm based search procedure. It deals with the design of a composite helicopter rotor blade for enhanced aerodynamic, structural, and dynamic performance. With composites, it is practical to fabricate non-rectangular blades with variations in twist distribution and airfoil sections along the blade span, thereby contributing to increased flexibility in aerodynamic design. Satisfactory aerodynamic design requires that the required horsepower for all flight conditions not exceed the available horsepower, that the rotor disk must retain lift performance to avoid blade stall, and that the vehicle remain in trim. Important factors in structural design include material strength considerations for both static and dynamic load conditions. A combination of flapwise, inplane, torsion, and centrifugal forces typically comprise the static loading. Another important consideration that encompasses both structural and aerodynamic design, is the autorotation capability. The autorotation requirement pertains to maintaining the mass moment of inertia of the rotor in the rotational plane at an acceptable level. This is a function of the vehicle gross weight, rotor aerodynamic performance, and the rotor system mass moment of inertia. Finally, dynamic design considerations of the rotor blade pertain to the vibratory response of the blade under the applied loads; this design limits the dynamic excitation of the fuselage by reducing the forces and moments transmitted to the fuselage.

A finite element in time and space formulation as developed by Bauchau and Kang (1993) was used to model the dynamics of a helicopter rotor blade. This formulation was based on a multibody representation of flexible structures undergoing large displacements and finite rotations, and requires that the equations of motion be explicitly integrated in time. An unsteady aerodynamic model is used to obtain the induced flow and to calculate the aerodynamic forces and moments in hover and forward flight. In addition to the geometric nonlinearities that are inherent in this problem, the loading on the blade varies as it moves around the azimuth - on the advancing side the flow velocity over the blade is additive to its tangential speed; on the retreating side, these velocities subtract. Consequently, in order to maintain force and moment equilibrium, the pitch of the blade is continuously changed as it rotates around the azimuth, and the hub loads are a function of the blade rotational frequency. There is a transient period during which equilibrium of the vehicle is established (trim), and this is generally obtained in 8-10 revolutions of the blade. This process of obtaining trim condition prior to load calculation introduces additional computational costs. The nature of the solution process used here is such that designs that correspond to aeroelastically unstable configurations will simply yield a divergent response in time. As described in a later section, this requires that the neural network based approximation procedure be treated somewhat differently.

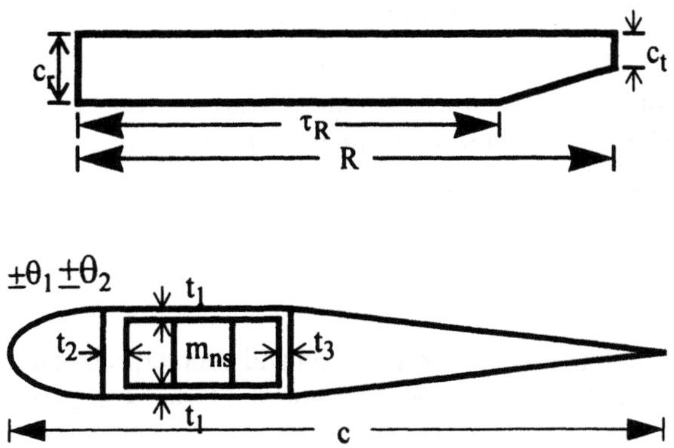


**Figure 6.7a and 6.7b** Planform and profile of rotor blade.

**Table 6.2** Design variables and bounding value for the rotor design problem.

| design variable | X | min | max |
|---|---|---|---|
| tuning mass | $m^i$ | 0.0 | 4.0 |
| horizontal flange thickness | $t_1^i$ | 0.25 | 0.35 |
| left vertical flange thickness | $t_2^i$ | 0.20 | 0.30 |
| right vertical flange thickness | $t_3^i$ | 0.20 | 0.30 |
| blade twist [deg] | $\theta_t$ | 4.0 | 8.0 |
| twist shape parameter | $\delta$ | 0.3 | 1.0 |
| taper inception point | $\tau_R$ | 0.50 | 0.80 |
| chord ratio | $\lambda_c$ | 1.0 | 2.0 |
| rotational speed [rad/sec] | $\Omega$ | 24.0 | 30.0 |
| ply angle of inner flange [deg] | $\theta_1$ | 30 | 90 |
| ply angle of outer flange [deg] | $\theta_2$ | 30 | 90 |

The objective of the design problem was to design the blade geometry and internal structure to minimize a weighted sum of hub shear force and bending moments for a hingeless rotor blade in forward flight; aerodynamic, performance and structural design requirements are considered as constraints, and dynamic requirements constitute a multicriterion objective function. The premise behind the approach is that a minimization of the hub loads and moments translates into lower vibrations that are transmitted to the fuselage structure.

The planform geometry of the blade is shown in Fig. 6.7a. The blade is tapered in both the chord and the depth, and all loads are assumed to be carried by the structural box shown in Fig.

6.7b. The design variables for this problem are also shown on the figures and summarized in Table 6.2. The blade planform geometry is defined by the chord ratio $\lambda_c$ and the point of inception of taper along the span denoted by $\tau_R$.The blade twist varies from $\theta_t$ at the root to zero at the tip; this variation maybe linear or nonlinear, and is controlled by the twist shape parameter $\delta$. Both $\theta_t$ and $\delta$ were design variables in the problem. Two nonstructural masses were positioned along the span of the blade, and the magnitude of these masses ($m_1$ and $m_2$) as well as their locations along the span ($d_{m1}$ and $d_{m2}$) were considered as design variables. The rotational speed of the rotor $\Omega$ was also selected during the design process. The remaining variables, although descriptive of the internal structure of the blade, have a strong influence on both aerodynamic and dynamic performance. These were the horizontal flange thickness $t_1$, and the left and right vertical sections of the box beam denoted by $t_2$ and $t_3$, respectively. For the Graphite/Epoxy rotor blade, the horizontal flanges are symmetric +/-45 deg laminates; this layup is also present on the outer half of both vertical sections of the box beam. The inner half of the vertical walls are divided into two segments, with a layup of +/-$\theta_1$ and +/-$\theta_2$ deg, respectively. This accounts for a total of 14 design variables for the problem. Note that it is relatively easy to increase the problem dimensionality by simply varying the thickness and orientations of plies in discrete segments along the span. The mathematical statement of the optimization problem can be written as in (Eqn. 6.21).

In this problem statement, $F_z$, $M_y$ and $M_z$ are the maximum peak-to-peak values of the scaled shear force, flap bending moment, and lead-lag bending moment, respectively; $HP_h$ and $HP_f$ denote the horsepower required during hover and forward flight, respectively, and $HP_a$ is the available horsepower; $C_T$ and $\sigma$ are the rotor thrust coefficient and solidity, respectively, and the subscripts L and U denote the lower and upper bounds of this ratio. These bounds are necessary to limit the lift performance of the rotor disk to avoid blade stall. The parameter FM is a Figure-of-Merit for the helicopter performance out of the ground effect, and subscipt L is used to denote its lower bound; the blade must have sufficient autorotation (AI) capacity to safely descend in an out-of-power configuration, and this is represented in constraint $g_7$; W is the sum of the structural and nonstructural weights and $W_U$ denotes its upper bound; $\overline{R}$ is used to denote the structural failure criterion, and a Tsai-Wu measure was used for the composite structure in this problem; $\sigma_{all}$ is the allowable stress for the material and $\sigma_{buck}$ are the static stresses due to buckling. The constraint set represents response quantities that are traditionally associated with structures, dynamics, and aerodynamic performance of the helicopter. For one given set of design variables, the analysis time required for the evaluation of the objective and constraint functions is substantial (about 18 CPU minutes on a SPARC station) and clearly not amenable to integration with in an iterative optimization environment. Both the BP and CP neural networks were used to generate the mapping between the design variables and the response quantities of interest. Some of these mappings are quite nonlinear, and require careful consideration of the choice of network architecture and of the number of training patterns. The range of design variable variations used in this training are shown in Table 6.3. A number of training patterns was generated, in the range of design variable variation, and this included both stable and unstable designs. A BP network with 14 input layer neurons, 10 hidden layer neurons, and 5 output layer neurons (a 14-10-5 network), corresponding to 5 output quantities, viz., vertical hub shear, flap moment, lag moment, rotor thrust, and the failure criterion index $\bar{R}$, was established and trained with 550 training patterns. This training presented problems in that it was difficult to reduce the training error to below 3%.

$$Minimize \rightarrow F = c_1 F_z + c_2 M_y + c_3 M_z$$

$$Subject \cdot to \cdot constraints$$

$$g_1 \equiv \frac{HP_h}{HP_a} - 1 \le 0 \qquad g_2 \equiv \frac{HP_f}{HP_a} - 1 \le 0 \qquad g_3 \equiv 1 - \frac{\frac{C_T}{\sigma}}{\left(\frac{C_T}{\sigma}\right)_L} \le 0$$

$$g_4 \equiv \frac{\frac{C_T}{\sigma}}{\left(\frac{C_T}{\sigma}\right)_U} - 1 \le 0 \qquad g_5 \equiv 1 - \frac{FM}{(FM)_L} \le 0 \qquad g_6 \equiv \frac{FM}{(FM)_U} - 1 \le 0 \tag{6.21}$$

$$g_7 \equiv 1 - \frac{AI}{(AI)_L} \le 0 \qquad g_8 \equiv \frac{W}{(W)_U} - 1 \le 0 \qquad g_9 \equiv \bar{R} - 1 \le 0 \qquad g_{10} \equiv \frac{\sigma_{buck}}{\sigma_{all}} - 1 \le 0$$

**Table 6.3** Lower/upper bounds on design variables.

| design variable | | symbol | min | max | accuracy |
|---|---|---|---|---|---|
| $d_1$ | mass 1 location | $m_1$ | 1 | 5 | 1 |
| $d_2$ | mass 2 location | $m_2$ | 6 | 10 | 1 |
| $d_3$ | tuning mass 1 (kg) | $d_{m1}$ | 0.0 | 5.0 | 0.1 |
| $d_4$ | tuning mass 2 (kg) | $d_{m2}$ | 0.0 | 5.0 | 0.1 |
| $d_5$ | horizontal flange thickness ratio | $t_1$ | 0.25 | 0.35 | 0.01 |
| $d_6$ | left vertical flange thickness ratio | $t_2$ | 0.20 | 0.30 | 0.01 |
| $d_7$ | right vertical flange thickness ratio | $t_3$ | 0.20 | 0.30 | 0.01 |
| $d_8$ | blade twist (deg) | $\theta_t$ | 4.0 | 8.0 | 0.1 |
| $d_9$ | twist shape parameter | $\delta$ | 0.3 | 1.0 | 0.1 |
| $d_{10}$ | taper inception point | $\tau_R$ | 0.50 | 0.80 | 0.01 |
| $d_{11}$ | chord ratio | $\lambda_c$ | 1.0 | 2.0 | 0.1 |
| $d_{12}$ | rotational speed (rad/sec) | $\Omega$ | 24 | 34 | 1 |
| $d_{13}$ | layup angle of inner vertical flange (deg) | $\theta_1$ | 30 | 90 | 1 |
| $d_{14}$ | layup angle of outer vertical flange (deg) | $\theta_2$ | 30 | 90 | 1 |

The training data was then sorted to separate the designs that yielded stable and unstable responses; it was considered expedient at this stage to establish 5 networks, each mapping the design variables into one output only (14-10-1 networks for all but the lag moment, where a 14-10-8-1 network was used). The training of each of these networks could be done in parallel. The training using these separated patterns proceeded well, converging to an error of less than 1%, with the exception of the lag moment (error was 2.6%). The observed pattern of time variation for the lag moment was quite nonlinear, and provides an explanation for the discrepancy in training.

**Table 6.4** Testing of the BP trained network.

| shear force (N) | | | flap moment (N m) | | | lag moment (Nm) | | |
|---|---|---|---|---|---|---|---|---|
| NN | actual | % | NN | actual | % | NN | actual | % |
| 13695 | 13546 | 1.10 | 11403 | 11711 | 2.08 | 5278 | 5266 | 0.24 |
| 8666 | 8649 | 0.19 | 6295 | 6891 | 0.50 | 4059 | 3994 | 1.65 |
| 6864 | 6790 | 1.09 | 6700 | 6745 | 0.67 | 4036 | 3880 | 4.72 |
| 9160 | 9373 | 2.27 | 8409 | 8406 | 0.04 | 3917 | 3772 | 3.84 |
| 14952 | 15028 | 0.51 | 12905 | 12984 | 0.61 | 4082 | 4086 | 0.10 |
| 17002 | 17001 | 0.03 | 12768 | 12736 | 0.25 | 5974 | 5899 | 1.27 |
| 7284 | 7339 | 0.75 | 6934 | 6957 | 0.33 | 11606 | 11559 | 0.40 |
| 6754 | 6591 | 2.47 | 7542 | 7569 | 0.35 | 8217 | 7980 | 2.97 |
| 13568 | 13352 | 1.62 | 9436 | 9211 | 2.45 | 7365 | 7104 | 3.67 |
| 8652 | 8671 | 0.22 | 8347 | 8521 | 2.04 | 5221 | 4921 | 6.08 |

**Table 6.4** Continued.

| thrust in cruise (N) | | | failure criterion | | |
|---|---|---|---|---|---|
| NN | actual | % | NN | actual | % |
| 778 | 784 | 0.76 | 0.261 | 0.261 | 0.01 |
| 517 | 506 | 2.17 | 0.243 | 0.241 | 0.83 |
| 369 | 374 | 1.33 | 0.247 | 0.249 | 0.84 |
| 498 | 502 | 0.79 | 0.313 | 0.308 | 1.62 |
| 857 | 873 | 1.83 | 0.250 | 0.250 | 0.02 |
| 926 | 921 | 0.54 | 0.319 | 0.316 | 0.94 |
| 435 | 427 | 1.87 | 0.238 | 0.244 | 2.45 |
| 348 | 352 | 1.13 | 0.306 | 0.307 | 0.65 |
| 722 | 712 | 1.40 | 0.325 | 0.328 | 0.91 |
| 450 | 453 | 0.66 | 0.235 | 0.239 | 1.67 |

Table 6.4 shows the results of testing these networks for generalization performance using 10 sets of design variables that were not part of the original training set. Similar generalization performance was obtained from networks trained with the unstable data. The difficulty in training the networks for the combination of stable and unstable data can be attributed to either

a) a completely different input-output relationship in one or more components or b) excessive data for the number of weights and bias constants in the network that could be varied to fit the data.

**Table 6.5** Testing of the trained CP network.

| shear force (N) | | | flap moment (N m) | | | lag moment (N m) | | |
|---|---|---|---|---|---|---|---|---|
| NN | actual | % | NN | actual | % | NN | actual | % |
| 8942 | 8687 | 2.93 | 7316 | 7504 | 2.53 | 3663 | 3705 | 1.13 |
| 14963 | 14272 | 4.84 | 10056 | 10264 | 2.02 | 8273 | 7892 | 4.82 |
| 12384 | 12579 | 1.55 | 10568 | 10458 | 1.05 | 5894 | 5758 | 2.36 |
| 8926 | 8919 | 0.08 | 7045 | 7058 | 0.18 | 5685 | 5713 | 0.49 |
| 10253 | 10156 | 0.96 | 13093 | 13185 | 0.69 | 11796 | 11183 | 5.48 |
| 8113 | 8046 | 0.84 | 8936 | 8841 | 1.01 | 12529 | 11932 | 5.00 |
| 9735 | 9653 | 0.85 | 9618 | 9685 | 0.69 | 22940 | 23695 | 3.18 |
| 14885 | 14100 | 5.56 | 10856 | 10547 | 2.93 | 36633 | 31888 | 8.10 |
| 13883 | 14016 | 0.96 | 10567 | 10648 | 0.76 | 29205 | 28742 | 1.61 |
| 21050 | 20975 | 0.36 | 16889 | 16205 | 4.22 | 29445 | 30113 | 2.21 |

**Table 6.5** Continued.

| thrust in cruise (N) | | | failure criterion | | |
|---|---|---|---|---|---|
| NN | actual | % | NN | actual | % |
| 472 | 483 | 2.28 | 0.25 | 0.258 | 0.39 |
| 766 | 763 | 3.93 | 0.312 | 0.312 | 0.01 |
| 763 | 754 | 1.19 | 0.235 | 0.237 | 0.84 |
| 509 | 502 | 1.39 | 0.264 | 0.266 | 0.75 |
| 933 | 907 | 2.86 | 0.329 | 0.333 | 1.20 |
| 376 | 371 | 1.35 | 0.347 | 0.349 | 0.57 |
| 467 | 454 | 2.86 | 0.394 | 0.390 | 1.02 |
| 650 | 642 | 1.25 | 0.510 | 0.507 | 0.59 |
| 690 | 694 | 0.58 | 0.412 | 0.405 | 1.73 |
| 1134 | 1093 | 3.76 | 0.334 | 0.333 | 0.31 |

Similar experiments were also performed using the full CP network, that generates an identity mapping of the type [X,Y] -> [X',Y']. This network architecture requires a much larger number of input patterns, and the quality of generalization depends upon the number of Kohonen layer neurons that are permitted (indirectly a measure of maximum cluster radius). If the cluster radius were set to zero, the number of Kohonen neurons will be equal to the number of training patterns, and the network would simply "memorize" all training data. The results of numerical experiments designed to test this network are shown in Table 6.5. A BP network was also trained to differentiate between stable and unstable patterns. In the present work, stable patterns were denoted by a 0 and unstable patterns by a 1. The purpose of this network was to act as a filter during optimization to discard the inclusion of unstable patterns in the genetic population. A testing of this BP network showed that it was able to distinguish between stable and unstable patterns correctly 95.4% of the time.

### 6.3.3 Fuzzy Sets in Modeling

In some engineering systems, it is not possible to express functions in precise mathematical form because such forms simply don't exist. In other systems, it is often preferable to define functions in a fuzzy manner, as this allows for a more realistic statement of the design problem. As an example, a constraint in a structural design problem may require that the structure withstand a stress of "about" 20,000 psi. Traditionally, this statement would be incorporated in the design problem as a requirement that the structure withstand stress levels of "at least" 20,000 psi; clearly, the original intent of the problem requirement is lost in such an interpretation. This type of fuzzy information can be effectively captured and expressed in the design problem statement using the theory of fuzzy sets.

Unlike ordinary classical sets which either wholly include or exclude an element, fuzzy sets are bounded by non-crisp boundaries which allow an element to be partially included in a set.

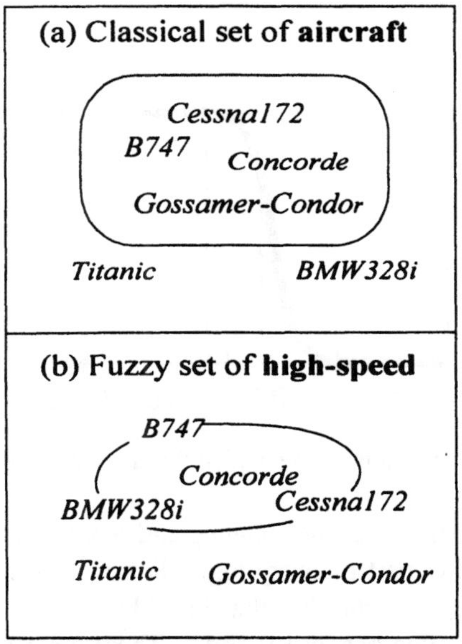


**Figure 6.8** Classical set vs. Fuzzy set.

In Fig. 6.8a, two different sets called **aircraft** and **high-speed** are shown and they can either include or exclude each of 6 given elements, namely, *Cessna172, B747, Concorde, BMW328i, Titanic, Gossamer-Condor*. While all elements describe vehicles, the set **aircraft** describes known aircraft which clearly includes *Cessna172, B747, Concorde, Gossamer-Condor,* and

excludes *BMW328i and Titanic.* In this case, there is a well defined boundary which separates those elements describing aircraft from those describing automobiles and ships. Therefore, the set **aircraft** is a classical set. However, if a set **high-speed** is to include only those elements that move at high speeds, then defining a vehicle with high speed from the given elements would, most definitely, include the element *Concorde* and exclude the elements *Titanic* and *Gossamer-Condor* given the known speeds of these vehicles. However, for the elements *Cessna172, B747,* and *BMW328i,* it would be difficult to determine whether to include or exclude these from this set. Some will consider a jet such as the *B747* to be high speed while others would not. Hence, it will have to sit on the fence as shown in Fig. 6.8b, indicating that it is partially included in the set. While classical sets can not include elements partially, fuzzy sets allow a partial inclusion of elements by assigning them degrees of membership denoted as $\mu$. The degree of membership indicates how much an element is included in a specific set and is usually defined by a utility function called membership function. A membership function of a fuzzy set **high-speed** is shown in Fig. 6.9. In this figure, the membership function indicates that the degree of membership of the element *B747* in fuzzy set **high-speed** is 0.7. Although a sigmoidal function was used describe the set **high-speed,** in this case, other types of membership functions such as linear, triangular, and trapezoidal are also available. Associated with each are a number of parameters that must be specified; they also offer a variety of smooth or nonsmooth functions to model variations of a quantity of interest.
"

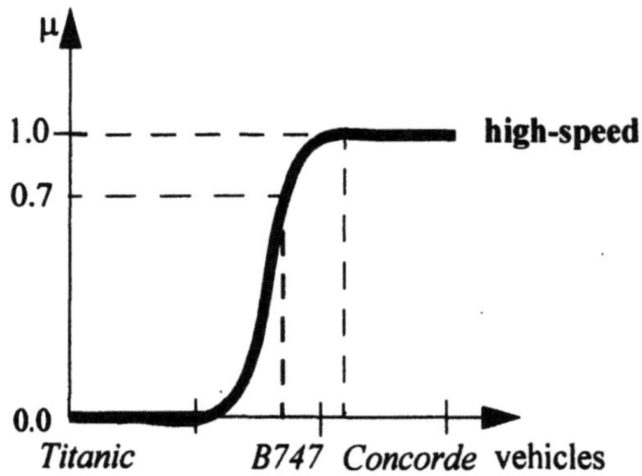


**Figure 6.9** Membership function for fuzzy set "**high-speed.**

While the above discussion pertains to the meaning and representation of fuzziness, not enough has been said about the 'logic' part of fuzzy logic. Fuzzy logic based reasoning is much like standard boolean logic except for the fact that in addition to 0 and 1, fuzzy logic can also operate with any numerical value between 0 and 1. There are a number of logical operations associated with the intersection and union of fuzzy sets. The intersection of fuzzy sets A and B is specified by T-norm operators which is expressed as follows:

$$\mu_{A \cap B}(x) = T(\mu_A(x), \mu_B(x)) = \mu_A(x) \otimes \mu_B(x) \tag{6.22}$$

where $\otimes$ is a binary operator for the function $T$; $\mu_A(x), \mu_B(x)$ are membership functions of these fuzzy sets. Among the common T-norm operators are minimum, algebraic product, bounded product and drastic product. The minimum operator, $T_{min}(a, b) = min(a, b)$, is typically used to determine the intersection of fuzzy sets. The parameters, *a* and *b*, vary between 0 and 1. Similar to fuzzy intersection, the union of fuzzy sets A and B is specified by T-conorm (or S-norm) operator and is defined as

$$\mu_{A \cup B}(x) = S(\mu_A(x), \mu_B(x)) = \mu_A(x) \oplus \mu_B(x) \tag{6.23}$$

where $\oplus$ is a binary operator for the function *S*. Among the generally used T-conorm operators are maximum, algebraic sum, bounded sum, and drastic sum. Again, the maximum operator $S_{min}(a, b) = max(a, b)$ is the most commonly encountered operator in determining the union of fuzzy sets.

Using fuzzy sets, linguistic values can be defined for a given set of input and output variables. In multidisciplinary design, it is often important to map input and output variables are mapped against each other. Towards this end, fuzzy rules are employed, which use conditional statements in the form of if-then rules. For example, a fuzzy if-then rule assumes the form "if *x* is *A* then *y* is *B*", where *A* and *B* are the linguistic values of the input variable *x* and the output variable *y*, respectively. The if-part of the rule "*x* is *A*" is called the antecedent or premise and the then-part of the rule "*y* is *B*" is called the consequent or conclusion. An example of such a rule could be of the form "if structural stress is high then reduce the load", where "high" is a linguistic value of variable "structural stress". In the case of boolean logic, if the antecedent part of the if-then rule is true then the consequent part of the rule is also true. However, fuzzy if-then rules do not operate in the same manner. Instead, in fuzzy if-then rules, if the antecedent is partially true to some degree, then the consequent is also partially true to that same degree. If the if-then rules in both boolean and fuzzy logic are written in a simpler manner by using a symbol, $\rightarrow$ , then they will be denoted as follows:

In boolean logic: $A \rightarrow B$
In fuzzy logic: $0.5A \rightarrow 0.5B$

If-then rules can also have more than one part in the antecedent and the consequence. In this case, using a *T*-norm operator, two antecedent parts are evaluated simultaneously, generating a single output value. This value is then used to compute the outcome of the consequent parts using an implication function. As stated above, of the various types of implication functions, the simplest and most commonly used is the truncation using the *min* operation. In Fig. 6.10, an example of fuzzy implication is shown for the following fuzzy rule.

If *x1* is **A1** or *x2* is **A2**, then *y* is **B**

where $x1 = 3$ and $x2 = 8$. In this figure, *x1* and *x2* are the fuzzy input variables in the if-then rule and *y* is the fuzzy output variable. First, the most common *T*-conorm operator, maximum, is implemented on the two parts of antecedent and a single value $\mu = 0.5$ is obtained. Then, the implication function modifies the output fuzzy set by chopping it off to the degree specified by the antecedent which is $\mu = 0.5$ in this case.

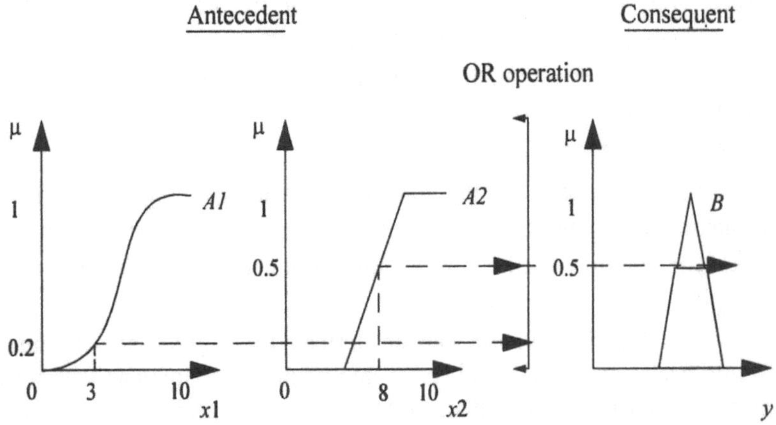


**Figure 6.10** Illustration of fuzzy implication.

In the event that more than one fuzzy rule is to be considered, the output fuzzy sets from each of the rules are aggregated into a single output fuzzy set by applying one of the available types of aggregation operations. In Fig. 6.11, an implementation of the most widely accepted aggregation process, the maximum method, is shown for two different output fuzzy sets.

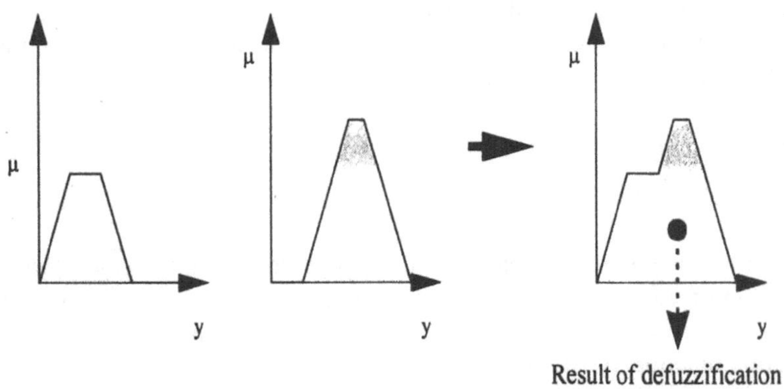


**Figure 6.11** Aggregation process of fuzzy output sets.

Since the maximum operator is commutative, the order in which the rules are aggregated in the case of multiple output fuzzy sets is unimportant. The aggregation of two output fuzzy sets returned by the implication process generates another fuzzy set. In order to extract a useful information from this aggregated fuzzy set, it must be defuzzified to generate a single number. Although converting a fuzzy set into a single crisp number can be done in several different ways, the most popular defuzzification method is the centroid calculation which returns the center of shaded area under the curve representing the aggregated output fuzzy set. In Fig. 6.11, the center of shaded region is denoted by a solid dot which is then projected onto the abscissa to determine a crisp value of $y$.

To demonstrate how fuzzy logic can be used to generate approximate models for analysis, consider a problem in which an approximation to the function $y = \sin(\theta), 0 \le \theta \le \pi$ is to be developed. In order for the fuzzy logic model to work properly, a set of fuzzy rules and the type and number of membership functions for the input and output must be first defined. Assuming that minimal knowledge about the half sine wave function is available, 3 gaussian ("small", "medium", "large") and 2 triangle ("small", "large") membership functions were selected to describe values of input and output of the function, respectively. Using these membership functions, the following three rules were generated.

Rule 1:If θ is **small** then $y$ is **small**.
Rule 2:If θ is **medium** then $y$ is **large**.
Rule 3:If v is **large** then $y$ is **small**.

These rules are developed based on the known fact that for θ close to 0 (small) and π (large), $y$ is near 0 (small), and for θ in the region of π/2 (medium), $y$ is close to 1 (large). In this study, a fuzzy implication is operated on each rules using truncation function as shown in Fig. 6.12. In this figure, three output fuzzy sets are produced from fuzzy rules for a known input value $x1$ and are aggregated into a single fuzzy set using the maximum operation. The aggregated fuzzy output set is then defuzzified to obtain $y1$ using the centroid method.

Building an optimal and robust fuzzy model requires that the membership functions be chosen appropriately. In addition to the choice of the functions, it is also possible to tune the parameters of the membership functions to build an optimal model on the basis of the available parameters. The gaussian membership function requires two parameters {$c$, s} to establish its shape, where {c} denotes the center of the gaussian and is located between 0 and π. The location of 3 vertex points {a,b,c} determine the shape of each triangular function. The location of the middle vertex in the triangular function {b} were bounded by -0.8 and 1.8. The remaining parameters such as {s} in gaussian and {a, c} in triangular membership functions were determined with respect to the location of {c} or {b} in gaussian and triangular membership functions, respectively. In order to tune a fuzzy model, an optimization algorithm can be employed. Evolutionary fuzzy modeling employs genetic algorithm based optimization to evolve fuzzy rules and membership function parameters. Genetic algorithm based optimization techniques have proven to be effective for synthesizing fuzzy models as described in Cordon and Herrera (1995), Karr and Gentry (1993), Krishna Kumar and Satyadas (1995), and Satyadas and Krishna Kumar (1994). The GA evolution procedure allows for selection of the type of membership functions and also the total number of rules. In a more limited exercise, when the only tuning parameters were the parameters which define the shape of each membership function, the optimal membership functions were obtained as shown in Fig. 6.13. In this exercise, only three input-output patterns were used for the purposes of training.

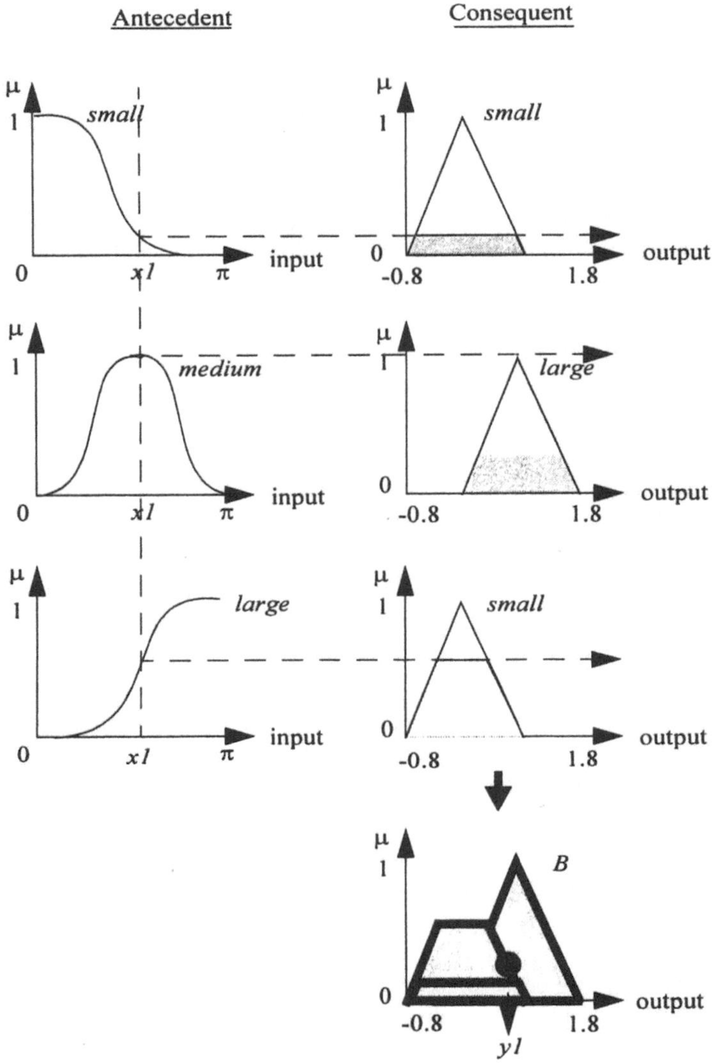


**Figure 6.12** Fuzzy inference process.

When compared against the neural network approach as a tool to generate surrogate models for optimization, the following general conclusions can be drawn. First, the performance of these approaches is dependent upon the behavior of the function to be approximated, and the amount of information available about the system. For a smoothly varying function, both the neural networks and fuzzy logic models perform well; neural networks, however, are less efficient at modeling non-smooth functions. In those cases where little data and a limited mathematical formulation is available, fuzzy logic provides a convenient approach to incorporate a user's knowledge into modeling the system. When significant number of input-output patterns are available, neural networks can be trained to generate the mapping with little user interaction. Fuzzy logic, on the other hand requires user input about the general form of input and output membership functions, and of the fuzzy rules to use in the model.

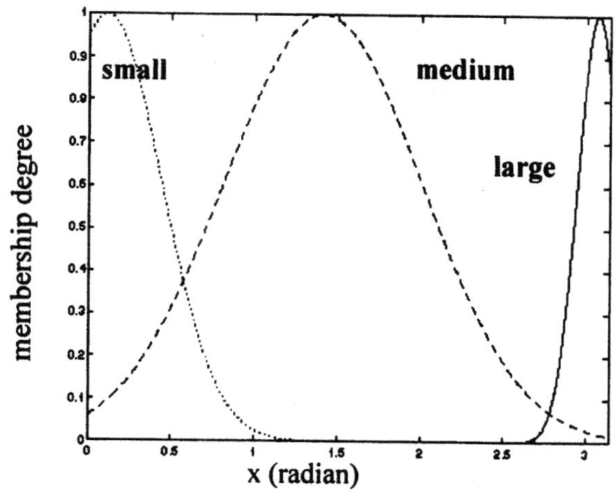


input membership functions

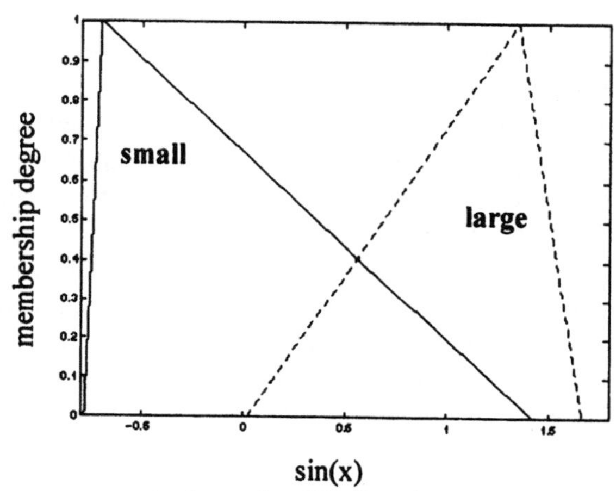


output membership functions

**Figure 6.13** Optimally tuned membership functions.

## 6.4 GA Based Modeling of the Immune System

This section describes a GA based modeling of another biological process, the immune system, and of its implementation in problems of design optimization. A principal advantage of the immune system simulation is in its seamless integration into a GA based search for an optimal design. In biological immune systems, foreign cells and molecules, denoted as antigens, are recognized and eliminated by type-specific antibodies. The task of recognizing antigens is formidable due to the very large number of possible antigens; it is estimated (Smith et. al. 1992) that the immune system has been able to recognize at least $10^{16}$ antigens. This pattern recognition capability is impressive, given that the genome contains about $10^5$ genes, and the immune system must use segments of these to construct antibodies for all possible antigens that are likely to be encountered; in a typical mammal, there are between $10^7$ and $10^8$ different antibodies. In biological systems, this recognition problem translates into a complex geometry matching process. The antibody molecule region contains a specialized portion referred to as the paratope which is used for identifying other molecules. The amino acids from which the paratope is constructed determine its shape and hence also the shape of the antigen molecules that can attach to the paratope. The antibody, therefore, can have a geometry that is specific to a particular antigen (specialist) or is capable of partial match and capturing of a broader group of antigens (generalist).

The genetic information for an antibody is encoded by separate gene segments, each of which is drawn from a library of gene segments. For this combinatorial mechanism to function effectively, the contents of the libraries must act cooperatively and partition the antigen recognition task evenly among the segments. This cooperative partitioning requires some organization, and an evolutionary approach has been used for this purpose. The evolutionary approach is similar to that used in GAs; the principal difference is in the choice of the fitness function. In immune systems, the fitness of an individual would be determined by its ability to recognize either a specific or a broader group of antigens.

The immune system model described herein is based on the use of binary string structures to represent both the antigen and antibody molecules. While this is a gross simplification of the biological model where genes are specified by a four-letter DNA alphabet, and where antibody-antigen pairings are based on complex matching of three-dimensional shapes, it fits more naturally into a GA based design optimization strategy. The simulation is initiated by first defining a scalar measure of the degree of match between binary bit string representations of the antibodies and antigens. The match can be measured either in terms of similarity or complimentarity in the string structures. Different levels of complexity can be built into the development of such a matching function, including the number of matches on a bit-by-bit basis and/or the length of contiguously matched substrings. If antibodies are to be developed that are a match of the antigens, a simple numerical measure that quantifies similarity for use in immune system modeling is to count the number of matching bits between a pair of strings as,

$$Z = \sum t_i \tag{6.24}$$

where $t_i$=1 if there is a match at the i-th location of the two strings, and is 0 otherwise. A larger value of $Z$ indicates a higher degree of match between the two strings. If a randomly distributed population of antibodies is exposed to a specific antigen, maximization of $Z$ using a traditional GA simulation would result in an antibody population that matches the antigen. Reversing the definition of $t_i$ would result in the evolution of an antibody population that is complimentary to the antigen.

The more interesting problem is in modeling the dynamics of the immune system when there are multiple antigens for which covering antibodies must be generated. The antibody pool can evolve to produce multiple copies of strings that match each of the present antigens. These antibodies would be termed as specialists, designed specifically for a particular antigen. Alternatively, it is also possible to evolve the antibody population into a generalist, which maximally covers all of the antigens. The manner in which the antigens are selected and exposed to the antibodies has a dominant effect on whether specialist or generalist antibodies are produced (Smith et. al. 1992) in the immune system simulation. Both of these antibody types are relevant in the context of design optimization, and the simulation strategies for their generation are summarized here for completeness.

Consider first a scenario in which a number of different antigens exist, and an antibody population must be evolved to provide generalist antibodies that provide maximum coverage of the antigens with the least number of antibodies; in the limit, this would result in a single antibody that maximally covers all of the antigens. A numerical process that would produce the desired effect is initiated by randomly initializing the population of antibodies and by specifying an antigen pool. The following stepwise procedure is adopted.

Step 1: Initialize the fitness of all antibodies to zero.

Step 2: Compute the fitness of the antibody pool based on similarity (or complimentarity) to the antigens, and this requires the following specific steps:

a) An antigen is selected at random.

b) A sample of antibodies of size $\mu$ is selected from the antibody pool without replacement.

c) The match score of each antibody is computed by comparing against the selected antigen, and the antibody with the highest score has the match score added to its fitness value; the fitness of the other antibodies is unchanged.

d) The antibodies are then returned to the antibody population, and the process repeated a number of times (typically two to three times the antibody population size).

Step 3: Based on the fitness computed in Step 2, a GA simulation is conducted with prescribed probabilities of crossover and mutation to evolve the antibody population through one generation of evolution.

Step 4: The process is then repeated from Step 1 till convergence in the antibody population is attained.

It turns out that the ability of the approach to develop either generalists, specialists, or both, is dependent upon the selection of the control parameter $\mu$ which denotes the sample size. A large sample size $\mu$ will facilitate the exposure of most antibodies to all antigens, particularly if the rate of sampling of the antigens does not contain a bias. In this case, the match scores of prospective antibodies that are specific to the sampled antigens will be higher, and these antibodies would evolve in the immune network simulation. On the other hand, the use of smaller sample size decreases the probability of a prospective antibody from being exposed to the different antigens. In this case, the tendency of the immune network simulation is to develop a generalist antibody, as the match scores of such antibodies will dominate the evolutionary process. In fact, increasing sample size would result in the appearance of a few specialists until ultimately, with a large enough sample size, the specialists would completely dominate the antibody evolution.

### 6.4.1 Immune Networks in GA Based Search

We next turn our attention to the use of immune network simulation in accelerating the convergence characteristics of the genetic algorithm by biasing the search in those regions which are identified to be promising at the early stages of the search process.

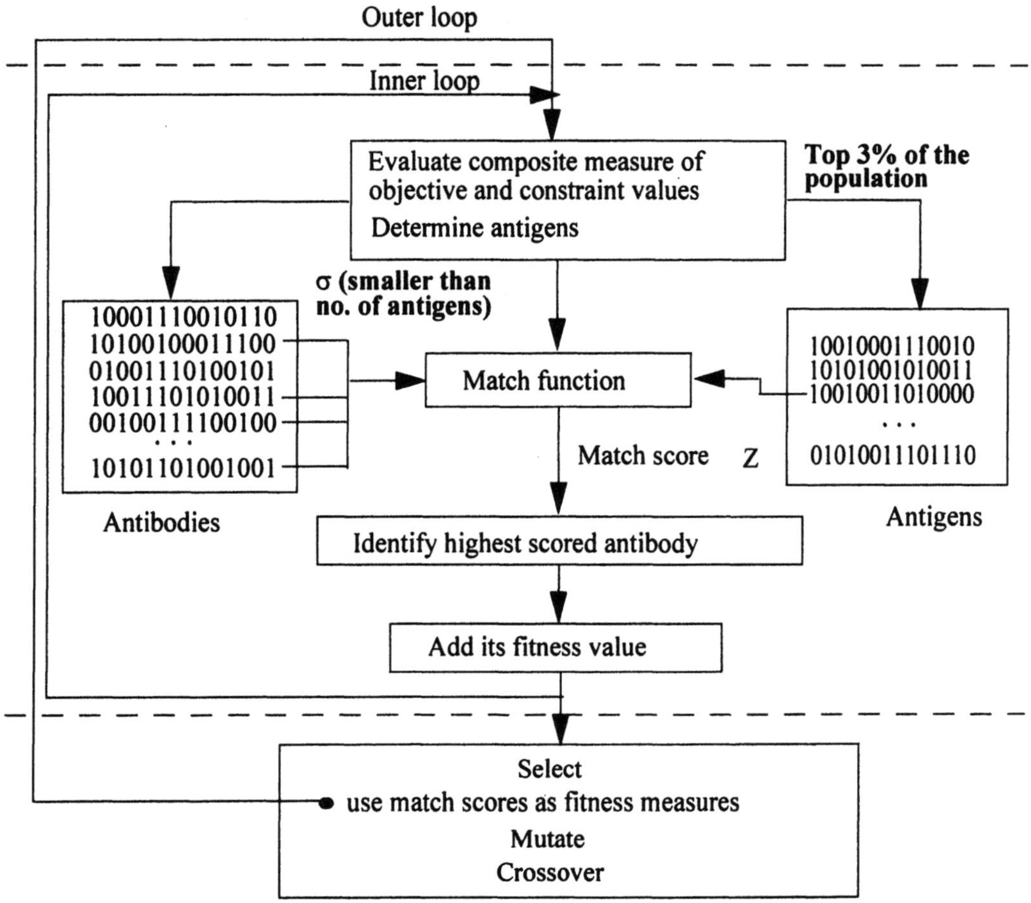


**Figure 6.14** Schematic of immune network based enhancement of GA search.

A stepwise procedure detailing this adaptation (the schematic flow of which is shown in Fig. 6.14) is as follows.

Step 1: A population of designs is randomly generated.

Step 2: The fitness function, which is a composite measure of the objective function and an appended penalty associated with constraint violation, is obtained for the entire population.

Step 3: Members within the top 3% of the population obtained at the end of Step 2 are designated as the antigens, and the entire population (including the antigens) is defined as the starting population of antibodies.

Step 4: Using an antibody sample size $\mu$ smaller than the number of antigens, the degree of match $Z$ is obtained for each member of the population according to the steps described in the previous chapter. This promotes the development of generalist antibodies.

Step 5: The match score of each design is used as a fitness measure in a traditional selection or reproduction operation. During this reproduction operation, the size of the population is unchanged.

Step 6: The crossover and mutation operations are performed on the new population of antibodies formed in Step 5.

Step 7: The process is then repeated from Step 2 with an intent of evolving the population to maximize the $Z$ function and cycled to convergence.

It is important to note that this process (shown schematically in Fig. 6.14) is somewhat different from the traditional GA implementation. In the latter, the fitness function is the composite measure of objective function value and the design constraints, and this is used in the selection or reproduction stage of the GA evolution. In the present work, this composite measure is only used to identify the antigens to which the population must adapt. The immune network simulation is conducted to determine a measure of schema similarity between all members of the population and the selected antigens, and a GA used to evolve the population with the intent of maximizing this similarity function. It stands to reason that the sample size $\mu$ be chosen to be smaller than the number of antigens so the solution obtained at the end of the immune network simulation will be a generalist.

The optimal sizing of a 10-bar truss, the geometry and applied loads for which are shown in Fig. 6.15, is used to illustrate the use of immune network modeling in enhancing the convergence characteristics of the genetic algorithm. Each cross-sectional area of the truss elements was considered as an independent design variable and these were sized to obtain a minimum weight structure. A load $P = 100$ *kips* was applied at nodes 3 and 5 in the directions of displacements $d_6$ and $d_8$ and allowable constraints on maximum stresses in the bar elements in both tension and compression of $\pm 25$ *kpsi,* and allowable limits on selected nodal displacements were specified in this design problem. The material was aluminum with a density $\rho = 0.1$ $lb/in^3$, Young's modulus of $10 \times 10^6$ $lb/in^2$, and a length $l = 360$ *in.*

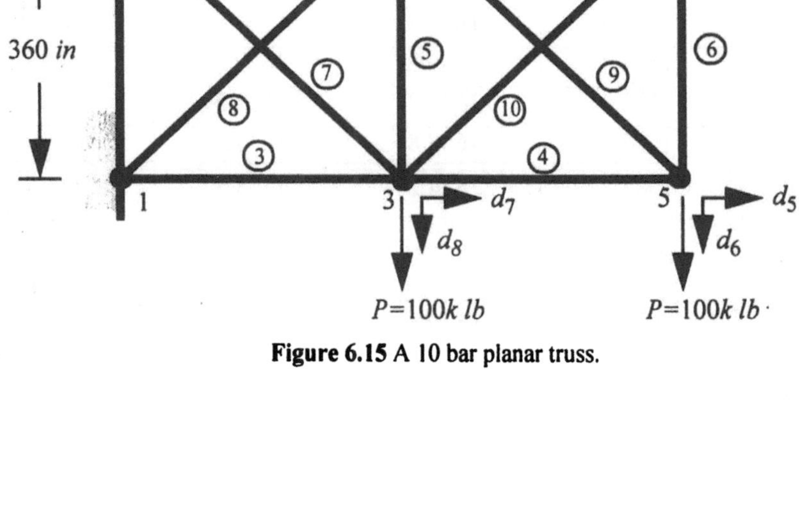


**Figure 6.15** A 10 bar planar truss.

The performance of the immune network based convergence-enhancement strategy was assessed against results obtained from a plain GA implementation and from use of efficiency enhancing strategies such as directed crossover and the multistage search. Each cross-sectional area, with a lower and upper bound of 0.01 $in^2$ and 12.0 $in^2$, respectively, was represented by a 10-digit binary string. With these lower and upper bounds, a precision of representation of 0.01173 $in^2$ was obtained for each design variable. A population size of 200 was used in the GA simulation. Probabilities of crossover and mutation of 0.8 and 0.02 were specified for plain GA, directed crossover, and multistage approaches while the immune network simulation approach used 0.8 and 0.005. In the directed crossover strategy, the crossover gain was recorded for the first 8 generations iterations and a genetic search with 50% regular crossover and 50% directed crossover was performed. In the multistage search method, four stages of genetic search were implemented using precisions of design variable representation of 1.0, 0.2, 0.05, and 0.01173 $in^2$, respectively. Also, the best 25% of the population at the ned of each stage was used to seed the population for the next stage. An antibody sample size of 4 and antigen population of 6 were chosen for the immune network simulation which allowed the population to converge to a generalist. A total of 80,000 function evaluations were allowed in each simulation and the results obtained in these experiments are summarized in Table 6.6. Results for 8 different simulations using different random number seeds are presented in this table.

**Table 6.6** Comparison of solutions from different strategies.

| Trial number | Plain GA | Directed Crossover | Multistage | Immune |
|---|---|---|---|---|
| | Optimal objective | Optimal objective | Optimal objective | Optimal objective |
| 1 | 1617.6 | 1620.8 | 1599.2 | 1636.8 |
| 2 | 1780.8 | 1772.8 | 1608.0 | 1612.8 |
| 3 | 1704.0 | 1712.0 | 1699.2 | 1660.8 |
| 4 | 1699.2 | 1700.8 | 1808.0 | 1686.4 |
| 5 | 1656.0 | 1667.2 | 1704.0 | 1608.0 |
| 6 | 1670.4 | 1659.2 | 1659.2 | 1664.0 |
| 7 | 1712.0 | 1672.0 | 1603.2 | 1608.0 |
| 8 | 1676.8 | 1668.8 | 1643.2 | 1659.2 |
| Mean | 1689.6 | 1684.2 | 1665.5 | 1642.0 |
| Standard deviation | 46.7556 | 45.0681 | 70.7555 | 29.9912 |
| Best results | 1617.6 | 1620.8 | 1599.2 | 1608.0 |

On average, the advanced strategies produce an improvement over the plain GA, even for a relatively small-sized problem. The multistage search produces the best overall objective but also shows the largest spread (standard deviation) in the optimized solutions from the mean. The immune network modeling process generated good results consistently. The best overall objective was similar to the one obtained in the multistage search. Both the mean and the deviation of optimal objective values were the lowest for the immune network approach which suggests that it is not only effective in locating the optimal solution but also less dependent on the random number seed. When the immune network approach is implemented in design problems of higher dimensionality, the improvement in the convergence rate is further amplified as the method is designed to identify the gross schema-patterns that correspond to best design strings early in the search process.

### 6.4.2 Immune Networks in Multicriterion Design

The immune network simulation has been shown to be a useful approach to generate fronts of Pareto optimal solutions in multicriterion design. A general mathematical statement description of the multicriterion design problem can be written as follows,

$$\min_{(X \in \Omega)} f(X) , \tag{6.25}$$

where $f(X)$ and $\Omega$ denote the objective function and the feasible set, respectively, given as follows.

$$f(X) = \{f_1(X), f_2(X), \ldots, f_m(X)\} \tag{6.26}$$

$$\Omega = \{X \in R^n | g(X) \le 0, \quad h(X) = 0\} \tag{6.27}$$

In the above, $m$ is the number of objective criteria that are to be minimized, and $n$ is the number of design variables. The vector function $g(X)$ and $h(X)$ define the inequality and equality constraints, respectively. An ideal solution for each criteria would be obtained by minimizing each separately and without regard to the other criteria. In general, however, criterion must be traded-off against each other, and this brings one to the subject of constructing a weighted composite objective function as follows.

$$f(X) = w_1 f_1 + w_2 f_2 + \quad \cdot \quad \cdot \quad + w_m f_m$$
$$\sum w_i = 1 \tag{6.28}$$

Since the weighting schedule is rarely known precisely, it is always advantageous to generate a front of compromise designs, with each point on the front representing a particular weighting choice. One simple approach to generate the front of Pareto optimal solutions to multicriterion optimization problems is to minimize the weighted composite objective function of Equation (6.28) for different sets of weights. Another intuitive approach is to minimize the difference between the individual objective functions and their corresponding ideal solutions. The ideal solution of each objective function can be determined by optimizing each of the objective functions independently. If the ideal solutions are defined as $f_1^*, f_2^*, \ldots, f_m^*$, then for an arbitrary design, $x$, the normalized distance of each component of the objective functions from its ideal optimum is given as follows.

$$d_i(x) = \frac{f_i(x) - f_i^*}{f_i^*} \qquad i = 1, \ldots, m \tag{6.29}$$

Now, the problem can be solved by minimizing the composite of all deviations as follows,

$$\text{Minimize} \quad \max_{i=1,\ldots,m} [w_i d_i(x)], \tag{6.30}$$

for multiple sets of weighting coefficients; the optimal solutions obtained in this manner would constitute the front of Pareto optimum solutions. In using the immune network simulation to develop a front of Pareto optimal designs, the basic idea is to treat good designs corresponding to different weighting criteria coefficients as antigens, and to evolve sub-populations of designs that cover each of these antigens. In other words, the feature of immune system simulation that encourages the growth and stabilization of specialist antibodies is required in this problem. A summary of how this is accomplished is as follows.

Step 1: Randomly generate a population of designs

Step 2: Formulate a number of different weight combinations that can be expected to span the Pareto optimal front.

Step 3: Evaluate each design in the population for its objective values corresponding to each of the chosen weighting combinations.

Step 4: Combine the objective and constraint values to obtain a composite measure of each design, and for each of the weight combinations.

Step 5: For each combination of weights, identify the designs with the best composite measure.

Step 6: Defining the identified designs of Step 5 as antigens and the whole population as a pool for antibodies (including the antigens), implement the immune system simulation as described earlier to generate specialist antibodies.

An implementation of this approach to the design of a simply supported I beam is presented next. The beam, shown in Fig. 6.16 is to be designed to minimize both the structural weight *f1*, and the static deflection at the mid-span *f2*.

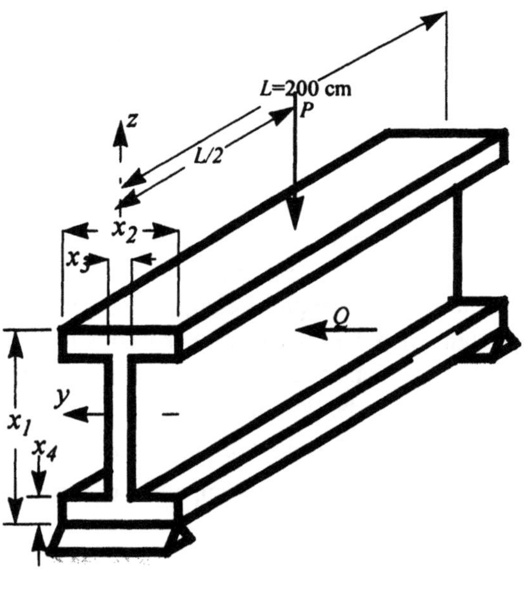


**Figure 6.16** An I-beam structure.

The beam is subjected to a concentrated central load $P = 600\ kN$ in the vertical direction, and a load $Q = 50kN$ in the transverse direction. The beam material has a Young's modulus $E = 2\text{x}10^4$ $kN/cm^2$ and can sustain a maximum bending stress $k_g = 16\ kN/cm^2$. The dimensions of the structure, $x_i$, $i = 1,4$ are the design variables for the problem. The mathematical statement of the optimization problem is as follows:

$$Minimize\ f = W1 \cdot f1 + W2 \cdot f2 \tag{6.31}$$

subject to

$$g1 \equiv \frac{\left[\frac{M_y}{W_y} + \frac{M_z}{W_z}\right]}{k_g} - 1 \le 0 \tag{6.32}$$

where, $M_y$ is the maximal bending moment in $y$ direction, $M_z$ is the maximal bending moment in $z$ direction, and $W_y$ and $W_z$ are the section moduli in $y$ and $z$ directions, respectively. From the known magnitude of forces $P$ and $Q$, $M_y$ and $M_z$ are determined to be 30,000 $kNcm$ and 2,500 $kNcm$ and the section moduli, $W_y$ and $W_z$, are expressed as follows.

$$W_y = \frac{x_2 x_1^3 - (x_2 - x_3)(x_1 - 2x_4)^3}{6x_1} \tag{6.33}$$

$$W_z = \frac{2x_4 x_2^3 + (x_3^3)(x_1 - 2x_4)}{6x_2} \tag{6.34}$$

The function $f1$ is the cross sectional area given as

$$f1 = 2x_2 x_4 + x_3(x_1 - 2x_4) \tag{6.35}$$

and $f2$ is the static deflection obtained as

$$f2 = \frac{PL^3}{48EI} \tag{6.36}$$

where

$$I = \frac{x_2 x_1^3 - (x_2 - x_3)(x_1 - 2x_4)^3}{12} \tag{6.37}$$

and the total length of the beam, $L$ is 200 cm.

For this problem, the upper and lower bounds, and the precision of design variables are as defined in Table 6.7; these definitions result in a total design string length of 30 bits.

**Table 6.7** Design variable specifications for I-beam problem.

| Design Variables | Upper Bound | Lower Bound | Precision |
|---|---|---|---|
| x1 | 10.00 | 80 | 0.13698 |
| x2 | 10.00 | 50 | 0.31496 |
| x3 | 0.9004 | 5 | 0.03228 |
| x4 | 0.9004 | 5 | 0.03228 |

A population size of 100 designs was used and a maximum of 20,000 function evaluations were allowed in each simulation. The probabilities of crossover and mutation were chosen as 0.6 and 0.01, respectively. To account for the random nature of the GA search, the process was repeated for eight different times using different random number seeds. Optimal designs obtained using the traditional approach and the immune network simulation are presented in Tables 6.8 and 6.9, respectively. There is excellent agreement between results from both approaches. Results in Table 6.9 correspond to the best designs obtained using the immune network simulation.

The immune network approach can also be extended to problems where the objective and constraint functions are imprecise or fuzzy. For example, if a loaded structure is to withstand a stress of 20,000 psi, the following constraint function may be applied in a classical design sense.

$$\sigma \leq 20,000 psi \tag{6.38}$$

This suggests that the design having the stress of 20,000 psi is acceptable whereas one having the stress of 20,001 psi is no, even if the objective value of the latter design may be far better than the former. Therefore, it would make more sense to define constraints such that a smooth transition is made from an acceptable level to an unacceptable one. For example, here is an inequality constraint, $g_j$ which is bounded by a prescribed value of $b_j$ as shown below.

$$g_j \leq b_j \tag{6.39}$$

If an allowable interval $\Delta b_j$ is applied to a design constraint $b_j$ in Eqn. (6.39) then a feasible fuzzy set $G_j$ can be defined as

$$G_j = [-\infty, b_j + \Delta b_j]. \tag{6.40}$$

**Table 6.7** Pareto designs for I-beam problem using traditional approach.

| Weights | | Design Variables | | | | Area (cm$^2$) | Deflection (cm) | Scaled Objective |
|---|---|---|---|---|---|---|---|---|
| w1 | w2 | x1 | x2 | x3 | x4 | | | |
| 0.0 | 1.0 | 80.0000 | 50.0000 | 5.0000 | 5.0000 | 850.00 | 0.0059 | 0.9999 |
| 0.1 | 0.9 | 80.0000 | 50.0000 | 0.9004 | 5.0000 | 563.03 | 0.0069 | 1.4862 |
| 0.2 | 0.8 | 80.0000 | 50.0000 | 0.9004 | 4.9677 | 559.86 | 0.0069 | 1.8119 |
| 0.3 | 0.7 | 80.0000 | 50.0000 | 0.9004 | 3.7734 | 442.57 | 0.0087 | 2.0692 |
| 0.4 | 0.6 | 80.0000 | 50.0000 | 0.9004 | 2.9664 | 363.32 | 0.0106 | 2.2199 |
| 0.5 | 0.5 | 80.0000 | 50.0000 | 0.9004 | 2.3853 | 306.27 | 0.0128 | 2.2839 |
| 0.6 | 0.4 | 80.0000 | 50.0000 | 0.9004 | 1.9334 | 261.89 | 0.0153 | 2.2668 |
| 0.7 | 0.3 | 80.0000 | 50.0000 | 0.9004 | 1.4815 | 217.51 | 0.0190 | 2.1621 |
| 0.8 | 0.2 | 80.0000 | 50.0000 | 0.9004 | 1.0618 | 176.30 | 0.0249 | 1.9501 |
| 0.9 | 0.1 | 80.0000 | 35.8268 | 0.9004 | 0.9004 | 134.93 | 0.0365 | 1.5720 |
| 1.0 | 0.0 | 63.6994 | 39.9213 | 0.9004 | 0.9004 | 127.62 | 0.0564 | 1.0015 |

**Table 6.8** Pareto designs for the I-beam problem using the immune network simulation.

| Weights | | Design Variables | | | | Area ($cm^2$) | Deflection (cm) | Scaled Objective |
|---|---|---|---|---|---|---|---|---|
| w1 | w2 | x1 | x2 | x3 | x4 | | | |
| 0.0 | 1.0 | 80.00 | 50.00 | 5.00 | 5.00 | 850.00 | 0.0059 | 0.9999 |
| 0.1 | 0.9 | 80.00 | 50.00 | 0.90 | 5.00 | 563.03 | 0.0069 | 1.4862 |
| 0.2 | 0.8 | 80.00 | 50.00 | 0.90 | 4.97 | 559.86 | 0.0069 | 1.8119 |
| 0.3 | 0.7 | 80.00 | 50.00 | 0.90 | 3.77 | 442.57 | 0.0087 | 2.0692 |
| 0.4 | 0.6 | 80.00 | 50.00 | 0.90 | 2.96 | 363.32 | 0.0106 | 2.2199 |
| 0.5 | 0.5 | 80.00 | 50.00 | 0.90 | 2.38 | 306.27 | 0.0128 | 2.2839 |
| 0.6 | 0.4 | 80.00 | 50.00 | 0.90 | 1.90 | 258.72 | 0.0155 | 2.2665 |
| 0.7 | 0.3 | 80.00 | 50.00 | 0.90 | 1.47 | 217.11 | 0.0191 | 2.1622 |
| 0.8 | 0.2 | 80.00 | 50.00 | 0.90 | 1.06 | 176.30 | 0.0249 | 1.9501 |
| 0.9 | 0.1 | 80.00 | 34.56 | 0.90 | 0.92 | 134.09 | 0.0369 | 1.5725 |
| 1.0 | 0.0 | 62.84 | 40.23 | 0.90 | 0.90 | 127.73 | 0.0576 | 1.0023 |

Using this fuzzy set, a fuzzy constraint function can be constructed as

$$g_j \in G_j,\ j = 1, \ldots, m\,. \tag{6.41}$$

Eqn. (6.41) indicates that if $g_j$ is a member of a fuzzy set $G_j$ then the design is feasible. In other words, the degree of membership of a feasible design is always $\mu_{g_j} \geq 0$.

$$\mu_{g_j} = \begin{cases} 1, & if\ g_j - b_j \leq 0 \\ 1 - \left\{\dfrac{g_j - b_j}{\Delta b_j}\right\}, & if\ b_j \leq g_j \leq b_j + \Delta b_j,\ j = 1, \ldots, m \\ 0, & if\ g_j - b_j > \Delta b_j \end{cases} \tag{6.42}$$

If the value of $g_j - b_j$ lies between 0 and $\Delta b_j$, a membership function is used to determine $\mu_{g_j}$ whose value varies between 0 and 1. In Eqn. (6.42), a linear membership function is used to describe the level of constraint violation between $b_j$ and $b_j + \Delta b_j$.

Similarly, the degree of membership for individual objective functions is obtained in the following manner. First, ideal design points, $x_i$* which corresponds to the minimum value of $f_i$ (denoted as $f_i^{min}$) are found for all criterion functions $f_i$, $i = 1, k$. All criterion functions $f_j$, $j = 1, k$, $j \neq i$ are also evaluated for each of the ideal design points $x_i$*, and the worst value for the $i$-th objective criteria obtained in this manner is denoted as $f_i^{max}$. Using these solutions as fuzzy boundaries, the following membership function, similar to Eqn. (6.42), can be written for each objective criteria.

$$\mu_{f_i} = \begin{cases} 0, & if\ f_i > f_i^{max} \\ \dfrac{f_i^{max} - f_i}{f_i^{max} - f_i^{min}}, & if\ f_i^{min} < f_i \le f_i^{max}, \quad i = 1, \ldots, k \\ 1, & if\ f_i \le f_i^{min} \end{cases} \tag{6.43}$$

The optimal solution of the fuzzy multicriterion design problem can be obtained by maximizing both $\mu_{g_j}$ and $\mu_{f_i}$ simultaneously. Therefore, an objective function for the fuzzy multicriterion problem can be formulated as follows.

$$\max_{x} \min_{i,j} \{\mu_{f_i}(x), \mu_{g_j}(x)\} \qquad i = 1, \ldots, k, \ j = 1, \ldots, m \tag{6.44}$$

As explained earlier, it is useful to construct a front of Pareto optimal solutions for the multicriterion design problem as it clearly demonstrates the required trade-offs among different solutions. In order to generate a solution which is a front of Pareto optimal solutions for fuzzy multicriterion design, an additional bias term $\Delta\mu_{f_i}$ is appended to $\mu_{f_i}$ in Eqn. (6.44) as follows.

$$\max_{x} \min_{i,j} \{\mu_{f_i}(x) + \Delta\mu_{f_i}, \mu_{g_j}(x)\} \qquad i = 1, \ldots, k, \ j = 1, \ldots, m \tag{6.45}$$

In principle, this accomplishes the same effect as the variation of weighting coefficients $w_i$ described in the previous section. Note that the value of $\Delta\mu_{f_i}$ can vary from 0 to 1, and different combinations of $\Delta\mu_{f_i}$ bias the search process to locate different points on the front of Pareto optimal solutions. As the value of $\Delta\mu_{f_i}$ imposed on a particular objective criteria $\mu_{f_i}$ increases, that criterion is emphasized to a lesser extent; a value of $\Delta\mu_{f_i} = 1$ would be akin to setting $w_i = 0$.

In using the immune network to account for fuzzy multicriterion design, and to develop a front of Pareto optimal designs, the basic idea is to treat good designs corresponding to different combinations of $\Delta\mu_{f_i}$ as antigens, and to evolve the population of designs in a manner of antibody evolution to cover these antigens. Hence, the feature of immune system simulation that encourages the growth and stabilization of specialist antibodies is adopted in this work. A summary of how this is accomplished is as follows.

(i) Obtain an ideal design point, $x_i^*$ for each individual objective criteria $f_i$ where $i = 1,\ldots,k$ by solving single criterion design problems separately.

(ii) By substituting these ideal solutions $x_i^*$ in each objective function $f_i$, evaluate its $f_i^{min}$ and $f_i^{max}$.

(iii) Formulate a fuzzy objective function $\Delta\mu_{f_i}$ for each individual criteria using its $f_i^{min}$ and $f_i^{max}$.

(iv) With predefined allowable intervals of $\Delta b_j$, formulate fuzzy constraint function $\mu_{g_j}$ for all $j = 1, m$.
(v) Randomly generate a population of designs.
(vi) Evaluate each design in the population for its objective values defined as $\mu_{f_i} + \Delta\mu_{f_i}$ for a number of different combinations of $\Delta\mu_{f_i}$.
(vii) For each combination of $\Delta\mu_{f_i}$, identify a design whose maximum value among $\mu_{f_i} + \Delta\mu_{f_i}$ and $\mu_{g_j}$ is the minimum.
(viii) Defining the identified designs of step (vii) as antigens and the entire population as a pool for antibodies (including the antigens), implement the immune system simulation as summarized in the following steps.
- A single antigen is randomly selected from the antigen population.
- From the population of N antibodies, a randomly selected sample of size S is taken without replacement (typically three times the number of antigens).
- Each antibody in the sample is matched against the selected antigen, and the match score is computed (similarity of match).
- Identify the antibody in the sample population with the highest match score. Ties are broken at random.
- Add the match score of the winning antibody to its fitness value.
- Repeat this process for typically three times the number of antibodies.

The following additional steps are required to complete the antibody evolution.
(ix) Perform selection on the basis of computed fitness value.
(x) Crossover and mutation is performed on the selected population using crossover and mutation probabilities of 0.6 and 0.01, respectively.
(xi) Above steps from (v) to (x) are repeated to convergence.

As an example, consider the design of a three bar truss structure. The objective of this optimization problem is to minimize two conflicting fuzzy criteria subject to some prescribed fuzzy constraints. The immune network simulation approach allows for the simultaneous generation of Pareto optimal designs for a number of different combinations of $\Delta\mu_{f_i}$. Results from these simulations were compared with results from a traditional approach, where, in the latter, the optimization had to be performed separately for each combinations of $\Delta\mu_{f_i}$. The geometry and loading of the 3-bar truss structure are shown in Fig. 6.17. The design variables for the problem are the cross-sectional areas of each bar element of the structure. However, symmetry in loading and geometry dictates that the cross-sectional areas of elements 1 and 3 be equal, thereby reducing the number of design variables to two ($x_i$, $i = 1,2$). Values of $P = 20$ *lb*, structure material density, r = 1 *lbm/in*$^3$, $E = 1$ *lb/in*$^2$, and length, $l = 1$ *in* were used in the computations. The objective of the optimization problem is to minimize both the structural weight $f_1$, and the vertical displacement $f_2$ of the loaded joint. As constraints for this problem, a maximum allowable stress of $\sigma^u = 20$ *psi* was permitted in elements 1 and 2, and a minimum allowable stress of $\sigma^l = -15$ *psi* was prescribed for element 3.

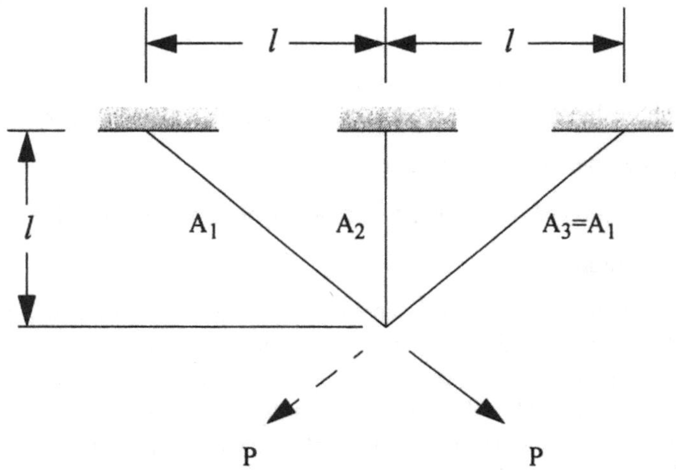


**Figure 6.17** A planar 3 bar truss structure.

The mathematical statement of the optimization problem is as follows:

$$minimize\ f(x) = \{f_1(x), f_2(x)\} \tag{6.46}$$

subject to

$$g_1 \equiv \frac{\sigma_1}{\sigma^u} - 1 \le 0 \tag{6.47}$$

$$g_2 \equiv \frac{\sigma_2}{\sigma^u} - 1 \le 0 \tag{6.48}$$

$$g_3 \equiv 1 - \frac{\sigma_3}{\sigma^l} \le 0 \tag{6.49}$$

where

$$\sigma_1 = p\left(\frac{x_2 + \sqrt{2}x_1}{\sqrt{2}x_1^2 + 2x_1x_2}\right) \tag{6.50}$$

$$\sigma_2 = p\left(\frac{1}{x_1 + \sqrt{2}x_2}\right) \tag{6.51}$$

$$\sigma_3 = -p\left(\frac{x_2}{\sqrt{2}x_1^2 + 2x_1x_2}\right) \tag{6.52}$$

The structural weight $f_1$, and the displacement $f_2$, can be expressed as

$$f_1 = \rho l(2\sqrt{2}x_1 + x_2) \tag{6.53}$$

$$f_2 = \frac{Pl}{E}\left[\frac{1}{x_1 + \sqrt{2}x_2}\right]. \tag{6.54}$$

The lower and upper bounds of each design variables were prescribed as 0.1 $in^2$ and 5 $in^2$, respectively. In order to obtain an ideal solution for each individual objective criteria, the above design problem was solved twice using a traditional genetic algorithm. The results show ideal designs to be $I_1$* =(0.7894, 0.4061) and $I_2$* =(5.000, 5.000)

The fuzzy problem statement for this problem can be written as follows. The two ideal design points, $I_1$* and $I_2$* were obtained by solving two different optimization problems with scalar objective functions $f_1$ and $f_2$, respectively. Substitution of these design points to evaluate both $f_1$ and $f_2$ yielded the minimum and maximum values for these criterion functions as follows.

$$\begin{aligned} f_1^{min} &= f_1(I_1^*) = 2.6389 \ lb \\ f_2^{min} &= f_2(I_2^*) = 1.6568 \ in \\ f_1^{max} &= f_1(I_2^*) = 19.1421 \ lb \\ f_2^{max} &= f_2(I_1^*) = 14.6652 \ in \end{aligned} \tag{6.55}$$

By allowing a 50% relaxation on all $\sigma_j$, $j$=1,3 in Eqns.(6.47), (6.48), and (6.49), a set of crisp constraints is converted to a set of fuzzy constraints as

$$\begin{aligned} &g_1 \in G_1, \qquad G_1 \equiv [-\infty, \sigma^u + \Delta\sigma^u] \\ &g_2 \in G_2, \qquad G_2 \equiv [-\infty, \sigma^u + \Delta\sigma^u] \\ &g_3 \in G_3, \qquad G_3 \equiv [\sigma^l + \Delta\sigma^l, \infty] \end{aligned} \tag{6.56}$$

where $\Delta\sigma^u = 10$, and $\Delta\sigma^l = -7.5$. Similar relaxations of $\Delta x_i^u = 2.5$ and $\Delta x_i^l = -0.05$ were also applied to the upper and lower bounds of design variables. Based on these fuzzy boundaries, a modified fuzzy formulation can be written as follows.

$$\max_x \min_{i,j} \{\mu_{f_i}(x) + \Delta\mu_{f_i}, \mu_{g_j}(x)\} \qquad i = 1, \ldots, 2, \ j = 1, \ldots, 7 \tag{6.57}$$

where

$$\mu_{f_1} = \begin{cases} 0, & if\ f_1 > 19.1421 \\ \dfrac{19.1421 - f_1}{16.5032}, & if\ 2.6389 < f_1 \le 19.1421 \\ 1, & if\ f_1 \le 2.6389 \end{cases} \tag{6.58}$$

$$\mu_{f_2} = \begin{cases} 0, & if\ f_2 > 14.6652 \\ \dfrac{14.6652 - f_2}{13.0084}, & if\ 1.6568 < f_2 \le 14.6652 \\ 1, & if\ f_2 \le 1.6568 \end{cases} \tag{6.59}$$

$$\mu_{g_i} = \begin{cases} 1, & if\ \sigma_i \le 20 \\ 1-\left(\frac{\sigma_i - 20}{10}\right), & if\ 20 < \sigma_i < 30 \\ 0, & if\ \sigma_i \ge 30 \end{cases} \qquad i = 1,2 \tag{6.60}$$

$$\mu_{g_3} = \begin{cases} 1, & if\ \sigma_3 \ge -15 \\ \frac{\sigma_3 + 22.5}{7.5}, & if\ -22.5 < \sigma_3 < -15 \\ 0, & if\ \sigma_3 \le -22.5 \end{cases} \tag{6.61}$$

$$\mu_{g_{3+j}} = \begin{cases} 1, & if\ x_j \le 5 \\ 1-\left(\frac{x_j - 5}{2.5}\right), & if\ 5 < x_j < 7.5 \\ 0, & if\ x_j \ge 7.5 \end{cases} \qquad j = 1,2 \tag{6.62}$$

$$\mu_{g_{5+k}} = \begin{cases} 1, & if\ x_k \ge 0.1 \\ \frac{x_k - 0.05}{0.05}, & if\ 0.05 < x_k < 0.1 \\ 0, & if\ x_k \le 0.05 \end{cases} \qquad k = 1,2 \tag{6.63}$$

The fuzzy multicriterion problem can be solved for different combinations of $\Delta\mu_{f_i}$ using both the traditional genetic algorithm and the proposed approach based on the immune network simulation. The traditional genetic algorithm was run several times to obtain optimal solutions corresponding to each combinations of $\Delta\mu_{f_i}$, whereas, the immune network approach obtained a set of solutions for all $\Delta\mu_{f_i}$ combinations in a single execution of GA. Relaxed upper and lower bounds of 7.5 $in^2$ and 0.05 $in^2$ were prescribed. Discrete representation of continuous design variables $x_1$ and $x_2$ were generated using the precision of 0.000007105 $in^2$, resulting to the binary string of 20 bits for each design. In each simulation, a population of 100 designs was evaluated for 30,000 function evaluations. In traditional genetic search, probabilities of crossover and mutation of 0.8 and 0.02 were used, while the immune network simulation was performed using probabilities of 0.6 and 0.01.

As shown in Table 6.9 and Table 6.10, 11 different combinations of $\Delta\mu_{f_i}$ were considered, which resulted in a total of 11 antigens. In the immune network simulation, 33 antibody designs were sampled from the population pool and matched against selected antigens. Since the sample size is significantly larger than the number of antigens, the antibody population converged to a set of specialists. The design variable values, structural weight, vertical displacement of the loaded joint, and minimum degree of membership for each combination of $\Delta\mu_{f_i}$ are displayed in Tables 6.9 and 6.10. Results from both approaches are also plotted in Fig. 6.18.

Both the figures and tables show excellent agreement in the results of the two approaches.

We know that a value of $\mu_{g_j} = 1$ indicates that the constraint is satisfied within the crisp boundary (either on the crisp boundary or with a larger margin of constraint satisfaction). Looking at Table 6.9, we note that in all cases except for Case 1 and 11, one or more constraints has a value of $\mu_{g_j}$ which is less than unity, indicating that the search process exploited the fuzzy definition of feasibility to improve the objective function value. This is more clearly shown in Table 6.10 which compares the objective function values obtained from problem formulations with and without fuzzy constraint representations. The departure outside of the crisp constraint boundary is largest in Case 6 (a minimum value of $\mu_{g_j}$ of 0.849 for traditional GA, and 0.850 for immune network approach), where no bias was imposed on either of the objective criterion.

**Table 6.9** Results for the 3 bar truss using traditional methods.

| ID | Bias terms | | Design variables | | Str. wght (*lbm*) | Disp l. (*in*) | $\mu_{f1}+\Delta\mu_{f1}$ | $\mu_{f2}+\Delta\mu_{f2}$ |
|---|---|---|---|---|---|---|---|---|
| | $\Delta\mu_{f1}$ | $\Delta\mu_{f2}$ | x1 | x2 | | | | |
| 1 | 1.0 | 0.0 | 4.99 | 5.00 | 19.1 | 1.66 | 1.00 | 0.99 |
| 2 | 0.8 | 0.0 | 3.89 | 5.03 | 16.0 | 1.82 | 0.98 | 0.98 |
| 3 | 0.6 | 0.0 | 2.79 | 5.06 | 12.9 | 2.01 | 0.97 | 0.97 |
| 4 | 0.4 | 0.0 | 1.71 | 5.11 | 9.97 | 2.23 | 0.95 | 0.95 |
| 5 | 0.2 | 0.0 | 0.65 | 5.16 | 7.02 | 2.51 | 0.93 | 0.93 |
| 6 | 0.0 | 0.0 | 0.55 | 3.53 | 5.10 | 3.60 | 0.85 | 0.85 |
| 7 | 0.0 | 0.2 | 0.55 | 2.31 | 3.87 | 5.23 | 0.93 | 0.92 |
| 8 | 0.0 | 0.4 | 0.59 | 1.51 | 3.20 | 7.30 | 0.96 | 0.96 |
| 9 | 0.0 | 0.6 | 0.65 | 1.01 | 2.85 | 9.63 | 0.98 | 0.98 |
| 10 | 0.0 | 0.8 | 0.71 | 0.66 | 2.68 | 12.0 | 0.99 | 0.99 |
| 11 | 0.0 | 1.0 | 0.79 | 0.39 | 2.63 | 14.7 | 0.99 | 1.00 |

**Table 6.9** Continued.

| ID | $\mu_{g1}$ | $\mu_{g2}$ | $\mu_{g3}$ | $\mu_{g4}$ | $\mu_{g5}$ | $\mu_{g6}$ | $\mu_{g7}$ | Min $\cdot\mu_j$ |
|---|---|---|---|---|---|---|---|---|
| 1 | 1.00 | 1.00 | 1.00 | 1.00 | 1.00 | 1.00 | 0.99 | 0.99 |
| 2 | 1.00 | 1.00 | 1.00 | 1.00 | 1.00 | 1.00 | 0.98 | 0.98 |
| 3 | 1.00 | 1.00 | 1.00 | 1.00 | 1.00 | 1.00 | 0.97 | 0.97 |
| 4 | 1.00 | 1.00 | 1.00 | 1.00 | 1.00 | 1.00 | 0.95 | 0.95 |
| 5 | 1.00 | 1.00 | 1.00 | 1.00 | 1.00 | 1.00 | 0.93 | 0.93 |
| 6 | 1.00 | 1.00 | 0.85 | 1.00 | 1.00 | 1.00 | 1.00 | 0.85 |
| 7 | 0.92 | 1.00 | 0.93 | 1.00 | 1.00 | 1.00 | 1.00 | 0.92 |
| 8 | 0.96 | 1.00 | 1.00 | 1.00 | 1.00 | 1.00 | 1.00 | 0.96 |
| 9 | 0.98 | 1.00 | 1.00 | 1.00 | 1.00 | 1.00 | 1.00 | 0.98 |
| 10 | 0.99 | 1.00 | 1.00 | 1.00 | 1.00 | 1.00 | 1.00 | 0.99 |
| 11 | 0.99 | 1.00 | 1.00 | 1.00 | 1.00 | 1.00 | 1.00 | 0.99 |

**Table 6.10** Results for the 3 bar truss using immune network approach.

| ID | Bias terms | | Design variables | | Str wgt (*lbm*) | Displ (*in*) | $\mu_{f1}+\Delta\mu_{f1}$ | $\mu_{f2}+\Delta\mu_{f2}$ |
|---|---|---|---|---|---|---|---|---|
| | $\Delta\mu_{f1}$ | $\Delta\mu_{f2}$ | x1 | x2 | | | | |
| 1 | 1.0 | 0.0 | 4.99 | 4.98 | 19.12 | 1.65 | 1.00 | 0.99 |
| 2 | 0.8 | 0.0 | 3.90 | 4.98 | 16.02 | 1.82 | 0.98 | 0.98 |
| 3 | 0.6 | 0.0 | 2.79 | 5.05 | 12.94 | 2.01 | 0.97 | 0.97 |
| 4 | 0.4 | 0.0 | 1.71 | 5.10 | 9.95 | 2.23 | 0.95 | 0.95 |
| 5 | 0.2 | 0.0 | 0.66 | 5.14 | 7.02 | 2.50 | 0.93 | 0.93 |
| 6 | 0.0 | 0.0 | 0.55 | 3.53 | 5.11 | 3.59 | 0.84 | 0.85 |
| 7 | 0.0 | 0.2 | 0.55 | 2.30 | 3.87 | 5.24 | 0.92 | 0.92 |
| 8 | 0.0 | 0.4 | 0.60 | 1.50 | 3.20 | 7.32 | 0.96 | 0.96 |
| 9 | 0.0 | 0.6 | 0.66 | 1.00 | 2.87 | 9.59 | 0.98 | 0.98 |
| 10 | 0.0 | 0.8 | 0.71 | 0.69 | 2.70 | 11.83 | 0.99 | 1.01 |
| 11 | 0.0 | 1.0 | 0.80 | 0.37 | 2.64 | 14.95 | 0.99 | 1.00 |

**Table 6.10** Continued.

| ID | $\mu_{g1}$ | $\mu_{g2}$ | $\mu_{g3}$ | $\mu_{g4}$ | $\mu_{g5}$ | $\mu_{g6}$ | $\mu_{g7}$ | Min $\cdot\mu_j$ |
|---|---|---|---|---|---|---|---|---|
| 1 | 1.00 | 1.00 | 1.00 | 1.00 | 1.00 | 1.00 | 1.00 | 0.99 |
| 2 | 1.00 | 1.00 | 1.00 | 1.00 | 1.00 | 1.00 | 1.00 | 0.98 |
| 3 | 1.00 | 1.00 | 1.00 | 1.00 | 1.00 | 1.00 | 0.97 | 0.97 |
| 4 | 1.00 | 1.00 | 1.00 | 1.00 | 1.00 | 1.00 | 0.95 | 0.95 |
| 5 | 1.00 | 1.00 | 1.00 | 1.00 | 1.00 | 1.00 | 0.94 | 0.93 |
| 6 | 1.00 | 1.00 | 0.85 | 1.00 | 1.00 | 1.00 | 1.00 | 0.84 |
| 7 | 0.93 | 1.00 | 0.95 | 1.00 | 1.00 | 1.00 | 1.00 | 0.92 |
| 8 | 0.97 | 1.00 | 1.00 | 1.00 | 1.00 | 1.00 | 1.00 | 0.96 |
| 9 | 1.00 | 1.00 | 1.00 | 1.00 | 1.00 | 1.00 | 1.00 | 0.98 |
| 10 | 1.00 | 1.00 | 1.00 | 1.00 | 1.00 | 1.00 | 1.00 | 0.99 |
| 11 | 1.00 | 1.00 | 1.00 | 1.00 | 1.00 | 1.00 | 1.00 | 0.99 |

The optimal solutions obtained from both traditional GA and the proposed approach are compared against the Pareto optimal solutions without relaxation (fuzzy representation) of the constraint boundaries. The front of Pareto optimal solutions without relaxation was obtained using the weighting method and is drawn as a dotted line in Fig. 6.18. It is worthwhile to point out that these results (obtained from the weighting method) are almost identical to those obtained using the fuzzy multicriterion approach with crisp constraints. Although, the optimal points obtained from the traditional GA and immune network simulation approaches appear to overlap with the Pareto optimal front without relaxation, the optimal points are in reality located slightly away from the Pareto front where objective values of $f_1$ and $f_2$ are less than those corresponding to the Pareto optimal front without relaxation. To reiterate, by allowing for some relaxation on constraint boundaries, a fuzzy multicriterion approach generates improvements in objective function values obtained from a crisp formulation; the maximum departure from the crisp constraint boundaries is indicated by a lower value of the constraint membership function.

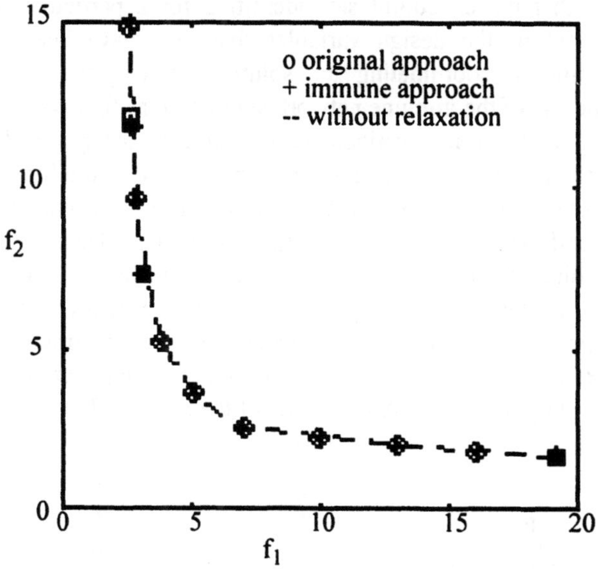


**Figure 6.18** Pareto front for the 3 bar truss problem.

### 6.4.3 Immune Networks in Decomposition-Based Design

An alternative approach to adapting GA search for large scale design problems (long string lengths) is based on partitioning the problem into an appropriate number of subproblems; a reasonable approach for partitioning is one where balanced subsets of design variables are assigned to different subproblems, and where each subproblem would be responsible for meeting the system level design objectives and for satisfying constraints most critically affected by the design variables of that subproblem.

Consider the design problem to be formulated in terms of a design variable vector ***6.*** Also, let the design constraints $g_j(X)$ belong to the global constraint set $G$. The vector $X$ and constraint set $G$ are said to define a system level problem that is formulated as follows:

$$\begin{aligned} &\mathit{Min\ or\ Max}\quad F(X), \\ &\mathit{subject\ to}\ \ G \equiv \{g_j(X), j = 1 \ldots NCON\} \leq 0 \end{aligned} \tag{6.64}$$

Assume further that the best topology for decomposing the problem domain resulted in three subproblems A, B, and C, and the design variables and constraints for each of these subproblems are denoted by $X_A$, $X_B$, $X_C$, and $g_A$, $g_B$, and $g_C$, respectively. The objective function F($X$) for each of the subproblems is the same, and is the system level objective function. The system level problem of eqn. (6.64) is now represented by the following three subproblems.

$$\begin{aligned} &Min\ or\ Max \quad F(X), \\ &subject\ to\ g_i(X_i) \leq 0, \quad X_k = const \\ &i, k = A, B,\ or\ C \qquad i \neq k \end{aligned} \tag{6.65}$$

The GA strategy in each subproblem works with shorter string lengths, and hence smaller population sizes are required in each subproblem. The principal challenge in this approach arises from the fact that the constraint sets identified for a particular subproblem, are not completely independent of the design variables that may have been assigned to another subproblem. This issue of coordinating the solutions among the several subproblems is addressed through the use of the immune network simulation as follows.

After partitioning of the design variable vector for each subsystem has been performed, genetic evolution is carried out in each subsystem in parallel, with the fitness function described in terms of the system level objective function. Only those subsections of the chromosome string which correspond to the design variables for a particular subsystem are subject to change in that subsystem evolution. This process can be carried out in each subsystem (in parallel if hardware permits) for a fixed number of generations. Next, a fraction of most fit strings from each subsystem are introduced as antigens for which a generalist immune system simulation is performed in all other subsystems. This serves as a correction step that introduces compatibility between the different subpopulations. A schematic of the process flow is shown in Fig. 6.19.

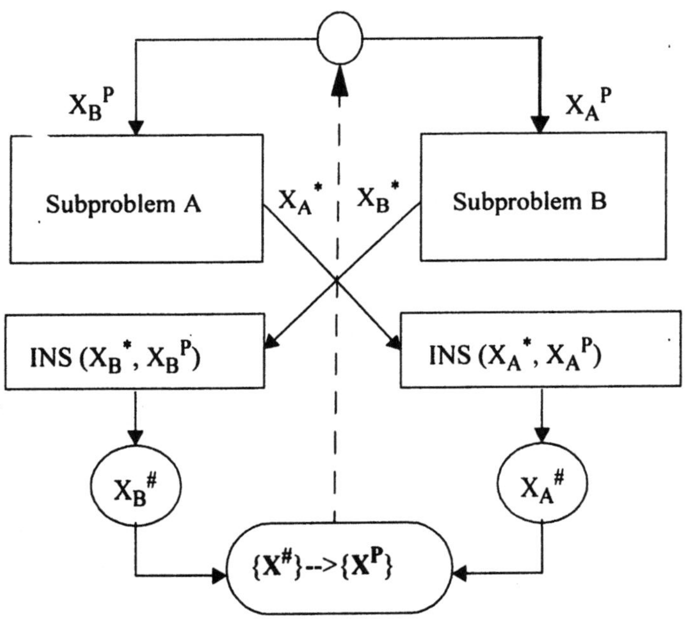


**Figure 6.19** Schematic of coordination using immune network simulations.

In using decomposition principles in multidisciplinary design optimization, an issue that has to be considered at the very outset is the search for a rational approach to decompose the problem. The trained BP network can be used to identify dependencies among design variables and design objectives, and used as a guideline for problem decomposition.

Consider the strengths of the interconnections between the neurons of the various layers of a trained multilayer feedforward network. This matrix of numbers can be viewed as a hologram representing the mapping between the desired input and output quantities, and offers little in terms of a physical interpretation of the problem domain. The flow of information between the input and output layers, however, is a strong function of these interconnection weights, and an analysis of the weights provides meaningful insight into the problem domain.

In order to assess the influence of the i-th input component on the k-th output component, all connecting paths between these neurons must be considered. For each neuron in the hidden layer, the contribution of a signal from the i-th input, i=1,n, can be computed as the ratio:

$$s_j = \frac{\left|w_{ji}^{(1)}\right|}{\sum_{r=1}^{n} \left|w_{jr}^{(1)}\right|} \qquad j = 1, M \tag{6.66}$$

Here M is the number of neurons in the hidden layer. The signal $s_j$ is amplified by the weights between the first and the second hidden layer (indicated by superscript '12'), and summed over all neurons in the first hidden layer,

$$s_k = \sum_{j=1}^{M} \left|w_{jk}^{(12)}\right| \bullet s_j \qquad k = 1, M_2 \tag{6.67}$$

where $M_2$ is the number of neurons in the second hidden layer. Finally, the product of these signals with the weights of the output layer is formed and added over all neurons in the second hidden layer as follows:

$$\tilde{t}_{ki} = \sum_{j=1}^{M} \left|w_{jk}^{(2)}\right| \bullet s_j \qquad k = 1, m \tag{6.68}$$

The quantities in eqn (18) can be normalized to yield elements of the transition matrix $[T]_{nxm}$ as follows.

$$[T] = \frac{\tilde{t}_{ki}}{\sum_{s=1}^{m} \tilde{t}_{ks}} \qquad i = 1, m \qquad k = 1, n \tag{6.69}$$

For an architecture with a single hidden layer, the above computation simplifies as

$$\tilde{t}_{ki} = \sum_{j=1}^{M} \frac{\left|w_{ji}^{(1)}\right|}{\sum_{r=1}^{n} \left|w_{jr}^{(1)}\right|} \bullet \left|w_{kj}^{(2)}\right| \qquad k = 1, m \tag{6.70}$$

where the subscripts '1' and '2' denote weights from input to hidden and hidden to output layers, respectively. The elements of the transition matrix show the contribution of the i-th input on the k-th output, and since each row of the matrix sums to unity after normalization, the contribution is available as a fractional quantity. This approach will be referred to as the ABS procedure for weight analysis.

An alternative approach (ALT), and one that preserves the sign of this dependency relationship, is to simply multiply the interconnection weight matrices of each layer in sequence, and normalizing the components by the maximum value in a particular row of the resulting matri6.

$$[\tilde{T}] = \frac{T_{ij}}{max_i |T_{ij}|} \tag{6.71}$$

where,

$$[T] = \prod_{k=1}^{NL-1} W^k \tag{6.72}$$

and NL and $W^k$ are the number of layers and the interconnection weight matrices of each layer, respectively.

Based on the information from the weights of a backpropagation neural network representing the mapping between global design variables and objective and constraint functions, the low-vibration rotor problem described in a previous section was decomposed into three subproblems. Each subproblem had a subset of the global design variables and constraints most influenced by those variables. The topology of problem decomposition is presented in Table 6.10.

**Table 6.10** Decomposition topology for the rotor blade design problem.

| | subproblem [1] | subproblem [2] | subproblem [3] |
|---|---|---|---|
| objective | $F(X) = c_1F_z + c_2M_y + c_3M_z$ | | |
| con-straints | $HP_h \leq HP_a$ <br> $\eta^L \leq \eta \leq \eta^U$ <br> $AI \geq AI^L$ | $HP_f \leq HP_a$ <br> $C_T/\sigma^L \leq C_T/\sigma \leq C_T/\sigma^U$ | $W_b \leq W_b^U$ <br> $\bar{R} \leq 1$ <br> $\sigma_{buck} \leq \sigma_{all}$ |
| design variables | $m_1, m_2, m_3, t_1^4, t_1^5, t_1^6, t_2^3, t_2^6, t_2^9, t_3^6, t_3^8, t_3^9, \tau, \pm\theta_1$ | $t_1^3, t_1^{10}, t_2^4, t_2^5, t_2^7, t_2^8, t_3^2, t_3^3, t_3^5, t_3^7, t_3^{10}, \theta_t, \delta, \pm\theta_2$ | $m_4, m_5, t_1^1, t_1^2, t_1^7, t_1^8, t_1^9, t_2^1, t_2^2, t_2^{10}, t_3^1, t_3^4, \lambda_c, \Omega$ |

The GA based optimization was performed in each subproblem in parallel, with the immune network simulation used to coordinate the solution process. Each subproblem had a population size of 100 which was selected in light of the string lengths for these three subproblems; string lengths of 63, 57, and 59 were used for subproblems A, B and C. Probabilities of crossover and mutation were kept at the same levels as in the previous problem.

For the immune network simulation, an antibody population size of 50 was chosen for each subproblem; antigen pool sizes of 5 or 10, corresponding to $\Gamma = 0.05$ and 0.1 were studied in the simulations. Consistent with the intent of letting the immune system develop specialists in proportion to the frequency of occurrence of antigens, a large antibody sample size corresponding to $\sigma = 5$ was considered. The results of the decomposition based approach were compared against those obtained from treating all design variables and constraints in a single group by an all-in-one approach. Fig. 6.20 shows this comparison in terms of the best objective function value and the required CPU time. Here, five different sets of results are presented, including four different implementations of the all-in-one approach and one set of results corresponding to the present decomposition approach. The all-in-one strategies include a straightforward GA implementation (the plain GA), GA with directed crossover, GA with multistage search, and a combination of the directed crossover and multistage search. Figs. 6.20a and 6.20b show results for three different initializations of the random number generator in the GA search. While there was an improvement in the best objective function value with the use of strategies like directed crossover and multistage search, the decomposition based approach provided the best result on average. Fig. 6.20b shows the relative CPU time required to reach the best objective function value obtained from the plain GA for each strategy. The decomposition based approach is clearly more efficient from a computational standpoint; this is attributed to the parallel implementation in decomposition based optimization.

## 6.5 Genetic Algorithms and Machine Learning

### 6.5.1 Classifier Systems

There has been significant recent interest in methods of computational intelligence and to bring them to bear on problems of multidisciplinary design. One recent effort has been in the use of classifier systems in improving response surface and neural network based approximations. Such systems may be considered as an evolved form of rule-based expert systems. In expert systems, an expert develops a number of rules in the IF (rule) THEN (action) format, where each rule may also be given some strength. The process of inference is then initiated with some conditions, and which then sets into motion a sequence of rule firings that initiate other action. Such an approach has been shown to require a fair degree of experience with the design domain under consideration, and successful implementations of this approach have been in tightly focussed, narrow-range problems. Among continuing problems with knowledge-based systems is the manner in which this information is collected, and its domain of applicability. With moderate change in operating conditions, the validity of the knowledge may be in question.

Classifier systems derive information from the computational domain to not only modify an initial set of rules prescribed by an user, but also evolve new rules by a process of genetic transformation. Early work describing the concept of classifier systems may be traced to a publication of John Holland (1962). Since that time, classifier systems have been studied extensively in the field of computer science, both from the standpoint of applications (Goldberg, 1983, Dorigo and Schnepf, 1991, Wilson, 1986), and development of the theoretical foundations (Forrest, 1985, Richards, 1995). The application of this approach to problems of engineering design have only recently received recognition (Goel and Hajela, 1997, Goel and Hajela, 1998), and a more widespread use in this area will hinge on further advancements in the basic technology that are directly responsive to this challenging application environment.

A classifier system can be divided into two parts, a set of rules or classifiers, and a message list. The message list contains at least one input from the external environment and also provides the framework for the rules to interact in that any rules generated internally are posted here as well. Hence, the message list is dynamic in nature, constantly evolving as the system changes. The classifier rules are made up of three distinct segments - conditions, actions and strength. The conditions allow the classifier to read the message list by searching for matches between the condition and the message list. If a match is found, the action is posted to the message list. A typical classifier system may thus be regarded as a massively parallel system in which rules are triggered either in parallel, or through a chaining mechanism in which the action of one classifier results in the firing of other classifiers. Some rules have the special role of affecting external action (effectors).

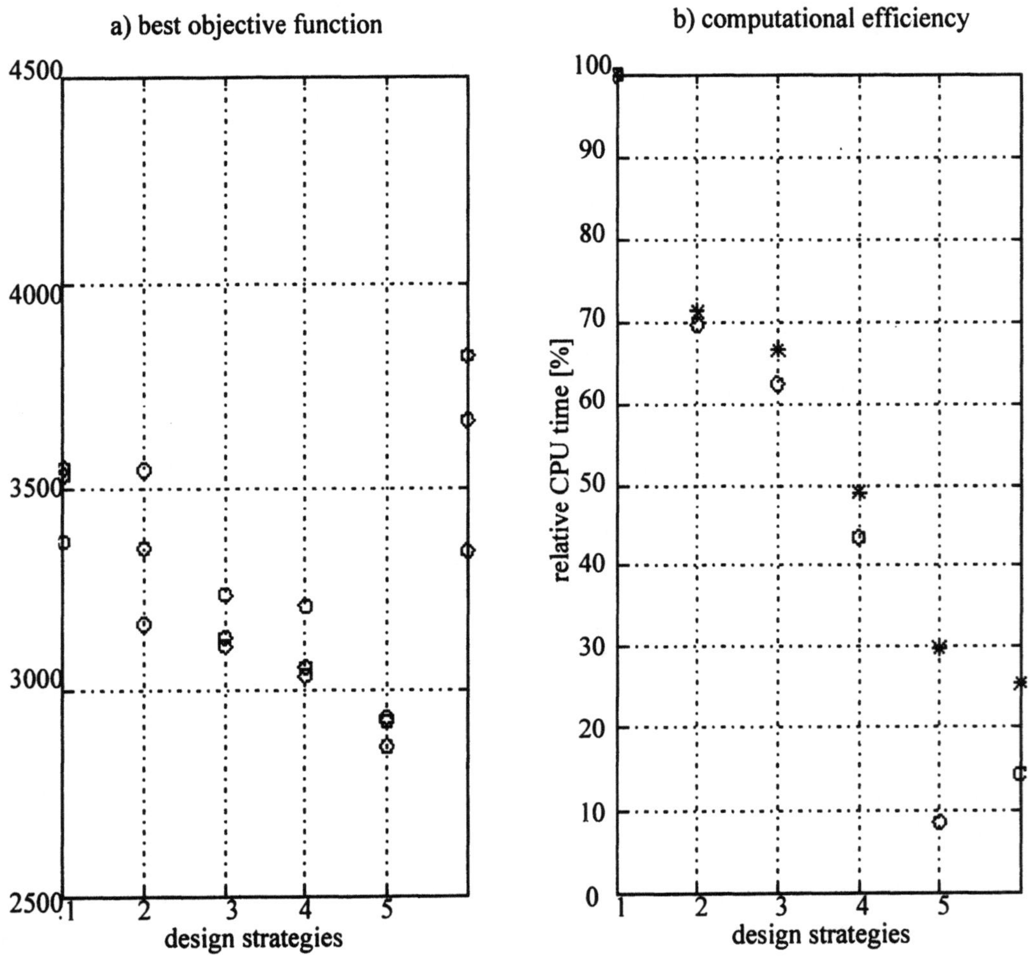


**Figure 6.20 a&b** Performance of large-scale design process; (1) plain GA, (2-4) advanced GA strategies (5-6) two immune network strategies.

As stated before, the strength is a number associated with each rule designed to indicate its value to the system, and forms the basis for learning. If a rule helps bring about useful responses, it gains strength. Similarly, an ineffective rule is weakened and perhaps ultimately purged from the system. The usefulness of a rule is measured by the performance of the system based on external outputs. When a useful output takes place, the responsible rules are *paid*, and their strength is proportionately increased. To facilitate the process of classifier evolution on the basis of their strength, they are generally coded in a chromosomal form (binary or another), and evolved using a genetic algorithm or another evolutionary strategy. A number of other features can be implemented in classifier systems. These include the introduction of *nonspecific rules* (in a binary rule this implies a number of locations occupied by a wild-card character) which allows for partial matches between rules and messages, tagging of information to distinguish between external and internally generated information, and taxing of rules so as to eliminate those that do not post messages. The latter is often done through the use of a *bucket-brigade* algorithm (Holland, 1985), in which not only is the classifier causing a favorable action rewarded, but all classifiers in the chain leading up to the firing of the effective classifier also receive a favorable score.

The issue of representation of classifiers is not trivial, as it has a direct bearing on the effectiveness of the computational process. If the classifier is coded by a very long binary string, the cardinality of the search space increases dramatically, making it difficult to establish exact matches between messages and classifier conditions. The specificity of the classifier has an important role in mitigating this problem. The specificity is defined as the fraction of the total number of bits of the classifier that are used for matching. Classifiers with low specificity have poor discrimination ability and are also called default rules since they get selected by default if other more specific rule matches are not available. In addition to default rules, it is possible to set up a default hierarchy in which exception rules are defined to cover for deficiencies in the default rules. In such a structure, the exception rules are somewhat more specific than the default rules and are placed in a hierarchical form under the default rule. In short, the specificity of the rule increases as one traverses the depth of the hierarchical tree structure. Additional research in this area is required if the classifier system approach is to be extended to more realistic engineering design problems.

This approach has been applied to improving function approximations using both response surfaces (Lee and Hajela, 2000), and backpropagation neural networks (Hajela and Kim, 1998). In the case of the former, classifers were used to discover the order of the polynomial that must be fitted over various regions of the design space in order to get good global function approximations. For functions where the extent of nonlinear behavior varied from region to region in the parameter space of interest, such a strategy was extremely helpful in selecting the appropriate polynomial order. For the backpropagation networks, this approach was helpful in identifying the required distribution of data to generate neural network approximations with acceptable levels of generalization performance.

### 6.5.2 Cellular Automata

Another area of computational intelligence where the genetic algorithm has received recent recognition is in the use of cellular automata computations for structural analysis. This focus has grown out of an awareness that traditional methods of analysis are ill-suited to a new generation of massively parallel computers, and new algorithms specifically designed for the new computer architectures are required. Cellular automata models are predicated on a logic that any system may be represented by a number of discrete particles or cells, and where the state of any particle is determined through interaction with the state of its immediate neighbors; the collective state of these particles is representative of the state of the system. The central requirement for a successful implementation of this approach is to develop the appropriate rules for interaction among neighboring cell sites, which when invoked, evolve the state of the cells to desired acceptable solutions. This model of computation has been adapted in a number of combinatorial mathematics and computer science applications, and in computation of fluid flow properties (Hardy et. al., 1976, Orszag and Yakhot, 1986).

To better understand how cellular automata function, consider a one-dimensional grid of cells to define a binary state system as shown in Fig. 6.21, where each cell can take on a value of 0 or 1.

| | |
|---|---|
| t=0 | 1 0 1 0 0 1 1 1 0 1 0 0 0 1 1 |
| t=1 | 0 1 0 0 0 1 1 1 1 0 0 0 0 1 0 |
| t=2 | 1 0 0 0 0 1 1 1 1 0 0 0 0 0 1 |

**Figure 6.21** One dimensional CA model.

Assume further that the neighborhood of any cell is defined as itself and the two adjacent cells. A simple local rule that can be used to evolve this binary state system from one discrete time-step to the next is that each cell state assumes a value that is a 'majority vote' of the 3 cells that define its neighborhood; boundary cells are flipped from a 0 to 1 or vice versa. This updating procedure is shown in Fig. 6.22.

| | | | | | | | | | |
|---|---|---|---|---|---|---|---|---|---|
| 000 | 001 | 010 | 011 | 100 | 101 | 110 | 111 | - | neighborhood |
| 0 | 0 | 0 | 1 | 0 | 1 | 1 | 1 | - | output |

**Figure 6.22** Majority vote rule.

Note that the boundary cells will have to be treated differently as the neighborhood of these cells is empty on one side. In the figure, the values of these cells are flipped at each iteration. Similar rules can be developed for cellular automata models in higher dimensions. Such a model is naturally amenable to implementation on massively parallel machines - the temporal updating of the state of different cell groups can be assigned to different processors. Since all such operations are performed simultaneously, proper load distribution among processors is not an issue of serious concern. Furthermore, minimal amount of information must be exchanged among processors at each time step.

The key to the success of this parallel distributed approach is that the proper rule for updating the states be available. In developing cellular automata based models for hydrodynamic flow analysis, the rules are based on an understanding of the physics of the problem. If cellular sites are characterized by fluid particles with a certain velocity, collisions with adjacent particles must follow the basic conservation of mass and momentum principles. Such simulations have been the subject of a number of recent studies. In problems where such rules are difficult to identify, or in situations where the physics of the problem is not properly understood, alternative strategies must be implemented. The role of genetic algorithms in discovering such rules of interaction has been established in recent research (Hajela and Kim, 1999, Hajela and Kim 2000).

To illustrate this concept, consider again the one-dimensional binary state cellular automata, and where the converged state of the system is known; let us further assume that the rule for evolution (majority vote rule) is not known. To determine a rule for evolution, the optimization procedure would function as follows.

- Step 1: A population of 8-digit binary bit strings (rules) would be generated at random. Each bit of this string corresponds to the output of one of the eight unique combinations of a 3-digit binary bit string representing the 3 neighborhood cell states.
- Step 2: Each rule in this population of rules would be applied to the same initial state (one at a time), and the state evolved to some converged final configuration. The similarity between the converged and known final state of the system would be the fitness assigned to the selected rule. This process is repeated for each rule in the population.
- Step 3: Using the fitness of rules in Step 2, the traditional selection, crossover, and mutation operations of a GA are performed.
- Step 4: The process is repeated from Step 2 until such times that the rule evolution process has converged.

The converged rule obtained from this simulation should be the 'majority vote rule' which was responsible for generating the desired state. This rule learning process can be extended to a problem with higher dimensions. This approach has been applied in developing displacement, strain, and stress fields in 2-D elastic structures (Hajela and Kim, 1999, Hajela and Kim 2000). The rule derivation process for these applications used a genetic algorithm in conjunction with equilibrium and minimal strain energy principles.

## 6.6 Closing Remarks

This chapter describes the applicability of recently emergent computational paradigms in problems of multidisciplinary engineering design, with a special focus on how such methods alleviate difficulties associated with the formal use of design optimization algorithms. The computational models considered in this context include genetic algorithms, neural networks, and a modeling of the biological immune system. Specific concerns addressed in this discussion include the need for global search methods such as the genetic algorithm which offer an increased probability of identifying a globally optimal solution. These methods need special attention when applied to large-scale problems typical of multidisciplinary design. Strategies by which to enhance their efficiency are presented in the discussion, including approximation techniques based on neural networks. Methods for multicriterion design are also included in the discussion. The population based search in GA allows for a simultaneous generation of the Pareto optimal front, and the immune network simulation is shown to be useful in this regard. The approach for multicriterion design also includes problems where the problem formulation may be fuzzy or imprecise. Special emphasis is also placed on decomposition methods for the multidisciplinary design problem, and how the genetic algorithm based search methods are adapted for this task. The chapter briefly introduces the role of genetic algorithms in machine learning and in the cellular automata based methods for analysis.

## 6.7 References

Abdi, F., Ide, H., Levine, M., and Austel, L., (1988). The Art of Spacecraft Design: A Multidisciplinary Challenge. *2nd NASA/Air Force Symposium on Recent Advances in Multidisciplinary Analysis and Optimization*, NASA CP-3031.

Barron, A.R., (1992). Neural Network Approximation. In *Proceedings of the Seventh Yale Workshop on Adaptive and Learning Systems*, Yale University, New Haven, CT, 69-72.

Bauchau, O.A., and Kang, N.K., (1993). A Multibody Formulation for Helicopter Structural Dynamic Analysis. *Journal of American Helicopter Society*, 38:3-14.

Cordon, O., and Herrera, F., (1995). A General Study on Genetic Fuzzy Systems. In Periaux J., and Winter G., eds., *Genetic Algorithms in Engineering and Computer Science* John Wiley and Sons Limited, England.

Dorigo, M., and Schnepf, U., (1991). Organization of Robot Behaviour Through Genetic Learning Process. In *Proceedings of the 5th International Conference on Advanced Robotics*, The Institution of Electrical and Electronics Engineering, Pisa, Italy.

Fogel, L.J., Owens, A.J., and Walsh, M.J., (1966). *Artificial Intelligence Through Simulated Evolution*. Wiley Publishing, New York.

Forrest, S., (1985). Implementing Semantic Network of Structures Using the Classifier System. In *Proceedings of the 1st International Conference on Genetic Algorithms* Hillsdale, New Jersey, Lawrence Erlbaum Associates, 80-92.

Goel, S., and Hajela, P., (1997). Adaptive Optimization Technique Using Classifiers Based Machine Learning Paradigm. In *Proceedings of the 38th AIAA/ASME/ASCE/AHS Structures, Structural Dynamics and Materials Conference* Kissimmee, Florida.

Goel, S., and Hajela, P., (1998). Turbine Aerodynamic Design Using Reinforcement Learning Optimization. *The 7th AIAA/NASA/ISSMO.USAF Multidisciplinary Analysis and Optimization Meeting* St. Louis, Missouri.

Goldberg, D.E., (1983). Computer-Aided Pipeline Operation Using Genetic Algorithms and Rule-Learning. *Ph.D. Dissertation*. University of Michigan, Ann Arbor, Michigan.

Haftka, R.T., and Gurdal, Z., (1993). *Elements of Structural Optimization*. Kluwer Academic Publishers, Dordrecht.

Hajela, P., (1981). Further Developments in the Controlled Growth Approach for Optimum Structural Synthesis. In *Proceedings of the 12th Design Automation Conference* September 12-15, Arlington, Virginia, American Society of Mechanical Engineers, Paper 82-DET-62.

Hajela, P., (1997). Stochastic Search in Discrete Structural Optimization – Simulated Annealing, Genetic Algorithms and Neural Networks. In Gutkowski W., ed., *Discrete Structural Optimization* Springer, New York, 55-134.

Hajela, P., and Kim, B., (1998). Classifier Systems for Enhancing Neural Network Based Global Function Approximations. In *Proceedings of the 7th AIAA/NASA/ISSMO/USAF Multidisciplinary Analysis and Optimization Meeting* St. Louis Missouri.

Hajela, P., and Kim, B., (1999). GA Based Learning in Cellular Automata Models for Structural Analysis. In *Proceedings of the 3rd World Congress on Structural and Multidisciplinary Optimization* Niagara Falls, New York.

Hajela, P., and Kim, B., (2000). On the Use of Energy Minimization for CA Based Analysis in Elasticity. In *Proceedings of the 41st AIAA/ASME/ASCE/AHS SDM Meeting*, April 1-3, Atlanta, Georgia.

Hajela, P., and Lin, C.-Y., (2000). Real Versus Binary Coding in Genetic Algorithms – A Comparative Study. In *Proceedings of the 5th International Conference on Computational Structures Technology*, September 6-8, Leuven, Belgium.

Hardy, J., De Pazzis, O., and Pomeau, Y., (1976). Molecular Dynamics of a Classical Lattice Gas: Transport Properties and time Correlation Functions. *Physics Review* A13:1949-1960.

Hecht-Nielsen, R., (1987). Counterpropagation Networks. *Journal of Applied Optics* 26:4979-84.

Holland, J.H., (1962). Outline for a Logical Theory of Adaptive Systems. *Journal of the Association of Computing Machinery* 3:297-314.

Holland, J.H., (1974). *Adaptation in Natural and Artificial Systems*. University of Michigan Press, Ann Arbor, Michigan.

Holland, J.H., (1985). Properties of the Bucket-Brigade Algorithm. In *Proceedings of the 1st International Conference on Genetic Algorithms* Hillsdale, New Jersey, Lawrence Erlbaum Associates, 1-7.

Karr, C.L., and Gentry, E.J., (1993). Fuzzy Control of pH Using Genetic Algorithms. *The Institution of Electrical and Electronics Engineering Transactions on Fuzzy Systems* 1:46-53.

Krishna Kumar, K., and Satyadas, A., (1995). Space Station Fuzzy Models and its application to an aircraft control problem. In Periaux J., and Winter G., eds., *Genetic Algorithms in Engineering and Computer Science* John Wiley and Sons Limited, England.

Lee, J., and Hajela, P., (2000). An Application of Classifier Systems in Improving Response Surface Based Approximations for Design Optimization. *Computers and Structures*, to appear.

LeRiche, R., and Haftka, R.T., (1993). Optimization of Laminate Stacking Sequence for Buckling Load Maximization by Genetic Algorithms. *American Institute of Aeronautics and Astronautics Journal* 31:951-956.

Lin, C.-Y., (1990). Genetic Search Methods for Multicriterion Optimal Design of Viscoelastically Damped Structures. *Ph.D. Dissertation*, University of Florida, Florida.

Lin, C.-Y., and Hajela, P., (1993). Genetic Search Strategies in Large Scale Optimization. In *Proceedings of the 34th AIAA/ASME/ASCE/AHS/ASC SDM Conference* La Jolla, California, 2437-2447.

Orszag, S., and Yakhot, V., (1986). Reynolds Numbers Scaling of Cellular-Automaton Hydrodynamics. *Physics Review Letters* 56:1691-1693.

Rechenberg, I., (1973). *Evolutionsstrategie: Optimierung Technischer System nach Prinzipien der Biologischen Evolution*. Frommann-Holzboog, Stuttgart.

Richards, R.A., (1995). Zeroth-Order Shape Optimization Utilizing A Learning Classifier System. *Ph.D. Dissertation* Stanford University, Stanford, California.

Rumelart, D.E., and McClelland, J.L., (1988). *Parallel Distributed Processing*. Volume 1, The MIT Press, Cambridge, Massachussets.

Rumelart, D.E., and McClelland, J.L., (1988). *Parallel Distributed Processing*. Volume 2, The MIT Press, Cambride, Massachussets.

Satyadas, A., and Krishna Kumar, K., (1994). Evolutionary Fuzzy Techniques for Fuzzy Controller Synthesis. In *Proceedings of the First Industry/University Symposium on Research for Future Supersonic and Hypersonic Vehicles* North Carolina, TSI Press, New Mexico, 148-155.

Schraudolph, N.N., and Belew, R.K., (1992). Dynamic Parameter Encoding for Genetic Algorithms. *Machine Learning* 9:9-21.

Smith, R.E., Forrest, S., and Perelson, A.S., (1992). Searching for Diverse Cooperative Populations with Genetic Algorithms. *Technical Report CS92-3* University of New Mexico, Department of Computer Science, Albuquerque, New Mexico.

Sobieszczanski-Sobieski, J., (1993). Multidisciplinary Design Optimization: An Emerging New Engineering Discipline. In *Proceedings World Congress on Optimal Design of Structural Systems* Rio de Janeiro, Brazil, August 2-6.

Szewczyk, Z., and Hajela, P., (1992). Feature Sensitive Neural Networks in Structural Response Estimation. In *Proceedings of the ANNIE'92, Artificial Neural Networks in Engineering Conference* November.

Tolson, R.H., and Sobieszczanski-Sobieski, J., (1985). Multidisciplinary Analysis and Synthesis: Needs and Opportunities. *American Institute of Aeronautics and Astronautics* Paper No. 85-0584.

Wilson, S.W., (1986). Classifier System Learning of a Boolean Function. *Research Memo RIS No 27r* The Rowland Institute of Science, Cambridge, Massachussets.

MIX
Papier aus verantwortungsvollen Quellen
Paper from responsible sources
FSC® C105338

If you have any concerns about our products,
you can contact us on
**ProductSafety@springernature.com**

In case Publisher is established outside the EU,
the EU authorized representative is:
**Springer Nature Customer Service Center GmbH**
**Europaplatz 3, 69115 Heidelberg, Germany**

Printed by Libri Plureos GmbH
in Hamburg, Germany